MODERN MEASUREMENTS

MODERN MEASUREMENTS
Fundamentals and Applications

Edited by

ALESSANDRO FERRERO
DARIO PETRI
PAOLO CARBONE
MARCANTONIO CATELANI

IEEE
IEEE PRESS

WILEY

Published by John Wiley & Sons, Inc., Hoboken, New Jersey.
Published simultaneously in Canada

Library of Congress Cataloging-in-Publication Data is available.

ISBN: 978-1-118-17131-8

10 9 8 7 6 5 4 3 2 1

CONTENTS

PREFACE

There are several books related to instrumentation and measurement that explain how measurement methods work and how instruments can be designed, selected, and used to implement the different measurement methods needed to obtain the desired measurement result with the desired accuracy. But no books are available, to our knowledge, that explain what "measuring" means and which essential activities must be performed to accomplish a measurement.

This is somehow surprising, because measurement has always been an important, and sometimes critical, activity that has developed with the humankind, becoming a significant part of its culture. Today everything is measured, and we are surrounded by instruments, often without being aware of their silent presence. Despite this pervasive presence of every kind of instruments and the many kinds of measurements we perform in our everyday life, measurement is probably the only relevant technical activity that is rarely grounded on a well-defined and widely known models. It is rather considered as a mere tool, not as an independent branch of science and technique, so that it is often confined to the role of appendix, sometimes not so relevant, of other subjects.

This book is aimed at filling this gap, and providing a unique framework, valid for any kind of measurement. So, do not expect too many technical details on how a specific instrument works, or a specific measurement method can be implemented. Many excellent books already cover these topics. Here you will find those fundamental bricks that are present in every modern instrument and are part of every modern measurement activity. You will also discover the importance of some "hidden" bricks, such as the primary standards, without which no universally recognized measurement would be possible.

You will discover the importance of digital signal processing in modern measurements, and how critical it may become if some specific mathematical issues are disregarded. Do not expect to find how modern oscilloscopes or multimeters are designed and developed, since this is the aim of many dedicated books and handbooks. On the contrary, you will find how the aforementioned fundamental bricks can be combined to implement two of the most popular and used instruments.

Most importantly, and maybe surprisingly for those readers who are not yet familiar with this field, you will discover that every measured value is "wrong" in the sense that, due to a number of different factors that influence any measuring activity, we are never able to know the so-called "true value" of the measurand. But you will also discover that we know how to quantify how "wrong" our measurement result is. This is achieved by providing the so-called measurement uncertainty, which tells us how good and reliable the obtained measurement result is. This information is essential when we use a measurement result as the input element of a decision-making process and we wish to quantify the risk of a wrong decision.

To accomplish our goal, we have organized this book into two parts. Part I (Fundamentals) presents a model of the modern measurement activity and the already recalled fundamental bricks. It starts (Chapter 1) with a general model that introduces these bricks and the uncertainty concept, provides an overview of these bricks in the following chapters, and finishes (Chapter 7) with a more general and complex model that encompasses both traditional (*hard*) measurements and *soft* measurements, aimed at quantifying nonphysical concepts, such as quality, satisfaction, and comfort.

Part II (Applications) is aimed at showing how the concepts presented in Part I can be usefully applied to design and implement measurements in some very important and broad fields. We cover system identification (Chapter 8), reliability (Chapter 9), and electromagnetic compatibility (EMC) (Chapter 10) not only for their importance in many application areas, from manufacturing to health and safety, but also because their intrinsic complexity is the perfect test bench to prove the usefulness of the concepts introduced in Part I.

We sincerely hope that this book may provide a new, systematic insight into such an important field as instrumentation and measurement, and can help both experts and beginners, with a useful analysis tool, to understand the essential meaning of the measuring activity, regardless of its specific application.

The Editors

ACRONYMS

AC	Alternate Current
ACF	Antenna Calibration Factor
ADC	Analog to Digital Converter
AMN	Artificial Mains Network
BAAS	British Association for the Advancement of Science
BCI	Bulk Current Injection
BIPM	Bureau International des Poids et Mesures
BLA	Best Linear Approximation
BW	Bandwidth
CF	Crest Factor
CGPM	Conférence Générale des Poids et Mesures
CIPM	Comité International des Poids et Mesures
CM	Common Mode
CODATA	Committee on Data for Science and Technology
DAC	Digital to Analog Converter
DC	Direct Coupling or Direct Current, depending on the specific context
DFT	Discrete Fourier Transform
DM	Differential Mode
DMM	Digital Multimeter
DNL	Differential Non-linearity
DSP	Digital Signal Processing
DTFT	Discrete-Time Fourier Transform
EFT	Electrical Fast Transient
EM	Electromagnetic
EMC	Electromagnetic Compatibility

EMI	Electromagnetic Interference
EUT	Equipment Under Test
EV	Errors in Variables
FFT	Fast Fourier Transform (algorithm implementation of the DFT)
FIR	Finite Impulse Response (digital filter)
FRF	Frequency Response Function
FSR	Full Scale Range
GUM	Guide to the Expression of Uncertainty in Measurement
I&M	Instrumentation and Measurements
IEC	International Electrotechnical Commission
IEV	International Electrotechnical Vocabulary
IF	Intermediate Frequency
IIR	Infinite Impulse Response (digital filter)
IL	Insertion Loss
INL	Integral Non-linearity
IV	Instrumental Variable
JCGM	Joint Committee for Guides in Metrology
LP	Local Polynomial
LS	Least Squares
LTI	Linear Time Invariant
MA	Moving Average (digital filter)
ML	Maximum Likelihood
NSA	Normalized Site Attenuation
pdf	Probability Density Function
PEM	Prediction Error Method
PRBS	Pseudo Random Binary Signal
PUMA	Procedure for Uncertainty Management
PWM	Pulse–Width Modulated
RF	Radio Frequency
RMS	Root Mean Square
SA	Site Attenuation
SFDR	Spurious-Free Dynamic Range
SHA	Sample and Hold Amplifier
SI	Système International d'unités
SINAD	Signal to Noise and Distortion Ratio
SML	Sample Maximum Likelihood
SVSWR	Site Voltage Standing Wave Ratio
THD	Total Harmonic Distortion
UFA	Uniform Field Area
VIM	International Vocabulary of Metrology
VSWR	Voltage Standing Wave Ratio
WCSL	Worst-Case Scalloping Loss
ZOH	Zero-Order Hold

PART I

FUNDAMENTALS

CHAPTER 1

MEASUREMENT MODELS AND UNCERTAINTY

ALESSANDRO FERRERO[1] and DARIO PETRI[2]
[1]Dipartimento di Elettronica, Informazione e Bioingegneria, Politecnico di Milano, Milano, Italy
[2]Dipartimento di Ingegneria Industriale, Università degli Studi di Trento, Povo, Italy

1.1 INTRODUCTION

Nowadays, we are surrounded by measuring instruments and we use them several times a day, very often unconsciously and unaware of their complexity and accuracy. To realize how instruments have become a big part of our life, just think of how many times we read the speed indicator when we drive our car, the fuel indicator to know when we have to refill the tank, or, when we eventually refill it, the meter on the fuel pump.

Interestingly enough, we usually don't pay too much attention to the accuracy of the instruments we use, even if we rely on their indication to make important decisions, such as, for instance, driving safely or paying the right amount of money for the quantity of good we purchase. Even more strangely, the only instrument we generally adjust to a reference is our watch, which is probably the most accurate instrument we use in our everyday life: even the cheapest ones don't lag or lead for more than one second a day, which means that their relative accuracy is in the range of $1 \cdot 10^{-5}$!

The above examples give us clear evidence that we do use and read instruments, but they still leave an important question open: are we also making a measurement? Trying to answer this question opens also another fundamental question: which is the difference between reading an instrument and making a measurement?

This chapter is aimed at providing an answer to this question.

Modern Measurements: Fundamentals and Applications, First Edition.
Edited by Alessandro Ferrero, Dario Petri, Paolo Carbone, and Marcantonio Catelani.

1.2 MEASUREMENT AND METROLOGY

To understand what measuring means, let's start from the definition of measurement, taken from the International Vocabulary of Metrology (VIM) [1].

> *Measurement*
>
> Process of experimentally obtaining one or more *quantity values* that can reasonably be attributed to a *quantity*.

So, a measurement process provides, as a part of the *measurement result*, one or more quantity values that can be attributed to a quantity intended to be measured, that is also called, always according to the VIM [1], *measurand.*

To fully understand this definition, we have to refer to the definition of quantity. We can find it again in the VIM.

> *Quantity*
>
> Property of a phenomenon, body, or substance, where the property has a magnitude that can be expressed as a number and a reference.

The VIM states that a reference can be a *measurement unit*, a *measurement procedure*, a *reference material*, or a combination of such.

When physical properties are considered, the reference is generally a measurement unit, whilst, when chemical measurement are considered, the reference is quite often a reference material.

The quantity values provided by the measurement are therefore a number and a reference together expressing the magnitude of a quantity [1].

Is this the measurement result? Or, better, can a measurement result be expressed only by a number and a reference? As we will see later in section 1.5 of this chapter, a measurement procedure cannot provide the *"true"* value of a measurand, due to a number of factors that we will thoroughly discuss later. This means that a measurement result can only provide a *finite amount of information* about the measurand, and we must know if that amount is enough for the intended use of the measurement result. Otherwise, the measurement result would be meaningless.

Therefore, any measurement result has to be provided with an attribute capable of quantifying how close to the measurand's value the obtained quantity value is. This attribute is called *uncertainty*, and the correct definition of measurement result, as provided by the VIM, is as follows.

> *Measurement result*
>
> Set of *quantity values* being attributed to a *measurand* together with any other available relevant information.

In a note to this general definition, the VIM states that:

> A measurement result is generally expressed as a single *measured quantity value* and a *measurement uncertainty*.

The above general definitions have introduced a number of concepts (*quantity value, reference, relevant information, uncertainty*), that will be covered in the next Sections, and show that a measurement is a definitely more complex procedure than simply reading an instrument.

The science that includes all *theoretical* and *practical* aspects of measurement, regardless to the measurement uncertainty and *field of application*, is called *metrology* [1]. Its definition, as provided by the VIM, is as follows.

> *Metrology*
>
> Science of measurement and its application

As many other sciences that have a deep impact on human life, metrology finds its roots back into the ancient times, and the evolution of the human needs had a significant impact on its development, and its present formulation. Therefore, before analyzing the fundamental concepts of nowadays metrology, let us find an answer to two basic questions: why do we measure? And for what do we use our measurement results?

A quick glance into the past will help us to find the answers to these questions.

1.3 MEASUREMENT ALONG THE CENTURIES

At the beginning of the human adventure, the measurement concept did not exist and the experimental activity was confined to direct observations. Instruments did not exist, and our senses were the only available tools to observe and somehow quantify the reality.

The first physical quantity that was barely measurable with our senses and probably started metrology was time. We can only imagine that this interest was related to the impact that the night and day cycle, the lunar period and the rotation of seasons have on human life and the human ability of predict them to optimize all activities aimed at providing food, from hunting to agriculture and cattle breeding.

As a matter of fact, the archeological findings from prehistory show with little doubt that the first conceived measuring instruments were aimed at the measurement of time, through the observation of the stars or the sun. Although the most famous among these instruments is the Stonhenge Circle, its "operating principle" is not yet clear. On the other hand, many remains have been found from a later period, across prehistory and history, that proved how time was measured through the observation of the displacement of the shadow of a vertical device (the *gnomon*) created by the sun on the ground [2].

However, at that time, this activity was more of a religious and prophetic kind, rather than aimed at increasing knowledge. We had to wait until the organized social structures appeared to meet awareness of knowledge and a systematic discussion about its meaning.

1.3.1 Measurement in Ancient Greece

There is no doubt that ancient Greece was the cradle of philosophy, and philosophy was considered the most important means of knowledge. We have also evidence, from documents and archeological findings, that instruments were used, mainly for length and capacity measurements. However, the great philosophers of that time did not consider instruments as tools to advance knowledge.

The motivation can be found in the dichotomy between the world of philosophical abstractions and the empirical world of observations, clearly expressed by Socrates in the fifth century BC.

In Socrates' philosophy, we can build our knowledge by defining abstract models that allow us to explain, through suitable logical steps, what happens to us. In this way, we can build our own world of philosophical abstractions.

On the other hand, there is a physical world around us, that shows up through a number of occurrences, facts and events, and that, at Socrates' time, we could perceive only through our senses. This is the so-called empirical world and, according to Socrates, the abstractions were aimed at explaining the true essence of the empirical world. In his philosophy, abstract logical constructions were the only way to knowledge, and observations had the simple role of triggering speculation. The use of observation results as a way to validate abstract models was not part of Socrates' approach, and was not part of the scientific method at least until Galileo's time. The subordinate role assigned to observations with respect to logics has lasted for so many centuries that even now the word "empiricism" implies a negative connotation.

1.3.2 Measurement in the Roman Empire

We have to wait until the Roman Empire to discover, from the many available historical documents and texts, that they had a rather modern approach to measurement. Indeed, we know that they had standards (at least for length, mass, and capacity), these standards were approved, kept and maintained by the central government, and secondary standards were disseminated in the provinces of the empire.

Evidence of the importance assigned by Romans to measurement is given by the fact that the primary standards were kept in the temple of Iuno Moneta.[1] Only valuable and important objects were kept in the house of a God. And only money-related objects were kept by the God of Money!

According to these facts, we expect to find several traces, in the Latin literature and in the archeological sites, of the use of their instruments and standards. For instance,

[1] Moneta, in Latin, means money. We cannot probably find better evidence that the economical importance of measurement was quite clear to Romans.

we know that the length standard was the Roman *pes* (Roman foot), and we would expect to find clear evidence of the use of the pes in the remains of imperial Rome and evidence in the dimensions of the many perfectly preserved buildings we can find in the regions of the world that were dominated by the Romans. The big surprise is that this is not true. So, when and for what was the unified length unit used?

The answer to this question comes from the wonderful archeological site of Pompei. Pompei was an important town of the Empire, located South of Rome, not far from the present Naples, close to the Vesuvius volcano. It was destroyed by an eruption in 79 AD. Actually, the explosive eruption killed every form of life in Pompei, but preserved everything under a layer of ash and light pumice stones. When it erupted, it hurled 100,000 tons/s of superheated stones and ash into the stratosphere, and formed a cloud with the classic shape of an umbrella pine tree. The first thing that hit the ground was a hot wind that was estimated to hit Pompei at about 350 km/h and at a temperature of more than 450°C. This wave of hot air killed and dried every form of life in seconds. Then ash came down, and light pumice stones, that buried everything without causing any major damage.

For this reason, many centuries later, archeologists found Pompei almost intact, and have learned so many things about everyday life at that time, and found remains that solve our questions about the use of the unified length standard [3]. Despite the dimensions of the buildings cannot be related to the pes, its use was clear by analyzing some artifacts.

In the Roman houses, water was stored in reservoirs located under the ground floor. All rooms had small wells for access to water, and the opening was protected by small marble curbs. The diameter of the curbs can be traced to the pes quite perfectly. Similarly, the dimensions and the distance between the holes in the iron lockers can be traced to pes submultiples. What is the reason?

Marble artifacts were not produced locally; they came from a distant region. Iron locks were produced by a different craftsman than the one who made doors or coffers on which the locks were installed. The marble devices and the locks had to fit accurately with another part. Since they were not produced locally, their dimensions could not be measured with the same instrument, and therefore they had to be measured according to a unique standard. The Roman Empire was the first global market, and they needed to ensure that parts crafted in different regions of the Empire could be used in all other regions without creating major problems. A need that looks quite similar to our present industrial needs was solved, in principle, in a similar way as nowadays.

The roots of the modern industrial metrology can be found in the use of standards in the Roman Empire. It was not yet a science, but only an application-driven technology. It was not yet a way to knowledge, but it was the first step toward the development of a new science.

1.3.3 Measurement in the Renaissance Period

At the fall of the Roman Empire, the economy receded to a local, small scale, and therefore there was no need for universal standards. Many different systems of units flourished at that time, often according to the convenience of the local squire and the consequences of such a scientific disaster are still suffered in the present days.

Things changed dramatically during the Renaissance, when the modern concept of science began to be elaborated. The experimental activity had a big part in the development of this new concept, so that we can now state that the meaning we assign to the measurement activity today finds its true origin at Renaissance time.

Two scientists, above all, contributed to assert this new concept: Galileo Galilei and Evangelista Torricelli.

Galileo used the new telescope to peer into the sky and proved that we can use an instrument to extend our senses and discover things (planets and satellites, in his case) that are far beyond the range of our "human sensors".

Torricelli invented the barometer, and proved that the atmosphere has a weight, that is something that our senses cannot, once again, show us.

Galileo's observation of the fall of bodies and his observation of the motion of planets proved the flaws of the geocentrical theory and paved the way to the formulation of the general law of gravitation.

Socrates' paradigm received a first, important shock. Abstractions alone cannot build our base of knowledge and observations are not only the starting point of logical constructions. New observations and new experiments are needed to validate theories obtained as the result of a logical construction. Moreover, we cannot totally rely on our senses to observe the physical world: our senses are limited, and keep important phenomena hidden to us, thus preventing us from fully understand them or sometimes leading us to wrong conclusions. The modern science moved its first steps thanks to new instruments and new measurement procedures.

1.3.4 Measurement in the Modern Age

A couple of centuries later, another storm upset the traditional scientific belief. During the Enlightenment period, science was trusted as the only gate to knowledge, and the experimental activity was considered as the most important tool for investigating physical phenomena and understanding them.

In this new framework, in which science was, for the first time, considered as a global language for describing physical phenomena, a new issue arose. If science has to be a global language, it must also have a global, unique reference. Therefore, the measurement units must be universally recognized and invariant.

This is a twofold problem. On one side the system of units must be unique, that is universally accepted and adopted. As we saw in Section 1.3.2, this problem was already dealt with at the time of the Roman empire, where conventional standards were defined to solve practical and economical problems. However, the common belief was that scientific problems cannot be solved on the base of conventions. This is the second part of the problem: the unique system of units must also be derived by physical properties of phenomena that can be considered as invariant in space and time.

This was indeed a revolution both in the scientific and everyday habits of that time. Even though the scientific community reached a wide agreement on this point, putting it into practice required another big revolution: the French Revolution. Indeed, these ideas gained enough political consensus to have the local measurement units banned and replaced by new ones, that were defined as global as possible, according to the scientific knowledge of the eighteenth century.

It is worth reminding that a huge effort was put in place to measure the length of the Earth's meridian, on which the meter definition could be based.

It was another big step in the long process that turned observations into measurements. However, standards defined on properties of our planet, such as the unit of length and the unit of time, were not global and invariant enough. Moreover, a new science was moving its first steps: electromagnetism. It was soon clear that the result of the interactions of electromagnetic quantities was a force. New measurement units had to be considered to serve as a reference for the measurement of electromagnetic quantities, and they had to be coordinated to the already existing ones, so that the physical relationships between the different electromagnetic and mechanical quantities could be reflected into the same relationship between units.

Both problems were clear to James C. Maxwell, the father of electromagnetism, who, in 1871, wrote:

> If we wish to obtain standards of length, time and mass which shall be absolutely permanent, we must seek them not in the dimensions, or the motion, or the mass of our planet, but in the wavelength, the period of vibration, and the absolute mass of these imperishable and unalterable and perfectly similar molecules.

This is the conceptual basis of our present International System of Units, or SI, as it will be shown in the next chapter. It took almost another century, and the work of another great electrical engineer, Giovanni Giorgi, to see these outstanding ideas put into practice.

1.3.5 Measurement Today

The long journey that changed the simple activity of observing the empirical world into the measurement science is now accomplished. The experimental activity, of which measurements are the most important part, is now considered an important, essential tool for the advancement of knowledge. Today, any scientific or technical proposal needs to be supported by suitable measurement results.

We have now fully understood what Lord Kelvin stated in his famous speech of May 6, 1886:

> I often say that when you can measure what you are speaking about, and can express it in numbers, you know something about it; but when you cannot express it in numbers your knowledge about it is of meager and unsatisfactory kind; it may be the beginning of knowledge, but you have scarcely, in your thoughts, advanced to the stage of science, whatever the matter may be. So, therefore, if science is measurement, then without metrology there can be no science.

Nowadays, measurement results are also goods that are sold and bought. In a modern, developed country, the economical value of measurements has been estimated to be the 6% of the Gross Domestic Product. Electrical measurements in industry represent the 4% of sales, and biomedical analysis represent the 5% of the total cost of the health system.

Measurements are essential to determine the quantity of goods in economical transactions, to evaluate cost of goods and services, to control quality of goods and, in general, every other important issue involved in transactions. This application field of measurements is generally known as *legal metrology*.

When technical issues are involved, measurements are employed, for instance, in inspection tests—on materials, components or on the final products—in comparing product characteristics or performances and, in general, in every step of an industrial process. This application field of measurements is generally known as *industrial metrology*.

Last, but not least, measurements are the key point in the experimental validation of theories and in studying and implementing new and more accurate measurement methods and standards. This application field of measurements is generally known as *scientific metrology*.[2]

1.4 MEASUREMENT MODEL

The concepts that we have briefly recalled in the previous sections suggest that measurement is a complex activity, far more complex than connecting and reading an instrument. Developing a model for the measurement activity is therefore necessary, to fully understand its meaning and which are the requirements to perform a measurement correctly.

Socrates' model of knowledge, recalled in section 1.3.1, is a good starting point to build the measurement model. According to this approach, our knowledge is built around abstract models that explain phenomena through suitable logical steps. This approach is still valid today: we have models to explain the different phenomena (not only the physical ones), and these models are based on suitable mathematical equations, or logical inferences, in which we assign an abstract symbol to each considered quantity. In this respect, models represent the symbolical world.

On the other hand, as already noted by Socrates, we become aware of the different phenomena because they reveal themselves through occurrences, facts and events that define the empirical world.

The main questions are: how can we validate our abstract model? How can we ensure that it represents the phenomena it models correctly? If we consider that a model is supposed to provide the values taken by the output symbol (or symbols) for a given set of values assigned to the input quantities, we can conclude that a way to validate the model is to observe the phenomenon it represents, measure the quantities

[2]This definition might appear somehow tautological, since metrolgy means "science of measurement," and therefore a science is scientific by itself. However, this definition is generally used whenever we want to refer to the applications of metrology in support of other branches of science.

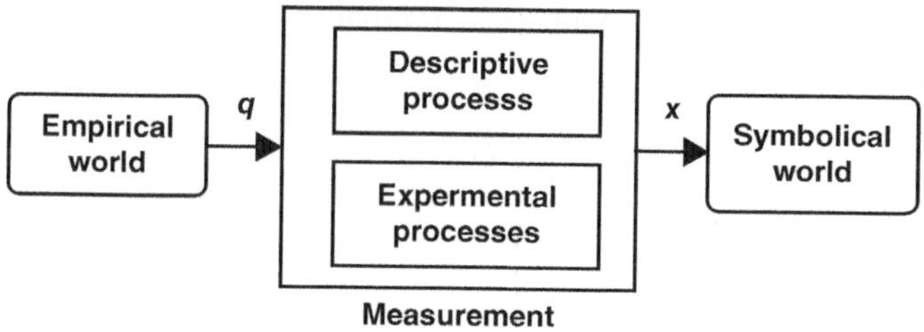

FIGURE 1.1 Model representing measurement as a bridge between the empirical and symbolical worlds.

that are supposed to affect this phenomenon, measure the quantity that represent the phenomenon itself, and verify whether the measurement result, for this quantity, corresponds to the value provided by the model.

For instance, if we want to validate the Ohm's law, we measure the resistance R of a resistor, we supply it with different voltage values V_k and check whether the corresponding measured values of current I_k flowing through the resistor are such that: $V_k = R \cdot I_k$, for all considered k measurements.

Measurement can be seen, therefore, as the bridge between the empirical world and the symbolical world. The measurement role is that of adding quantitative information to symbols.

1.4.1 A First Measurement Model

Defining measurement as a bridge between the empirical and symbolical worlds is a starting point to build a useful measurement model. So let us try to refine this idea, using the graphical representation shown in Figure 1.1.

According to this picture, the measurement process maps a single *manifestation* q of a *property* (for instance weight, length, ...) of an object (that is a phenomenon, a body, or a substance) belonging to the empirical world onto a symbol (x in Figure 1.1) that belongs to the symbolical world. In fact, measurement is an objective empirical operation that assigns a symbol x (called the *measured value*) to the manifestation q (called the *measurand*) of the considered property of an object [4, 8, 9]. If we consider again the above example where the current I flowing through a resistor was measured, the considered property of the resistor is the current and, if the measurement process returns "3.5 A", we assign this symbol, whose meaning is assumed to be known, to the current.

The above relationship between an empirical property and a symbol can be represented using a mathematical formalism. Consider a generic property and let $q_1, q_2, \ldots, q_n, \ldots$ represent single manifestations of that property, so that we can define a set of all possible manifestations as

$$Q = \{q_1, q_2, \ldots, q_n, \ldots\}. \tag{1.1}$$

Let X represent a class of symbols x. Formally, measurement is an objective empirical operation:

$$\mu : Q \rightarrow X, \tag{1.2}$$

or

$$x = \mu(q), \tag{1.3}$$

where q represents a generic manifestation of the considered property. It is worth noticing that μ is not a one-to-one mapping, since it maps different but indistinguishable property manifestations to the same symbol.

To better describe this mapping, we need to look more in details into the measurement block in Figure 1.1 and identify the internal processes that contribute to the generation of the measured value. Usually, the first process that comes to mind is the *experimental process*. The above example on the validation of the Ohm's law describes, actually, an experimental process. However, the experimental processes cannot alone realize the mapping structure (1.2). They can only put into practice the *descriptive processes* that allow one to identify and implement (1.2).

So, let us analyze the most important processes that must be considered.

1.4.1.1 *Measured Property Definition* The first required step is to properly define the manifestation of the property q that we intend to measure. This requires a model for the object that manifests the measurand.

For instance, let us assume that we want to measure the width of a plate which is almost rectangular. This implies that we are describing the plate itself as a rectangle, that is its width is assumed to be the same all along the other side, and hence leads us to measure a single value on a single position, and assign this value to its "width".

If we are not sure that the plate is almost rectangular, we can model it, for instance, with a trapezoid, and measure two width values, on the opposite edges of the plate. In this case, we are assuming that its "width" varies linearly among the two obtained values.

The above examples show that the definition of the manifestation of the property that we intend to measure has a direct impact on the experimental process, and hence cannot be disregarded. We will see later, in section 1.5.1, that this point has a significant impact also on the quantity of information that the measurement process provides about the measurand.

1.4.1.2 *Mapping Definition* Once the manifestation of the property q that we intend to measure has been suitably defined, the mapping relation μ in (1.2) has to be defined in such a way that all possible empirical relations between measurands can be mapped into abstract relations between symbols.

To understand this point, let us consider the class of the height of a set of people, and some empirical relations that can be established for property "height". We may have:

- *Unary relations*
  For instance: Charlie is *tall*, Snoopy is *not tall.*
- *Binary relations*
  For instance: Charlie is *taller* than Linus, Charlie is *much taller* than Snoopy.
- *Ternary relations*
  For instance: Snoopy is *taller* than Charlie *if* sitting on Linus' shoulders.

Mapping μ must preserve these relations on the returned symbols. This means that, for instance, the empirical relation "Charlie is taller than Linus" must be reflected into relation: μ(Charlie) > μ(Linus). Therefore, a measurement procedure that provides the following quantity values: μ(Charlie) = 1.55 m and μ(Linus) = 1.40 m is correct, while a measurement procedure that provides the following values: μ(Charlie) = 1.35 m and μ(Linus) = 1.43 m is not correct, because it does not preserve the empirical relation set between properties.

The above concept can be expressed using a mathematical formalism. Let us assume that a set $\mathcal{R}_Q$ of empirical relations $R_{Q1}, R_{Q2}, \ldots, R_{Qn}, \ldots$ exists on the set Q of all possible manifestations of the considered property. Let us express this as

$$\mathcal{R}_Q = \{R_{Q1}, R_{Q2}, \ldots, R_{Qn}, \ldots\}$$

and let

$$\mathcal{R}_X = \{R_{X1}, R_{X2}, \ldots, R_{Xn}, \ldots\}$$

be a set of relations defined on the set of symbols X.

Then, the property is formally represented by an *empirical relational system* [4]

$$\mathcal{Q} = \langle Q, \mathcal{R}_Q \rangle,$$

which is mapped by the measurement procedure onto the *numerical relational system* [4]

$$\mathcal{X} = \langle X, \mathcal{R}_X \rangle.$$

The so called *representation condition* requires measurement to be a correspondence between property manifestations and symbols in such a way that the relations between the property manifestations imply and are implied by the relations between their images in the symbol set. Thus, $\mathcal{R}_Q \rightarrow \mathcal{R}_X$ must be a one-to-one mapping.

When the above constraint is satisfied, mapping

$$\mu : \mathcal{Q} = \langle Q, \mathcal{R}_Q \rangle \rightarrow \mathcal{X} = \langle X, \mathcal{R}_X \rangle \tag{1.4}$$

is called *homomorphism*.

1.4.1.3 Scale Definition The representation condition defined in the previous section 1.4.1.2 is a required feature of mapping μ, but does not yet explain how a numerical value quantifying property manifestation q can be assigned to symbol x. To do so, a suitable *measurement scale* should be defined, through the following steps.

1. A set $\Omega_S = \{\omega_{S_n}\}$ of *standard objects* is chosen in such a way that it provides a set $Q_S = \{q_{S_n}\}$ of manifestations of the same property as the measurand.
2. A relation $\mu_S : \Omega_S \rightarrow Q_S$ is defined between the set of standard manifestations of the property and symbols, so that a set $X_S = \{x_{S_n}\}$ of symbols is obtained, corresponding to each standard manifestation, and the representation condition is satisfied.
3. The measurand is compared with the set of standard manifestations and a single manifestation is selected from set $Q_S = \{q_{S_n}\}$, as the result of the comparison.
4. Relation μ_S is applied to the selected standard manifestation and the associated symbol is provided as the measured value.

The above process can be formally described by the following set of relations.

$$\begin{array}{ccc} q & \stackrel{\mu}{\rightarrow} & x \\ \downarrow & & \uparrow \\ \{q_{S_n}\} & \stackrel{\mu_S}{\rightarrow} & \{x_{S_n}\} \end{array}$$

There are different possible ways to define a measurement scale, and their properties may limit the analysis that can be performed on the measurement results. Five different kinds of scales are generally employed, ordered according to an increasing level of richness in the information they convey. The ordering is such that all operations allowed at a given level are allowed also at the upper levels.

- *Nominal scales*
  When this kind of scale is adopted, each measured object is placed in a *class*, according to the value manifested by the considered property. Symbols are only *class tags* and the classes are not ordered.

  The only admissible operation on these scales is a one-to-one mapping, and the only statistical inference is the mode.[3]

[3]Value of the most commonly occurring item.

When this kind of scale is adopted, the quantity of information provided by the measured value (see Section 2) is associated to the probability of correctness of the assignment of the measurand to one of the possible classes.

- *Ordinal scales*
  When this kind of scale is adopted, each measured object is placed in an *ordered class*, always according to the values of the considered property. The classes are ordered with respect to the property. Properties for which an ordinal scale (or an upper level scale) is defined are called *quantities*, and the related symbols are called *quantity values*.

  An example of ordered scale is the *Mohs* scale of mineral hardness, and the ordering is based on the ability of a mineral to scratch a softer one.

  Increasing mappings (that preserve the order) are additional allowed operations on this scale, and median[4] and percentiles are additional allowed statistical inferences.
- *Interval scales*
  This kind of scale is an ordinal scale that conveys additional information about the *size* of the interval between *adjacent classes*.

  Examples of interval scales are the Celsius and Fahrenheit temperature scales, and the Richter seismic magnitude scale.

  Sums and subtractions are the only allowed mathematical operations. Affine mappings between scales that preserve the order and interval proportions[5] are also allowed. An example of affine mapping is the well-known transformation of a temperature value ϑ_C given on a Celsius scale into a temperature ϑ_F given on a Fahrenheit scale: $\vartheta_F = 1.8 \cdot \vartheta_C + 32$.

  Mean and standard deviation are allowed statistical inferences for these scales.
- *Ratio scales*
  This kind of scale is an interval scale to which a *zero* (or *absolute*) element is added, representing the lack of the considered property.

  This scale is the most useful and commonly used scale in physical sciences, since it preserves ordering, size of intervals, and ratio between properties.

  All arithmetic operations are allowed on this scale, and the proportional mapping ($y = a \cdot x, a > 0$) is also allowed.

  The geometrical mean and the coefficient of variation are also allowed statistical inferences.
- *Absolute scale*
  This kind of scale is used to count elements. In this case there is only one possible mapping: the element count. Therefore, the only meaningful transformation is *identity*.

Unless differently stated, in the following we will refer only to measurement ratio scales.

[4] Value of the middle ranked item.

[5] An affine mapping that preserves the order is mathematically expressed as: $y = a \cdot x + b, a > 0$.

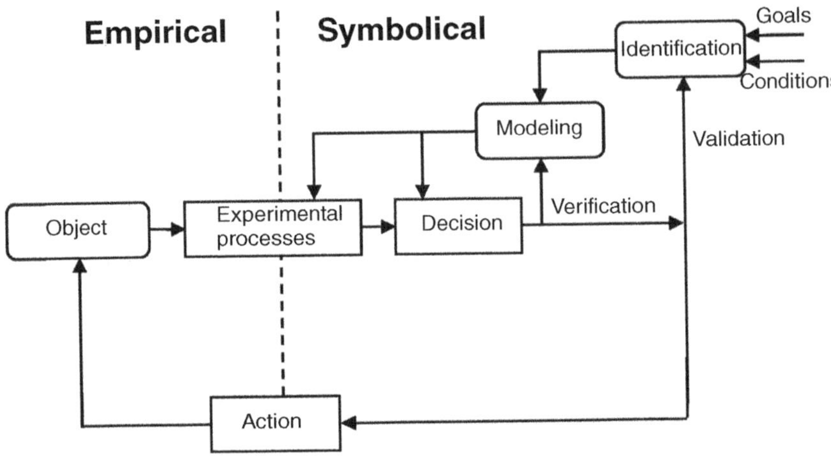

FIGURE 1.2 A measurement model including different kinds of involved descriptive processes.

1.4.2 A More Complex Measurement Model

The model presented in the previous section 1.4.1 describes how measurement activities accomplish the bridge function between the empirical and symbolical worlds by performing both descriptive and experimental processes. Indeed, the former processes are required to ensure a correct implementation of experimental processes and the interpretation of the obtained results.

Even though the model in Figure 1.1 is actually the core of the measurement activity, a more detailed model can be drawn, in which descriptive processes and feedback paths are identified, as shown in Figure 1.2.

The first point that must be considered is that, regardless of the field of application, usually measurement is not a self-motivating process, but is rather a goal-oriented activity. Indeed, the measurement results are quite always employed as relevant input elements in a *decision-making process* aimed at identifying the best *actions* needed to achieve established *goals* while satisfying given *conditions*.

For instance, the speed value provided by the speed indicator in a car is employed to decide whether the car speed has to be decreased or increased (the action) in order to reach destination in a given time, while driving safely (the goals) according to the weather conditions and the speed limits (the conditions).

Obviously, this model still represents measurement as a bridge between the empirical and symbolical worlds, as discussed in the previous section. Moreover, it can be readily verified that the model in Figure 1.2 refines the one in Figure 1.1, by explicitly showing the different descriptive processes that have to be performed to achieve the measurement goals. Let's analyze them.

Identification Usually, the result of any measurement activity does not depend only on the measurand, but is affected also by a number of other properties belonging to the empirical environment in which the measurement is performed. Therefore, the identification of the measurement context, that is the identification of all properties affecting the measurement result and the possible interactions among them, is the first step to be performed.

Four are the critical points in context identification.

- Measurand identification
  At first, it is necessary to identify which property (the measurand) of the object we are analyzing is the best suited one to achieve the goals, under the given conditions.[6] This means that we have to identify which property conveys the most relevant information needed to accomplish the goal.
- Measurement system identification
  Once the measurand has been identified, we need to identify a suitable measurement system. The selected measurement system should be the one that is expected to be the best one to achieve the goals, under the given conditions.
- Empirical environment identification
  During the measurement activity, measurand, measurement system and environment interact. For instance, a temperature probe changes the heat transfer of a body as well as its thermal capacity when it contacts it. Similarly, an electronic instrument dissipates some power in the environment, thus contributing to its temperature variations.

  These interactions manifest themselves through the effect that *variations* in one or more *environment properties* (i.e. temperature, pressure, humidity) and *time* have on the measurement result.

  It is therefore necessary to identify which environment properties, called *influence properties* (or *influence quantities*), affect the measurement result in a significant way,[7] so that they can be (and should be) controlled during the measurement process.

  Moreover, all significant mutual interactions between measurand, measurement system and environment must be identified, in order to achieve an accurate description of the whole measurement process.

  Usually, only a limited number of environmental properties may influence the measurand and the measurement system significantly. It is also worth noting that influence properties can be controlled and kept constant within a given tolerance, while time cannot be controlled or kept constant.

Modeling The identification of the aforementioned items is useless unless a mathematical model is defined that describes the measurement context accurately, that is the behavior of each identified interaction among the measurand, the measurement

[6]In order to achieve the goals, the identification of more than only one property might be required. However, in the following, we are considering the measurement of only one property for the sake of simplicity. All conclusions can be easily extended to the case of the measurement of multiple properties.

[7]The meaning of "significant" will be fully understood later, in section 1.5, when measurement uncertainty will be covered. Here, let us only anticipate that the measurement result is affected in a significant way if the uncertainty contribution associated with the considered influence property is non-negligible with respect to all other uncertainty contributions.

system and the environment. In that model, the identified properties are represented by *mathematical variables*, and the mutual interactions are represented by *mathematical relationships*.

In general, these variables and relationships are assumed to be deterministic. However, non-deterministic approaches are employed, such as probabilistic, fuzzy, neural, and neuro-fuzzy models, when either the measurand or the measurement systems, or both, are too complex to be described by a deterministic model in a satisfactory way.

The output of the modeling step is probably the most critical element in the whole measurement process, since it provides the symbolical description of the measurement process itself. All further steps are based on this model, that is therefore affecting the *quality*[8] of the measurement result.

As a matter of fact, this model provides always an *incomplete knowledge* of the measurement context, mainly because of the incomplete knowledge of the measurand, the fact that quantities and interactions that have, singularly taken, negligible effect have not been considered in the model, and the intrinsic approximation of the mathematical representation of variables and relationships.

This incomplete knowledge of the measurement context is reflected in the amount of information that the measurement result provides about the measurand. Due to all approximations performed in the description of the context, only a *limited* amount of information can be obtained by the measurement process, regardless of the accuracy of the employed instrumentation.

It can be stated that the limited amount of information provided by the measurement activity is useful in the subsequent decision step if and only if we can quantify the amount of missing information. This part of the modeling process is called *uncertainty expression and evaluation*. Although this is logically a part of the descriptive processes, it is such an important part of modern metrology that it will be covered as a separate point in Sections 1.5 and 1.6. It is nevertheless important to keep in mind that the need for considering, expressing, and evaluating uncertainty in measurement originates in this step of the measurement process.

Experimental processes The *experimental processes* block of the measurement model depicted in Figure 1.2 puts into practice the model defined in the previous step, by performing the processes mentioned in Section 1.4.1.

While the descriptive processes are quite general, the experimental processes are specific for each measurement application, and depend on the given goals and conditions. As it will be mentioned in the next chapters, the modern measurement systems are largely based on the conversion of the input signals into the discrete domain and the subsequent digital signal processing (DSP) of the acquired data. Therefore they apparently allow the measurement model to be implemented in a rather straightforward way.

[8]Here we prefer to refer to the general concept of *quality*, since it gives an immediate perception that the measurement result is not *ideal*, or *perfect*, and therefore its quality can be good or not. The next sections 1.5 and 1.6 will show how this "quality" can be expressed and evaluated

The contribution of such measurement systems to uncertainty is therefore mainly related to the additional approximations introduced by a digital implementation and to imperfections of the field interface (usually sensors and transducers), the electronic front-end and the sampling and analog-to-digital converting devices.

Decision As previously mentioned, measurement results are quite always employed as input data in a decision-making process. Generally, this process involves the comparison of the measurement result with one or more threshold values, and a decision is taken according to where it is located, with respect to the given threshold values.

This process is only apparently simple, because, due to measurement uncertainty, the measurement result cannot be represented with a single value. As it will be shown in Section 1.6, any measurement result does always represent an interval of values, with a stated coverage probability [1]. We do not want to enter the mathematical details of a decision-making process in the presence of uncertain values, because this is beyond the scope of this book. We simply wish to draw the readers' attention on the fact that the involved comparison process must be carefully considered, and the final decision does always imply a risk that the decision itself is wrong. The risk of wrong decision depends on measurement uncertainty, and is generally higher, the higher measurement uncertainty is.

It is worth noting that the role of the decision process is not confined to actions that follow the measurement process, but has some interesting implications within the measurement process itself. As shown in Figure 1.2, the normal consequence of the decision-making process is an action on the measured object. This is also, in general, the final goal of the measurement process: data are provided to act on the measured object so that the given goals can be achieved.

Figure 1.2 shows also other connections of the decision block toward the modeling and identification blocks. These connections show a feedback path, inside the measurement model, so that the model can be verified and the whole measurement process can be validated.

If the whole measurement context has been correctly modeled and no significant environmental properties and significant interactions have been neglected, the quality of the final measurement result should meet the expectation. If this is not the case, the model has to be refined, until the expected result quality is met.

Once the model has been verified, it is possible to validate also the whole measurement process, by checking that all goals have been achieved, without violating any of the given conditions. This step allows us also to verify that the whole measurement context has been correctly identified. If the identification is correct, the measurement goals are achieved, otherwise also the identification process has to be refined.

1.4.3 Final Remarks

The different processes that take part in any measurement activity have been briefly reported and discussed in this section. The obtained model shows the steps required to implement a measurement activity that is correct from a methodological point of view, and not only from an experimental (or instrumental) perspective.

The most important conclusion that can be drawn is that any measurement result does provide quantitative information about the measurand, thus achieving the original task of measurement to be a bridge between the empirical and symbolical worlds, but the amount of achieved information is always limited and, generally, never complete.

From a philosophical and ontological perspective, limited information should not be considered an advancement of knowledge, since knowledge itself implies complete and absolute cognition of the considered phenomenon. However, measurement is the only available tool to "know something" of what we "are speaking about", according to Lord Kelvin's famous speech, at least as far as scientific issues are concerned.

Therefore, from a more practical, engineering point of view, a way must be found to quantify also the completeness (or incompleteness) of the information conveyed by a measurement result. If and only if this additional, quantitative information is added to a measurement result, the result itself can be advantageously used in any subsequent decision making process. Otherwise, it is only a meaningless number.

1.5 UNCERTAINTY IN MEASUREMENT

The last section has clarified which is the role of the *other available relevant information* that, according to the definition of measurement result given by the VIM [1], must be specified together with the provided quantity values. This kind of knowledge conveyed by the measurement result is aimed at quantifying how complete the information associated with the quantity values provided by the measurement process is.

The problem of qualifying the measurement result with a quantitative attribute capable of characterizing the completeness of the information provided about the measurand has been considered since the beginning of metrology, and has undergone a significant revision during the last two decades of the twentieth century, when the modern concept of *uncertainty* has been introduced and adopted.

Understanding this concept is not immediate. As noted by the Guide to the Expression of Uncertainty in Measurement (GUM) [5], the word *uncertainty* means doubt, and thus in its broadest sense *uncertainty of measurement* means doubt about the information provided by the result of a measurement. The first question that we have to answer to, if we wish to fully understand the uncertainty concept, is where this doubt originates and why. Additional considerations on the model presented in Section 1.4 will be helpful in finding an answer to this question.

However, the concept of doubt is hardly quantifiable. Therefore, the use of the term *uncertainty* to quantify how much we should doubt about the obtained measurement result seems somehow contradictory and tautological, since we apparently try to quantify a doubt with a doubt!

Fortunately, this contradiction is only apparent and is due to a lexical problem. Because of the lack of different words for the *general concept* of uncertainty and the specific quantities that, as we will see in Section 1.6, provide *quantitative measure* of the concept, the same word "uncertainty" is used, in metrology, in these two different senses [5].

To avoid confusion, Section 1.5 is aimed at introducing and discussing the general concept, while Section 1.6 will consider the quantitative measure of the concept.

1.5.1 The Origin of the Doubt

The measurement activity, as noted in Section 1.4.2, is always framed in a context, and this context is generally complex. To identify the best actions required to achieve the measurement goals, according to the given context, a model is needed, as reported in Section 1.4.2.

It is well known that any model, even the best one, provides only an approximated description of the entity it is aimed at representing. The model of the measurement context makes no exception to this general rule, and therefore the origin of the doubt, that is, the sources of the different possible contributions to uncertainty, has to be looked for in this model.

According to the different activities modeled by the different blocks in the diagram of Figure 1.2, the following contributions to uncertainty can be defined.

Definitional uncertainty The *identification* and *modeling* blocks of Figure 1.2 have the tasks of identifying and modeling the measurand, the relevant environment properties, and their mutual interactions.

The resulting description of measurand and environment is imperfect. For instance, if we have to measure the external diameter of a pipe, we can model the pipe as a geometrical cylinder. This implies that we are assuming the diameter is the same all along the pipe, and leads to perform a single measurement in a randomly chosen point along the pipe. If the pipe surface is rough, the diameter is not the same along the pipe's length and therefore different measurement results can be obtained, depending on where the measurement is performed. This contribution to uncertainty is called *definitional*,[9] because it is originated by an imperfect definition of the measurand. The consequence of not considering it may lead to incorrect decisions. If, for instance, the diameter has to be used to size a hole into which the pipe has to go through, modeling the pipe as a cylinder makes it possible that the hole is too tight and the pipe cannot be inserted.

Likewise, an incorrect definition of the mutual interaction between measurand and environment may lead to similar problems. If, for instance, we have a metal pipe, and the interaction between temperature and metal thermal expansion is not considered, the same problem as the one mentioned above might be encountered.

Since the identification and modeling steps are the preliminary steps of any measurement activity, and affect all subsequent steps, definitional uncertainty represents the *lower bound* of measurement uncertainty. Its correct evaluation is therefore important, also from an economical point of view, because it makes no sense to reduce the

[9]The VIM definition states that the definitional uncertainty is the *component of measurement uncertainty resulting from the finite amount of detail in the definition of a measurand.*

other contributions to uncertainty at a much lower level than that of the definitional uncertainty. Indeed, the additional cost is not repaid by a lower uncertainty, since the dominant contribution is, in this case, the definitional one.

Interaction uncertainty In general, when a measurement system is connected to the measurand, it interacts with the measurand and this interaction may change the measurand value during the measurement process (loading effect). For instance, if a voltmeter is connected to a voltage source to measure its output value, and the input impedance of the voltmeter is of the same order of magnitude as the source internal impedance, the measured voltage will be significantly lower than the desired open-source value.

An imperfect or incomplete description of the interaction between measurand and measurement system originates a contribution to uncertainty that is called *interaction* uncertainty. Actually, the origin of this uncertainty contribution is still in the description and modeling steps, and therefore this contribution to uncertainty can be seen as a definitional uncertainty.[10] However, the interaction between measurand and measurement system is so important that it is generally useful to consider the resulting contribution to uncertainty separately from the other ones.

Instrumental uncertainty We call *measurement system* all instruments and procedures employed in the measurement activity. The imperfect behavior of this system and its interactions with the environment give rise to another uncertainty contribution, called *instrumental* uncertainty.[11]

It is worth noting that a significant part of this contribution can be originated outside the employed measurement devices and equipment. In Section 1.4.1 a number of descriptive processes have been discussed, as part of the measurement model. The practical implementation of the scale definition described in Section 1.4.1.3 is the most important of these processes, as far as instrumental uncertainty is concerned.

This process requires the definition of a set of standard objects, with which the measurand is compared during the measurement procedure. A symbol is associated with each standard object, so that the comparison procedure provides the symbol associated with the measurement result (as in Section 1.4.1.3). The implementation of this process requires the practical realization of the standard objects.[12] Thus, due to the unavoidable imperfections, the standard objects do always provide a value that is different from the one assigned to the associated symbols.

[10]For this reason, this contribution to uncertainty is not explicitly defined by the VIM.

[11]Instrumental uncertainty is defined by the VIM as the *component of measurement uncertainty arising from a measuring instrument or measuring system in use*.

[12]When physical quantities are measured, these standard objects are, in principle, the primary standards that realize the involved measurement unit. Since, in practice, primary standards cannot be used in every measurement process, local (or secondary) standards can be used, provided they are *traced* back to the primary ones. In this case, the instrumental uncertainty contains an additional contribution, called *calibration uncertainty*, that reflects the uncertainty of the calibration process employed to trace the local standards back to the primary ones.

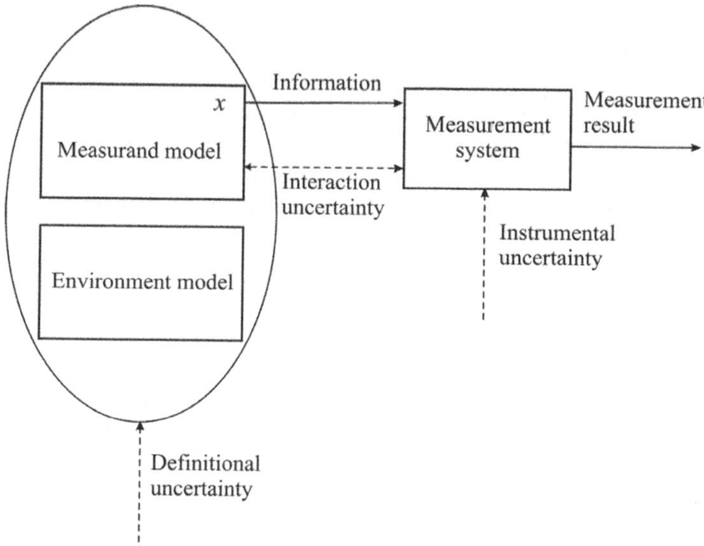

FIGURE 1.3 Different uncertainty sources affecting the information flow in the measurement procedure.

This is generally a lower bound to instrumental uncertainty. Indeed, taking into account that the calibration process is performed at reference environmental conditions, any deviation, in the actual operative conditions, from these reference conditions increases the instrumental uncertainty. In principle, if the mutual interactions between the instrument and environment are known, a correction can be applied, and the instrumental uncertainty can be reduced back to the calibration one. However, due to the intrinsic approximation of the mathematical model describing the instrument–environment interactions, an additional uncertainty contribution always arises.

The above considerations and the origin of the different uncertainty contributions are graphically sketched in Figure 1.3, where the information flow that originates the measurement result is also shown.

1.5.2 The Different Effects on the Measurement Result

In Section 1.5.1, the origin of the different uncertainty contributions has been analyzed and discussed. It can be synthetically concluded that the uncertainty contributions originate in some effects that cannot be properly taken into account by the considered mathematical and logical models.

It can be also intuitively concluded that the different effects generate uncertainty contributions that may affect the measurement result in different ways.

Of course, it is not possible to analyze, in this short survey chapter, all possible different effects. All of them, however, can be grouped into two main classes, each one characterized by a well-defined impact on the final measurement result: *systematic* effects and *random* effects.

Systematic effects Systematic effects are constant in time and reflect, generally, the effect of an influence quantity on the measurement result. Therefore, if the influence quantity does not change its value when repeating the measurement procedure, the resulting effect remains the same, and the measurement result does not change.

Systematic effects are therefore very dangerous, because their detection is very difficult. A way to detect them is to change the values of the influence quantities, or modify the measurement procedure, so that the systematic effect changes, manifesting itself in a variation of the measurement result. Another way is to compare the measurement results obtained by using two independent measuring systems, provided that one is known to give rise to negligible systematic effects with respect to the other.

Thus the systematic effect can be recognized and *corrected*.

Let us consider, for instance, a weighbridge, and let us suppose that the two beams have different lengths. This difference is directly reflected on the measurement result and a systematic effect arises, that cannot be corrected if the length difference is unknown, as it usually is. However, this effect can be compensated for, if a different measurement procedure is adopted. Let us call W_1 the obtained measurement result. If the measurement is repeated exchanging the position, on the two pans of the weighbridge, between the unknown weight (the measurand) and the standard weights, a different result W_2 is obtained. It is well known that the systematic effect due to the different beam lengths is compensated for if the value: $W = (W_1 + W_2)/2$ is taken as the measurement result.

Let us now consider a strain measurement performed by a resistive strain gauge. It is known that these devices are very sensitive to temperature. However, the resulting effect is fairly linear in a quite wide range of temperature variations. Therefore, if the same strain is measured at different temperatures, the law of variation can be easily obtained through a linear interpolation, thus enabling the correction of the measurement result when the environment temperature is known.

In general, the correction factors for the most significant influence quantities are provided by the instrument manufacturer or by the calibration certificates so that it is no longer necessary to repeat the measurement procedure under different values of the influence quantities. In this case, the systematic effect is identified and corrected by means of a priori information.

The correction of the systematic effects is such an important issue in measurement applications that the GUM [5], in its article 3.2.4, states: *It is assumed that the result of a measurement has been corrected for all recognized significant systematic effects and that every effort has been made to identify such effects.*

At last, it is important to consider that, due to the always present imperfections, the corrections can only reduce the impact of the systematic effects on the measurement result, but cannot cancel it totally. Therefore, a residual uncertainty component remains, due to incomplete knowledge of the systematic effects [5].

Random effects Random effects are due to unpredictable or stochastic temporal and spatial variations of influence quantities. They give rise to variations in repeated observations of the measurand.

Due to the unpredictable nature of these effects, they cannot be compensated. However, because of their stochastic nature, the statistical *expectation* of their effect on the measurement result is zero.

Therefore, their impact on the measurement result can be usually reduced by repeating the measurement procedure, under the same experimental conditions, a statistically significant number of times, and taking the average value of all measured values. Moreover, the statistical standard deviation of the measured values provides information about the measurement uncertainty, as it will be shown in Section 1.6.

1.5.3 The Final Effect

In general, a measurement procedure is affected by both systematic and random effects. Their origin and main characteristics have been discussed in Sections 1.5.1 and 1.5.2. We still have to analyze which is their ultimate effect on the measurement result.

To do this, let us suppose that the value of the measurand is represented by the center of a target, as shown in Figure 1.4. Let us now suppose to repeat the measurement procedure under the same experimental conditions. Each measured value can be graphically represented as a shot on the target.

According to the considerations in the previous sections, the random effects cause a dispersion of the measured values, similar to the effects, on real targets, of unpredictable wind gusts or unconscious movements of the shooter. If only random effects

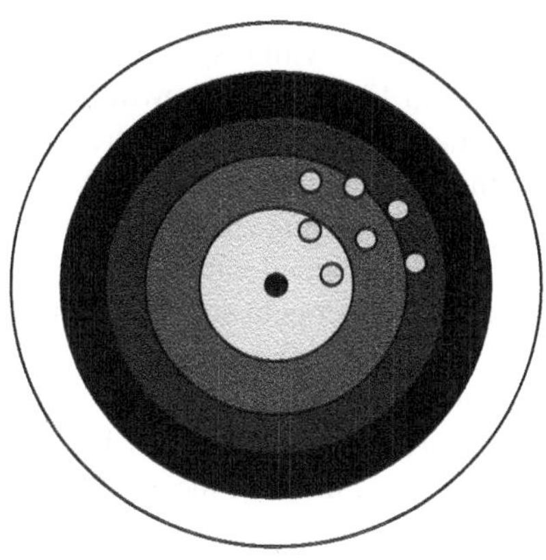

FIGURE 1.4 Graphical representation of the result of repeated measurements. The black center of the target represents the measurand value. Each small gray dot represents a single measured value. The dispersion of the measured values around their average is due to the random effects. The displacement of all values with respect to the center is due to the systematic effects.

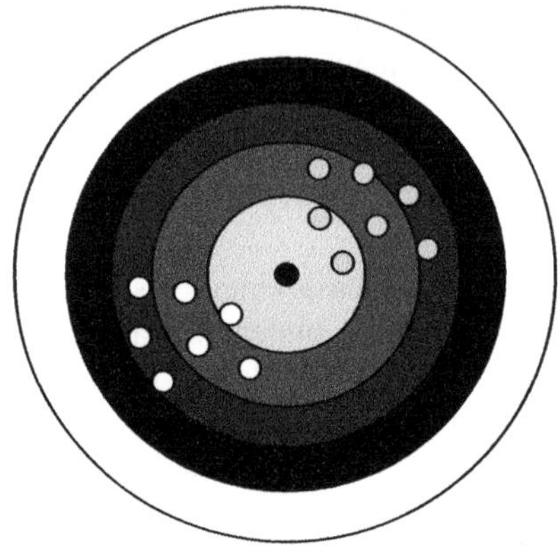

FIGURE 1.5 Graphical representation of the result of repeated measurements, for the same measurand as that in Figure 1.4, but under different measurement conditions. The new measured values are represented by the small white dots, while the gray dots are the same as in the case of Figure 1.4. The impact of the changed systematic effects is evident in the different position of the white dots with respect to the gray ones.

were present, all shots would be uniformly distributed around the center of the target. Conversely, systematic effects cause their average value (the center of gravity of the shot pattern) to be displaced with respect to the center. This is similar to the effect of an incorrect setting of the sight on the weapon used to shoot the target.

Let us now suppose to repeat the measurement procedure after a period of time, or in another place, or with different instruments, or under different environmental conditions. The results can be now graphically represented by Figure 1.5, where the previous results are shown as well.

The dispersion of the shot positions with respect to their center of gravity is quite similar to the one of the shots in Figure 1.4, showing that the random effects did not change. On the contrary, the center of gravity of the new shots is displaced, with respect to the previous one, thus showing that the systematic effects have changed, as expected, since we assumed that something changed in the measurement conditions. In our example, this effect can be explained with a different setting of the sight.

We can now draw a first significant conclusion. We cannot express a measurement result with a single value, because this is meaningless. This same value will be barely obtained by a new measurement.

Moreover, this single measurement value cannot be used in a comparison, and hence in a decision-making process. Since all measurement results are different, we would be forced to conclude that, for instance, the mass of an object measured in New York today is different from the mass measured in any other place, at the same latitude and height, tomorrow. There is no doubt that this is not acceptable.

However, if we observe again Figures 1.4 and 1.5, we understand that the measurement result is represented by the whole set of shots. Only the complete set is representative of the capabilities of the implemented measurement procedure, exactly as the whole set of shots is representative of the capabilities of a shooter. We will never judge the ability of a shooter from a single shot: it might be a very lucky shot of a terrible shooter!

It should now be clear that a measurement result is meaningful only if it expresses a distribution of values. If all significant systematic effects have been recognized and have been corrected for, we may expect that the measurand value falls within the distribution of values expressed by the measurement result, close to their average value.

It should also be clear, now, that measurement uncertainty should provide a quantitative measure of this dispersion.

1.6 UNCERTAINTY DEFINITION AND EVALUATION

According to the concepts presented in Sections 1.5 and 1.5.3, a measurement result cannot be expressed by a single measurement value. On the other hand, the measurand value should be expected to fall within the set of all possible measurement values provided by an experimental process. Also, it is possible to assume that these values distribute about the measurand value if all systematic effects are compensated for.

Therefore, we can now understand why, according to the definition given by the VIM [1], the result of a measurement cannot be expressed only by a single quantity value, but it requires also some additional information aimed at characterizing the distribution of values that could reasonably be attributed to the measurand.

Before discussing how this additional information can be expressed, it is important to understand the conditions it has to satisfy to be usefully employed in any practical situation. According to the GUM [5], the three following conditions must apply.

- The ideal method for evaluating and expressing the uncertainty of the result of a measurement should be *universal*. This means that the method should be applicable to all kinds of measurements and to all types of input data used in measurements.
- The actual quantity used to express uncertainty should be:
  - *Internally consistent*. This means that it should be directly derivable from the components that contribute to it, as well as independent of how these components are grouped and of the decomposition of the components into subcomponents.
  - *Transferable*. This means that it should be possible to use directly the uncertainty evaluated for one result as a component in evaluating the uncertainty of another measurement in which the first result is used.

The GUM [5] considers also an additional important requirement in many industrial and commercial applications, as well as in the areas of health and safety. In these areas, it is often necessary to provide an interval about the measurement result that may be expected to encompass a large fraction of the distribution of values that could reasonably be attributed to the quantity subject to measurement. Thus, the GUM states [5] that the ideal method for evaluating and expressing uncertainty in measurement should be capable of readily providing such an interval, in particular, one with a *coverage probability* or *level of confidence* that corresponds in a *realistic way* with that required.

Different approaches may be followed to satisfy the above requirements. Two of them have, until now, prevailed over all possible ones, and will be considered in this chapter.

The first one is the well known, traditional *error* approach, that has been followed until the theoretical considerations that brought to the first edition of the GUM, in 1995, proved that it was flawed from a strict ontological perspective. The second one is, of course, the one followed by the GUM in its current edition [5].

We are here considering both of them, including the old error approach, because its critical analysis is extremely useful to understand the more modern *uncertainty* approach followed by the GUM.

1.6.1 The Error Concept and Why it Should be Abandoned

The error approach is quite simple. It is based on the assumption that the true value x_t of measurand x can be somehow known. If a measured value x_m is returned by the measurement process, then the *measurement error* is defined as

$$\varepsilon = x_m - x_t \tag{1.5}$$

and the *relative measurement error* can be defined as

$$\varepsilon_r = \frac{x_m - x_t}{x_t}.$$

If the maximum error affecting the measurement result is denoted by ε_M, the interval

$$(x_m - \varepsilon_M, \; x_m + \varepsilon_M) \tag{1.6}$$

can be built about the measured value, within which the value that can be attributed to the measurand is supposed to lie with 100% confidence.

In many cases, a measurand y is not measured directly, but is determined from N other quantities $x_1, x_2, \ldots, x_N$ through a functional relationship

$$y = f(x_1, x_2, \ldots, x_N). \tag{1.7}$$

The expansion of (1.7) about the true values $x_{1_t}, x_{2_t}, \ldots, x_{N_t}$ in a first order Taylor series yields

$$\varepsilon_y = \frac{\partial f}{\partial x_1}\varepsilon_{x_1} + \frac{\partial f}{\partial x_2}\varepsilon_{x_2} + \cdots + \frac{\partial f}{\partial x_N}\varepsilon_{x_N}, \tag{1.8}$$

where ε_{y} is the measurement error affecting y, under the assumption that the higher order terms of the Taylor series expansion are negligible for small values of the measurement errors $\varepsilon_{\mathrm{x_i}} = x_{\mathrm{m_i}} - x_{\mathrm{t_i}}, i = 1, 2, \ldots, N$.

It can be readily checked that the error concept satisfies the three requirements given at the beginning of Section 1.6. In particular, (1.8) ensures that errors are transferable, and (1.6) ensures that the error can be employed to define an interval, about the measurement result that may be expected to encompass a large fraction of the distribution of values that could reasonably be attributed to the measurand.

The major flaw of the error approach lies in the fundamental assumption done to define the error itself in (1.5). In fact, this equation implies the knowledge of the measurand true value. However, all considerations in section 1.5 give clear evidence that no instrument and no measurement process can ever provide the true value of the measurand. Therefore, in practice, the true value of the measurand is always unknowable.

Moreover, the very basic concept of measurand can be lost when using a more refined model of the measured object. This is the case, for instance, with the concept of the length of an object at the atomic level, or with other physical quantities when considering the modern quantum physics.

The direct consequence of these statements is that the measurement error concept is meaningless also from a practical point of view, since the error itself cannot be evaluated, being its definition (1.5) based on an unknowable, and sometimes non-exactly definable quantity.

For this reason, the error concept has been abandoned, and replaced by the more modern concept of uncertainty.

1.6.2 Uncertainty Definition: The GUM Approach

As discussed in the previous section 1.6.1, the true value of the measurand is unknown and unknowable. Therefore, the task of *characterizing the distribution of values that could reasonably be attributed to the measurand* is made difficult by the fact that, not knowing the measurand, we cannot know whether the measured values belong to that distribution or not. This means that a more complex mathematical approach than the simple error concept is required to define and handle this problem.

In particular, the GUM [5] suggests to model the measured values as realizations of a random variable. Following this approach, the distribution of the measured variables is represented by a *probability density function* (pdf) and the related *standard deviation* provides a quantitative estimate of the dispersion of the possible measured values about their mean value, which is usually assumed to be the best estimate of the measurand value.

According to these considerations, the GUM considers the standard deviation of the pdf of the values that can be attributed to the measurand as a quantitative measure of uncertainty. This quantitative measure of uncertainty is called *standard uncertainty* and is defined as follows [5].

> *Standard uncertainty*
>
> Uncertainty of the result of a measurement expressed as a standard deviation.

Thus, standard uncertainty is defined as an estimate of the standard deviation of the distribution of values that can be reasonably assigned to the measurand.

The suggested notation for the standard uncertainty associated to a measured value x is $u(x)$.

Since we assumed to know the pdf of the distribution of measured values, a *coverage probability* (or *level of confidence*) can be assigned to an interval built about its mean value $\bar{x}$ as $[\bar{x} - u(x), \bar{x} + u(x)]$. Of course, the coverage probability depends on the pdf of the measured values. If, for instance, the pdf is normal, the coverage probability is the well known 68.27% value.

It might be then concluded that the additional GUM requirement mentioned at the beginning of Section 1.6 is satisfied, since standard uncertainty allows one to provide an interval [1] that encompasses a fraction of the distribution of values that could reasonably be attributed to the measurand with a known coverage probability.

However, the GUM requires that the coverage interval encompasses a *large* fraction of the distribution of values that could reasonably be attributed to the measurand, and that its coverage probability corresponds in a *realistic way* to that required.

The interval defined using standard uncertainty may not satisfy these requirements. Therefore, the GUM defines the *expanded uncertainty* as follows [5].

> *Expanded uncertainty*
>
> Quantity defining an interval about the result of a measurement that may be expected to encompass a large fraction of the distribution of values that could reasonably be attributed to the measurand.

The suggested notation for the expanded uncertainty associated to a measured value x is $U(x)$.

The GUM states that the fraction may be viewed as the coverage probability or level of confidence of the defined interval.

It is also extremely important to note that a specific level of confidence can be associated with the interval defined by the expanded uncertainty if and only if the pdf characterized by the measurement result and its standard uncertainty are known, or if *explicit or implicit assumptions* are made on the pdf itself.

Of course, the level of confidence that may be attributed to this interval can be known only to the extent to which such assumptions may be justified.

The expanded uncertainty can be obtained by multiplying the standard uncertainty by a suitable *coverage factor K*, defined by the GUM as follows [5].

> *Coverage factor*
>
> Numerical factor used as a multiplier of the standard uncertainty in order to obtain an expanded uncertainty.

TABLE 1.1 Value of the coverage factor K_p that produces an interval having level of confidence p when assuming a normal distribution

Level of confidence p (%)	Coverage factor K_p
68.27	1
90	1.645
95	1.960
95.45	2
99	2.576
99.73	3

If, for instance, a normal pdf is assumed for the distribution of values, the levels of confidence shown in Table 1.1 can be associated with the given coverage factors.

1.6.3 Evaluating Standard Uncertainty

The GUM identifies two methods for evaluating standard uncertainty, defined as follows [5].

Type A evaluation

Method of evaluation of uncertainty by the statistical analysis of a series of observations.

Type B evaluation

Method of evaluation of uncertainty by means other than the statistical analysis of series of observations.

1.6.3.1 Type A Evaluation Method of Uncertainty Under the assumption that the result of a measurement has been corrected for all recognized systematic effects (as done in article 3.2.4 of the GUM [5]), the distribution of the measured values can be characterized through the statistical analysis of repeated observations.

In most cases, the best available estimate of the *expectation* (or *expected value*) μ_X of a random variable X, for which n independent observations X_k have been obtained under the same measurement conditions, is the arithmetic mean of such observations, given by

$$\bar{X} = \frac{1}{n}\sum_{k=1}^{n} X_k. \tag{1.9}$$

Thus, the arithmetic mean $\bar{X}$ obtained from (1.9), can be taken as the numerical value of the measurement result.

The variability of the individual observations, or more specifically, their dispersion about their mean $\bar{X}$, can be characterized by the *standard deviation* σ of the underlying probability distribution of the random variable X, that is the positive square root of its variance σ^2.

An estimate of σ^2 is provided by the experimental variance of the observations, given by

$$s^2(X_k) = \frac{1}{n-1}\sum_{k=1}^{n}(X_k - \bar{X})^2. \tag{1.10}$$

It is well known that the variance of the mean value (1.9) is given by $\sigma^2(\bar{X}) = \sigma^2/n$. In many situations of practical interest, its best estimate is given by

$$s^2(\bar{X}) = \frac{s^2(X_k)}{n}. \tag{1.11}$$

The experimental standard deviation of the mean $s(\bar{X})$, equal to the positive square root of (1.11), quantifies how well $\bar{X}$ estimates the expectation μ_X of X, and may be used as a measure of the uncertainty of $\bar{X}$. It represents the standard uncertainty evaluated by a Type A evaluation method.

The major drawback of the Type A evaluation method is the need for independent repeated observations, whose number should be large enough to ensure that $\bar{X}$ provides a reliable estimate of the expectation μ_X of the random variable X. This condition cannot be always satisfied, mainly because of time constraints or difficulties in ensuring stable measurement conditions throughout the repeated observations.

Numerical example To show how standard uncertainty can be evaluated according to the Type A evaluation method, let us suppose to measure a dc voltage V with a five-digit multimeter, using the 10 V range. Let us also suppose that we repeat the measurement procedure several times, under the same measurement conditions.

In this example we consider only $n = 7$ repeated observations, for the sake of brevity. The obtained measured values are reported in Table 1.2.

The measured value is given by the arithmetic mean $\bar{V} = 7.2583$ V of the observations, and, according to (1.11), its standard uncertainty is: $u(\bar{V}) = 0.35$ mV.

TABLE 1.2 Measured voltage values for a Type A evaluation of standard uncertainty

Observation	Measured value [V]
1	7.2587
2	7.2595
3	7.2576
4	7.2568
5	7.2583
6	7.2592
7	7.2581

1.6.3.2 *Type B Evaluation Method of Uncertainty* In many practical situations, independent repeated observations of an input quantity X, as required by the Type A evaluation, are difficult to perform, for many different reasons, both technical and economical. For the same reasons, the systematic effects on the measurement result can be barely recognized and corrected. Nevertheless, it is still necessary to evaluate the standard uncertainty associated with the measured value. This can be done, as stated by the GUM [5], by scientific judgment based on all of the available information on the possible variability of X.

The pool of information may include [5]

- previous measurement data;
- experience with or general knowledge of the behavior and properties of relevant materials and instruments;
- manufacturer's specifications;
- data provided in calibration and other certificates;
- uncertainties assigned to reference data taken from handbooks.

In order to estimate a standard uncertainty value from the available information, this information should allow the operator to make reasonable assumptions on at least

- the shape of the probability density function that may be reasonably assumed to represent the variability of the considered input quantity X: normal, uniform, triangular, U-shaped, Weibull, Poisson, ...;
- an interval, about the measured value, that is expected to encompass a fraction of the distribution of values that could be reasonably attributed to the measurand with a given coverage probability.

If such assumptions can be made, then the provided interval defines the expanded uncertainty, and the coverage probability assigned to such interval, together with the assumed shape for the probability density function, allow one to compute the coverage factor and, consequently, the standard uncertainty.

Of course, the obtained uncertainty value is strongly dependent on the reliability of the available information, and the way it is used. It is worth noting [5] that the proper use of the pool of available information for a Type B evaluation of standard uncertainty calls for insight based on experience and general knowledge, and is a skill that can be learned with practice.

It is also worth noting that, when performed by an experienced operator, a Type B evaluation of standard uncertainty can be as reliable as a Type A evaluation, especially in a measurement situation where a Type A evaluation is based on a comparatively small number of statistically independent observations.

Numerical example To show how standard uncertainty can be evaluated according to the Type B evaluation method, let us consider again the example considered in Section 1.6.3.1 and let us suppose to measure a dc voltage V with a five-digit multimeter, using the 10 V range. In this case, only one measurement is performed, and let us suppose that the instrument reading is $V_r = 7.2587$ V.

Let us also suppose that the manufacturer accuracy specifications are available and that, for the considered range, they provide an interval of possible values, whose half-amplitude a is given by 0.02% of the reading + 6 digits.[13] Therefore, according to the reading and the selected range, it is:

$$a = 7.2587 \cdot 2 \cdot 10^{-4} + 6 \cdot 10^{-4} = 2.1 \cdot 10^{-3}\text{V}$$

According to the manufacturer, interval $V_r \pm a$ encompasses all values that can be reasonably attributed to the measurand, and its coverage probability is, therefore, 100%.

In order to evaluate the standard uncertainty, it is necessary to make proper assumptions about the way these values distribute. Since the manufacturer does not provide any additional information about the possible distribution of the measurement results, we can refer to the maximum entropy principle and assume that the probability distribution of the values that can be attributed to the measurand is uniform.

Under this assumption, the standard uncertainty is given by

$$u(V_r) = \frac{a}{\sqrt{3}} = 1.2 \cdot 10^{-3}\text{V}.$$

1.6.3.3 Combining Uncertainty Values Provided by Type A and Type B Evaluation Methods

In many practical situations, the two methods for standard uncertainty evaluation—Type A and Type B—quantify different contributions to uncertainty.

For instance, in the dc voltage measurement example considered in Sections 1.6.3.1 and 1.6.3.2, the Type B evaluation method, based on the manufacturer specifications, mainly considers the instrument sensitivity to differences between the values assumed by the influence factors during the actual measurement operations and instrument calibration respectively. Indeed, these differences remain usually constant when we repeat the measurement over a time interval of short duration and manifest themselves as systematic effects.

On the other hand, the Type A evaluation method is supposed to consider the effects, on the instrument reading, of variable noise originated inside the instrument or superimposed to the input signal, which manifests itself in terms of random effects during repeated observations.

Therefore, a better estimate of the standard uncertainty, in the given example, should consider both uncertainty contributions, $u_A(V)$ provided by the Type A evaluation method, and $u_B(V)$ provided by the Type B evaluation method.

[13] It is worth noting that this expression quantifies the effect, on the instrument reading, of variations in the influence factors (such as environmental temperature and instrument aging) with respect to calibration conditions. The coefficients in this expression depend on both the instrument sensitivity to the influence factors and the magnitude of their admissible variations.

Since the effects taken into account by the two evaluation methods are due to different physical phenomena, they can be assumed not correlated. Therefore, the two uncertainty contributions can be composed quadratically, and the final uncertainty value is given by

$$u(V) = \sqrt{u_A^2(V) + u_B^2(V)}. \tag{1.12}$$

In the case of the numerical example considered in Sections 1.6.3.1 and 1.6.3.2, the obtained standard uncertainty is: $u(V) = 1.2 \cdot 10^{-3}$ V.

It is worth noting that, in the considered example, the contribution $u_A(V)$ provided by the Type A evaluation method is negligible. This is usually the case in modern digital instrumentation when the input signal has a small amount of superimposed noise and the instrument is working correctly.

1.6.4 The Combined Standard Uncertainty

In most cases, a measurand Y is not measured directly, but is determined from N other quantities $X_1, X_2, \ldots, X_N$, through a functional relationship f:

$$Y = f(X_1, X_2, \ldots, X_N). \tag{1.13}$$

In general, the input quantities $X_1, X_2, \ldots, X_N$ may be themselves measurands, and may depend on other quantities. For the sake of clarity, and without losing in generality, these quantities will be considered, in the following, as measurands, but no further dependence on other quantities will be considered.[14]

An estimate y of measurand Y can be obtained by applying (1.13) to the estimates $x_1, x_2, \ldots, x_N$ for the input quantities $X_1, X_2, \ldots, X_N$. Thus, the measured value for Y is given by

$$y = f(x_1, x_2, \ldots, x_N). \tag{1.14}$$

Since the input estimates $x_1, x_2, \ldots, x_N$ in (1.14) are measurement results, each of them has an associated standard uncertainty $u(x_i)$, $i = 1, \ldots, n$, which is supposed to contribute to the standard uncertainty of the final measurement result y.

The problem that must be dealt with now is how to combine the standard uncertainties $u(x_i)$ in order to estimate a standard deviation associated to the measurement result y. This estimate is called *combined standard uncertainty* and is denoted by $u_c(y)$.

Let us suppose that function $f(\cdot)$ in (1.14) is fairly linear, about the measured value y, at least for small deviations of each input quantity X_i about their estimates x_i.

[14] In the case any of these quantities depends on other quantities, the method described in the following part of this section can be iteratively applied.

Under this assumption, function $f(\cdot)$ can be approximated by the first order terms of a Taylor series expansion about the expectations $E(X_i) = \mu_i$ of the input quantities, thus providing the following deviation about the expectation $E(Y) = \mu_y$ of the measurand

$$y - \mu_y = \sum_{i=1}^{N} \frac{\partial f}{\partial x_i}(x_i - \mu_i), \tag{1.15}$$

where $\mu_y = f(\mu_1, \mu_2, \ldots, \mu_N)$.

From a strict mathematical point of view, the derivatives in (1.15) should be evaluated in the expected values of the input quantities, which are generally unknown. Thus, in practice, they are approximated by the values obtained in the estimates $x_1, x_2, \ldots, x_N$ of the input quantities.

Moreover, if the magnitude of all derivatives in (1.15) is close to zero, or the non-linearity of $f(\cdot)$ is significant, higher-order terms must be considered in the Taylor series expansion. However, for the sake of simplicity, this case is not considered here.

Equation (1.15) can be squared, providing

$$(y - \mu_y)^2 = \left[\sum_{i=1}^{N} \frac{\partial f}{\partial x_i}(x_i - \mu_i)\right]^2, \tag{1.16}$$

which can be expanded as

$$(y - \mu_y)^2 = \sum_{i=1}^{N} \left(\frac{\partial f}{\partial x_i}\right)^2 (x_i - \mu_i)^2 + 2 \sum_{i=1}^{N-1} \sum_{j=i+1}^{N} \frac{\partial f}{\partial x_i}\frac{\partial f}{\partial x_j}(x_i - \mu_i)(x_j - \mu_j). \tag{1.17}$$

Equation (1.17) can be rewritten in terms of expectations, taking into account that

- $E[(y - \mu_y)^2] = \sigma_y^2$ is the variance of y;
- $E[(x_i - \mu_i)^2] = \sigma_i^2$ is the variance of x_i;
- $E[(x_i - \mu_i) \cdot (x_j - \mu_j)] = \sigma_{i,j} = \sigma_{j,i}$ is the covariance of x_i and x_j.

Taking also into account that standard uncertainty is a standard deviation by definition, we get

$$u_c^2(y) = \sum_{i=1}^{N} \left(\frac{\partial f}{\partial x_i}\right)^2 u^2(x_i) + 2 \sum_{i=1}^{N-1} \sum_{j=i+1}^{N} \frac{\partial f}{\partial x_i}\frac{\partial f}{\partial x_j} u(x_i, x_j), \tag{1.18}$$

where $u(x_i, x_j) = u(x_j, x_i)$ is the estimated covariance associated with x_i and x_j. The degree of correlation between x_i and x_j is characterized by the estimated *correlation coefficient*, defined as

$$r(x_i, x_j) = \frac{u(x_i, x_j)}{u(x_i)u(x_j)}, \tag{1.19}$$

where $r(x_i, x_j) = r(x_j, x_i)$, and $-1 \leq r(x_i, x_j) \leq +1$.

By replacing (1.19) into (1.18), we get

$$u_c^2(y) = \sum_{i=1}^{N} \left(\frac{\partial f}{\partial x_i}\right)^2 u^2(x_i) + 2 \sum_{i=1}^{N-1} \sum_{j=i+1}^{N} \frac{\partial f}{\partial x_i} \frac{\partial f}{\partial x_j} u(x_i)u(x_j)r(x_i, x_j) \tag{1.20}$$

that is also known as the general formulation of the *uncertainty propagation law* [5]. $u_c(y)$ is the *combined standard uncertainty* of y.

It can be immediately recognized that, if the estimates x_i and x_j are independent, we have $r(x_i, x_j) = 0$. If this holds for all input variables x_i, Equation (1.20) can be simplified as

$$u_c^2(y) = \sum_{i=1}^{N} \left(\frac{\partial f}{\partial x_i}\right)^2 u^2(x_i). \tag{1.21}$$

According to all the above mathematical derivations, Equations (1.20) and (1.21) allows one to evaluate the standard uncertainty of the result of a measurement, when the measurand is not measured directly, but is determined from a number of other quantities, and the standard uncertainties of their estimates are known.

It is worth noting that the uncertainty propagation law cannot provide any information about the distribution of probability associated with the measured value y, and hence the obtained standard deviation $u_c(y)$ cannot be used to identify intervals with given coverage probability about y.

To do so, we have to refer to the Central Limit Theorem. This theorem states that, if function $f(\cdot)$ in (1.13) is linear, none of the considered σ_i^2 variances dominates over the others, and $N \to \infty$, then the probability density function of Y exhibits a normal behavior, no matter on the shapes of the single probability density functions associated with X_i.

Of course, if N is a finite value, the probability density function of Y has only an approximated normal behavior, and the larger N, the better the approximation.

According to all above considerations, the following important conclusion can be drawn.

The uncertainty propagation law (1.20) (or its simplified version (1.21) in case of uncorrelated input quantities) can be used to evaluate the combined standard uncertainty of a measurement result when the measurand is determined from N other quantities, and the obtained combined standard uncertainty can be used to obtain an expanded uncertainty with given coverage probability if the following conditions are met.

1. Function $f(\cdot)$ in (1.14) is fairly linear about y, at least for small deviations of each input quantity X_i about its estimate x_i.
2. The partial derivatives of $f(\cdot)$ with respect to all input quantities exist and at least one of them is not nil.
3. The effect of none of the input variables prevails over the others.
4. The number N of considered input quantities is high enough (theoretically infinite, though, in practice, values of $N \geq 5$ are usually high enough).

If all above conditions are satisfied, the Central Limit Theorem can be applied and a normal probability distribution can be assigned to the estimate y of measurand Y.

The above conditions cannot be always fully satisfied. The application of the uncertainty propagation law may result in an incorrect estimate of the combined standard uncertainty and, more critically, in an incorrect estimate of the coverage probability assigned to the expanded uncertainty.

In order to overcome this problem, the BIPM issued, in 2008, a Supplement to the GUM [6] that recommends the use of Monte Carlo simulations to estimate the probability distribution of y starting from the probability distributions of the input quantities x_i whenever the uncertainty propagation law cannot be used.

Numerical examples Let us consider a dc power measurement, and let us suppose that we can measure voltage V and current I with two multimeters. Of course, the desired power value is obtained as: $P = V \cdot I$.

Let us also suppose that V_m is the measured value for voltage V and I_m is the measured value for current I, and that the only available information to evaluate standard uncertainty comes from the manufacturer specifications. Therefore, a Type B evaluation method has to be applied, as shown in the numerical example in Section 1.6.3.2.

Let us suppose that the readings and the associated standard uncertainty values (evaluated as shown in Section 1.6.3.2) are

- $V_m = 8.0125$ V (in the 10 V range), with $u(V) = 1.3 \cdot 10^{-3}$ V;
- $I_m = 50.105$ mA (in the 100 mA range), with $u(I) = 2.6 \cdot 10^{-5}$ A;

The measured power value is, therefore, $P_m = 0.4015$ W.

Let us now suppose that the two multimeters, used to measure voltage and current, are different and so their calibration histories. Under these assumptions, the two measured values V_m and I_m can be assumed to be uncorrelated, and Equation (1.21) can be used to evaluate the combined standard uncertainty.

It is

$$\frac{\partial P}{\partial V} = I = 50.105 \text{ mA}, \tag{1.22}$$

when evaluated for $I = I_m$, and

$$\frac{\partial P}{\partial I} = V = 8.0125 \text{ V}, \tag{1.23}$$

when evaluated for $V = V_m$.

By replacing the above evaluated standard uncertainties $u(V)$ and $u(I)$, and the values provided by (1.22) and (1.23) into (1.21), we get: $u_c(P) = 2.2 \cdot 10^{-4}$ W.

Let us now suppose to employ a single, two-channel multimeter to measure both V and I. In this case, a complete and positive correlation can be assumed between the measured values V_m and I_m, and hence equation (1.20) with $r(V, I) = 1$ shall be used to evaluate the combined standard uncertainty.

Considering, for the sake of simplicity, the same numerical values as above, we get: $u_c(P) = 2.7 \cdot 10^{-4}$ W. Therefore, in this example, using two independent instruments decreases uncertainty and, therefore, improves the *quality* of the measurement result.

However, this is not a general conclusion. To prove this, let us suppose to measure the resistance of the load connected to the measurement section considered for power measurement. The desired resistance value is: $R = V/I$.

Thus, the instrument readings V_m and I_m and the standard uncertainties $u(V)$ and $u(I)$ are the same as above. The measured resistance value is: $R_m = 159.91\ \Omega$.

The partial derivatives become

$$\frac{\partial R}{\partial V} = \frac{1}{I} = 19.96 \text{ A}^{-1}, \tag{1.24}$$

when evaluated for $I = I_m$, and

$$\frac{\partial R}{\partial I} = \frac{-V}{I^2} = -3191.58\ \Omega\text{A}^{-1}, \tag{1.25}$$

when evaluated for $V = V_m$ and $I = I_m$.

If we use two different, independently calibrated multimeters for voltage and current measurements, then applying (1.21) with the above determined values returns $u_c(R) = 86$ mΩ.

If, on the other hand, we use the same multimeter for both voltage and current measurements, then using (1.20) and the above determined values we have $u_c(R) =$ 58 mΩ. In this case, due to the different relationship between the measurand and the input quantities, a positive correlation produces an uncertainty reduction.

1.7 CONCLUSIONS

This chapter covered the very fundamentals of the measurement science. First of all it showed how measurement has been an important part of the knowledge processes, and how its importance has evolved along the centuries, as the scientific and human need for knowledge has evolved.

It showed how the measurement activity can be modeled in terms of different descriptive and experimental processes that concur to generate the measured value. A strict analysis of the defined model gave evidence of the different sources of imperfections that cause the measured values to differ from the measurand value. This same analysis proved that the measurand value remains unknown and unknowable, and proved that a quantitative estimate of the deviation between the measured and the measurand values is definitely necessary for every practical measurement application.

The modern uncertainty concept was analyzed as the presently most effective way to provide this quantitative estimate, and the standard recommendations on how to implement this concept were presented.

It is worth noting, as a final remark, that the uncertainty concept, as defined in Section 1.5, has general validity and provides a scientifically sound answer to the need for a quantitative estimate of how "good" a measurement result is. On the other hand, the practical implementation recommended by the presently available standard documents [5, 6] and presented in Section 1.6 has some limitations, mainly due to the considered mathematical framework: the theory of probability.

Different solutions have been recently proposed (and well summarized in Reference 7) to overcome these limitations, by referring to the more general mathematical framework of the theory of evidence. Though the discussion of new proposals is beyond the scope of this book, we wish to conclude this chapter as we started it: with a clear indication that the measurement science, as every other science, is always evolving, and today problems should be a motivation for searching new solutions, even if they may clash against well assessed theories, methods, and tools.

FURTHER READING

Several books and articles can be found that cover the topics that have been synthetically described in this chapter. In particular, the following books are related to the history of measurement.

Alder, K. *The Measure of All Thing*, Free Press, New York (2002).

Keithley, J.F. *The Story of Electrical and Magnetic Measurements*, IEEE Press, Piscataway, NJ (1999).

Zupko, R. *Revolution in Measurement: Western European Weights and Measures Since the Age of Science*, American Philosophical Society, Philadelphia, PA (1990).

Klein, H.A. *The Science of Measurement: A Historical Survey*, Dover Publications, Mineola, NY (1988).

Favre, A. *Les origines du système métrique*, Les Presses Universitaires de France, Paris (1931).

The following books and articles are related to the measurement model and uncertainty evaluation.

Rossi, G. B. *A Probabilistic Theory of Measurement with Applications*, Springer, New York (2014).

Sydenham, P. H. *Handbook of Measurement Science*, Vol.1, Theoretical fundamentals, John Wiley & Sons Inc, Hoboken, NJ (1992).

Sydenham, P. H., Hancock, N. H., Thorn, R. *Introduction to Measurement Science and Engineering*, John Wiley & Sons Inc, Hoboken, NJ, (1992).

Finkelstein, L. Measurement and instrumentation science – an analytical review, *Measurement*, **14** (1), 3–14 (1994).

Finkelstein, L. State and advances of general principles of measurement and instrumentation science, *Measurement*, **3** (1), 2–6 (1985).

Giordani, A., Mari, L. Measurement, models and uncertainty, *IEEE Transactions on Instrumentation and Measurement*, **61** (8), 2144–2152 (2012).

Ferrero, A., Salicone, S. Uncertainty: only one mathematical approach to its evaluation and expression?, *IEEE Transactions on Instrumentation and Measurement*, **61** (8), 2167–2178 (2012).

REFERENCES

1. OIML. *International Vocabulary of Metrology – Basic and General Concepts and Associated Terms (VIM)*, 3rd edition, (2007). Available at BIPM http://www.bipm.org/en/publications/guides/vim.html
2. Keithley, J.F. *The Story of Electrical and Magnetic Measurements*, IEEE Press, Piscataway, NJ (1999).
3. Di Pasquale, G., Marchis, V. Some considerations about the Roman Pes, *Nuncius*, **11**, 669–675 (1996) (in Italian).
4. Sydenham, P.H. *Handbook of Measurement Science, Vol.1: Theoretical Fundamentals*, John Wiley & Sons, Inc., Hoboken, NJ (1992).
5. BIPM JCGM 100:2008. *Evaluation of measurement data – Guide to the expression of uncertainty in measurement* (GUM), GUM 1995 with minor corrections (2008). Available at BIPM http://www.bipm.org/en/publications/guides/gum.html
6. BIPM JCGM 101:2008. *Evaluation of measurement data – Supplement 1 to the Guide to the expression of uncertainty in measurement – Propagation of distributions using a Monte Carlo method.* Available at BIPM http://www.bipm.org/en/publications/guides/gum.html
7. Salicone, S. *Measurement Uncertainty – An Approach via the Mathematical Theory of Evidence*, Springer, New York (2007).
8. Sydenham, P.H., Hancock, N.H., Thorn, R. *Introduction to Measurement Science and Engineering*, John Wiley & Sons, Inc., Hoboken, NJ (1992).
9. Luce, R.D., Krantz, D.H., Suppes, P., Tversky, A. *Foundations of Measurements*, Vol.3: Representation, Axiomatization, and Invariance, Dover Publications, Mineola, NY (2006).

EXERCISES

1.1 Why do we measure?

- **(a)** To provide a basis for the exchange of goods and services
- **(b)** To encourage cooperative activities
- **(c)** To control, manage, and improve processes
- **(d)** All of the above

1.2 Typically, any measurement result depends on

(a) The value of the measurand

(b) The measurand value and other properties pertaining to the measurement context

(c) The measurand value and the measurement unit

(d) The influence quantities

1.3 The main components of the measurement context are

(a) The environment and the operator

(b) The measurand and the measured system

(c) The measured system, the measurement system and the environment

(d) The measurement system and the influence quantities

1.4 The main types of measurement uncertainty sources are

(a) The definitional and interactional uncertainty

(b) The measurement system uncertainty and the operator inexperience

(c) The acquisition system uncertainty and instrumental noise sources

(d) The definitional, interactional and instrumental uncertainties

1.5 The measurement uncertainty mainly derive by

(a) The definition of the measurand

(b) The uncertainty of the employed instruments

(c) The uncertainty associated with the adopted model of the context

(d) The operator and the empirical procedures performed

1.6 An optimal model of the measurement context allows to achieve

(a) Meaningful information about the measurand with minimum complexity

(b) Useful information about the measurand

(c) The maximum information about the measurand

(d) Enough information about the measurand

1.7 If the input signal has negligible superimposed noise, the uncertainty of the result provided by an electronic instrument is typically

(a) Mainly of Type B

(b) Mainly of Type A

(c) Negligible if the instrument is correctly used

(d) Both of Type A and Type B

1.8 Generally, any measurement context model should contain

(a) The variable that model the measurand and its interactions with the main influence properties

(b) The variable that model the measurand, the set of variables modeling the main influence properties and the related allowed ranges, the relationships describing the associated interactions, the set of variables modeling the secondary influence properties and the related allowed ranges

(c) The model of the measurand and of the measurement system

(d) The variables that model the measurand and the main and secondary influence properties

1.9 Which of the following statements about measurement uncertainty evaluation is incorrect?

(a) Indications provided by the GUM should be disregarded

(b) Monte Carlo procedures can be a useful tool to study uncertainty propagation

(c) The measurement model inputs should take into account uncertainty arising from transducers

(d) The Law of Propagation of Uncertainty suggested by the GUM can be difficult to apply

1.10 Which of the following statements is true

(a) In nominal scales, the classes containing the measurement result must be labeled with numbers

(b) In ordinal scales, addition and subtraction are allowed

(c) In interval scales, the gap between two different classes is significant

(d) In ratio scales, the zero is defined by shared conventions

1.11 In any measurement, a lower bound to measurement uncertainty is given by

(a) Definitional uncertainty

(b) Combined uncertainty

(c) Instrumental and interaction uncertainty

(d) The effect of environmental factors

1.12 It is enough that any measurement result (explicitly or implicitly) provides the following information

(a) Measurement value and uncertainty

(b) The pdf of the measurement value

(c) Measurement value and measurement unit

(d) Measurement value, uncertainty, measurement scale and limits of validity of the information provided

1.13 The measurement error concept is obsolete and not used any longer to qualify a measurement result because

(a) Modern instruments provide the true value of the measurand

(b) We do not know how to process it when indirect measurements are considered

(c) It refers to the true value, and the true value of the measurand is always unknown and unknowable

(d) It does not take into account random effects

1.14 Standard uncertainty is defined as

(a) An interval about the true value of the measurand

(b) An estimator of the standard deviation of a probability distribution

(c) A confidence interval about the measured value with a given level of confidence

(d) An interval about the measured value, within which the value of the measurand is supposed to lie with full certainty

1.15 Type-B evaluation method of standard uncertainty is used when

(a) A systematic effect is present

(b) It is not possible to evaluate uncertainty by means of a statistical analysis of a series of observations

(c) A great number of experimental data is available

(d) A calibration certificate is not available

1.16 The expanded uncertainty provides a confidence interval about the measurement result with a given confidence level if

(a) Explicit or implicit assumptions about the probability distribution of the measurement result can be done

(b) A suitable coverage factor can be considered

(c) The probability distribution associated with the measurement result is normal

(d) The probability distribution associated with the measurement result is uniform

1.17 The measurement representational condition states that

(a) Each measurement result must contain the value of measurement uncertainty

(b) Each measurement result must be represented by a set of numbers

(c) Any empirical relations between measurands must be mapped to abstract relations between symbols

(d) Any empirical relations between measurands must be mapped to arithmetic operations

1.18 The final goal of a measurement process is

(a) Increase the personal knowledge of the measurement expert

(b) Provide a quantitative input to a decision-making process

(c) Define the experimental set-up

(d) Attain an uncertainty value as low as possible

1.19 Legal metrology covers

(a) That part of the measurement science that is used by lawyers in trials

(b) Technical expertise to issue laws related to measurements

(c) Measurement methods and instruments employed in economical transactions and the application of sanctions

(d) Quality measurements to evaluate the performance of courts

1.20 The coverage factor is

(a) A multiplier of standard uncertainty that provides expanded uncertainty

(b) The coverage probability associated with the interval defined by the expanded uncertainty

(c) A factor that takes always numerical value equal to 2 if the probability distribution assigned to the possible values of the measurand is normal

(d) A factor that takes always numerical value equal to $1/\sqrt{3}$ if the probability distribution assigned to the possible values of the measurand is normal

CHAPTER 2

THE SYSTEM OF UNITS AND THE MEASUREMENT STANDARDS

FRANCO CABIATI
National Institute of Metrological Research (INRIM), Turin, Italy

2.1 INTRODUCTION

The system of measurement units is a subject with a highly conventional content, susceptible to changes with the technological evolution, and with legislative implications, which compel to express relations and definitions in words. This usually leads to dealing with the subject from a descriptive and historical point of view.

In the approach followed here, the attention is rather driven to the logical and mathematical aspects of unit and unit system. Thus, it was deemed more useful to concentrate on the coherent system of units in its more recent form, as ultimate result of the development subsequent to the technical and scientific use of measurement units, which led to structuring them in systems.

Just limiting to the most important milestones in the development of such unit systems, one can mention:

- the CGS system (centimeter, gram, and second), in its three variants CGS_{es} (electrostatic), CGS_{sim} (symmetric) and CGS_{em} (electromagnetic) introduced in the attempt of including in different ways also the electromagnetic units;
- the MKS system (meter, kilogram, and second), introduced together with the *Convention du Mètre* in 1875, with sizes corresponding to those of the present units;
- the MKSA system (meter, kilogram, second, and ampere), which included also the practical unit of electric current, to extend the system to the electromagnetic derived units, followed later by the kelvin (for thermodynamic temperature) and the candela (for luminous intensity);

Modern Measurements: Fundamentals and Applications, First Edition.
Edited by Alessandro Ferrero, Dario Petri, Paolo Carbone, and Marcantonio Catelani.

- the SI (*Système International d'Unités*), named in 1960 and later completed with the inclusion of the mole as unit of amount of substance.

Further changes in SI unit definitions were made, before the system has reached the present form, for the kelvin, the candela and the meter.

It can not be ignored that, in spite of the official convention, non-SI units are still in use. In particular, CGS units sometime appear in scientific contexts, while imperial and US customary units are still widely used in technical environments. Of course, such unit systems and related structures are not taken into consideration in the following.

In Sections 2.1 to 2.3, the structure and general features of a coherent unit system, such as SI, are considered from a general theoretical point of view, In Sections 2.4 and 2.5, the SI is analyzed in detail, the basic units are considered in their formal and most substantial aspects, looking for the underlying mathematical relations beyond the wording of definitions, and the perspective of short-term evolution of the system are outlined. In the last section, the role of the standards and the problem of realizing a measurement unit is addressed, its solution is applied to the basic and other important units to obtain the primary standards and the unit dissemination is shortly described. Where relevant, the realization is dealt with both for the present and the perspective future definition of the unit.

The proposed exercises at the end of section, beyond their use as tests of understanding, are also conceived as occasions not only for applying the learned theoretical notions, but also for deepening some of them.

2.2 ROLE OF THE UNIT IN THE MEASUREMENT PROCESS

The measurement process dealt with in Chapter 1 is the most general one, as it has to cover the field of all the measurable properties or quantities. In a field restricted to quantities variable with an even extreme fineness, of general use, but also of technical and scientific relevance, a more sophisticated system of references has been developed. The feature of conventionality, essential for the intersubjectivity of measurement, has been transferred from the reference scale, still directly involved in the measurement process, to a potentially universal and reproducible entity. Such an entity is the measurement unit and, more specifically, the unit definition: a fully abstract entity which needs to be realized in a primary standard to take part in the empirical measurement process.

A unit definition identifies, more or less explicitly, a reference quantity and establishes its value and its relation with the unit, thus allowing the value of a primary standard to be determined.

In the case of measurements performed within a unit system, the measurement value x appearing in the most general model of Figure 1.1 of Chapter 1 is a composite expression including a rational number r and a measurement unit $U(q)$. That is

$$x(q) = r \cdot U(q). \tag{2.1}$$

In the same model, both the descriptive and experimental processes must include the unit realization, yielding a primary standard, and the chain of intermediate standards which make possible the reference to the measurement unit. Such passages, implicitly included in the model, are essential to complete the logical path from the measurand to the measurement result.

Following the most classical measurement process, the measurand is compared with a standard quantity of the same kind and this usually implies the generation of a scale. Of the different kinds of measurement scales identified in Section 1.3.1.3, only the ratio scale are here considered, as a unit system does not define scales, but only units, from which a ratio scale can be generated. In fact, a measurement scale can be built up as an ordered series of reference quantities (standards), such that for two generic subsequent element of the series $q_{S,n}$ and $q_{S,n-1}$ it is

$$q_{S,n} - q_{S,n-1} = \rho \cdot q_{PS}, \tag{2.2}$$

ρ being a fractional or integer number and q_{PS} a *primary standard* which realizes the definition of the unit $U(q)$ (see Section 2.6).

Thus the relation diagram in Section 1.3.1.3, with some modification, could be extended as in Figure 2.1 in order to show, together with Equation (2.2), the link with the measurement unit, which makes meaningful the expression of the measurement result as in (2.1). In the figure, x_{PS} is the value of q_{PS} established by the relation μ_{PS} following the definition of $U(q)$, while x_{Sn}, which express the values of the standard scale q_{Sn}, are only indirectly assigned through a chain of comparisons starting from the primary standard q_{PS}.

In a less traditional measurement process, yet most usual recently, the direct comparison between measurand and standard is performed automatically inside a measuring instrument. The comparison may be even previously transferred to a different quantity by applying a transducer to the measurand.

In a measuring instrument, the traceability to a primary standard is obtained through the calibration of the whole instrument. This could be represented with the relation diagram of Figure 2.2, where μ_M is the relation established by the measuring instrument, which yields a reading value for every applied measurand.

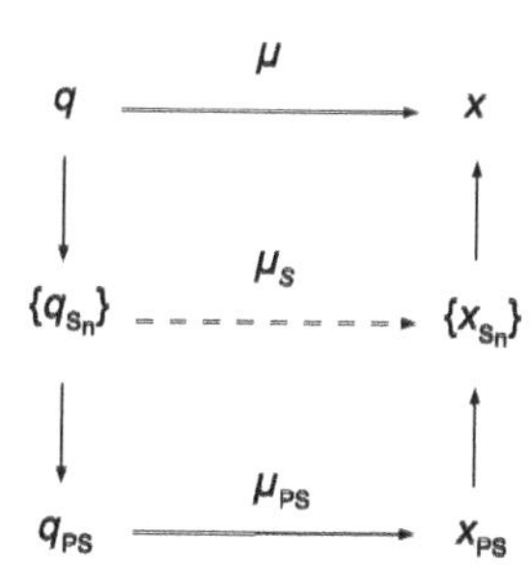

FIGURE 2.1 Relation diagram of the measurement process performed within a unit system. The previous model is modified to show the logical path linking the measurement result x to the measurement unit realized by the primary standard q_{PS}.

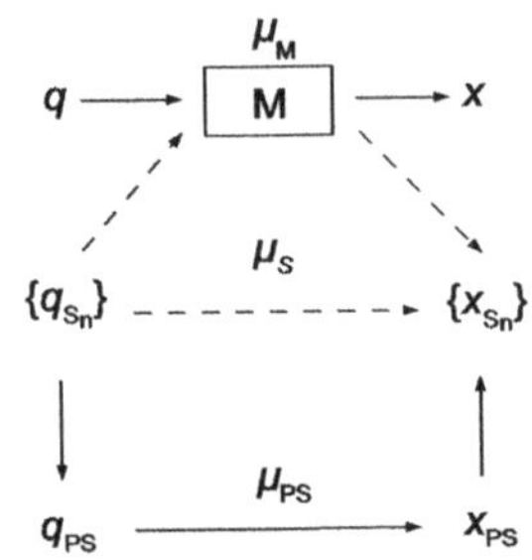

FIGURE 2.2 Relation diagram of the measurement process performed by a calibrated measuring instrument. The relation μ_M between measurand and measurement result is directly established by the instrument M, but that relation must be validated through a calibration procedure, where the same instrument is applied to a standard scale traceable to a primary standard realizing the unit.

In order to make meaningful the reading value, that is to allow it to be assumed as the measurement value, the instrument should be submitted to a calibration process, where the relation μ_M is verified applying to the instrument a convenient scale of standard quantities (see Section 2.6). That is, the comparison of the measurand with a standard referred to the unit through a primary standard, essential for the measurement process, is performed with a procedure equivalent to the substitution method, where the second step is delayed from the time of calibration to the time of measurement.

2.3 IDEAL STRUCTURE OF A UNIT SYSTEM

A system of measurement units in its archaic form is just a collection of independent units of different quantities with no relation among them (consider that, for quite a long time, the volume of liquids or loose materials was measured by means of volume standards with no relation with the length standards).

With the development of experimental sciences, the relations among quantities and among units could not have been disregarded, also because scientific experiments are largely based on indirect measurements, where the value of a quantity is obtained from two or more direct measurements of different quantities. In fact, taking into account those relations allowed many units to be expressed by reference to other units, thus reducing the number of reference quantities to a minimum.

Namely, if N is the number of the quantities whose units are to be defined, and M the number of independent relations, the number of independent reference quantities, for which a conventionally fixed value is necessary and sufficient to be assigned, is reduced to $R = N - M$. Each of the R definitions that relates a unit to the relevant reference quantity reciprocally fixes the value of that quantity.

The R units that include in their definition a physical reference quantity, and then link the unit system to the real world, are usually called *base units* and the others *derived units*. The definitions of the derived units express in words the relation between the unit to be defined and other units, which can be either derived or base units.

The relations among units reflect only consolidated and largely agreed scientific theories, expressed by relations among the respective quantities. The general expression of such relations for a generic quantity Q_i is

$$Q_i = \prod_j Q_j^{e_{i,j}}, \tag{2.3}$$

where $j \neq i$ and $e_{i,j}$ is an exponent equal to zero except for those quantities that are related to Q_i.

For reasoning about relations between units and quantities, it is useful to consider the current notation

$$Q = \{Q\} \cdot [Q], \tag{2.4}$$

where $\{Q\}$ is a number expressing the value[1] and $[Q]$ is a symbol for the unit, that is a quantity homogeneous with Q exactly complying with the relevant unit definition. Any algebraic operation is allowed on the entities in (2.4).

With (2.4), a generic relation $Q_i = k \prod_j Q_j^{e_{i,j}}$ used for an indirect measurement, k being a numerical coefficient, may be written as

$$\{Q_i\}\, [Q_i] = k \prod_j \{Q_j\}^{e_{i,j}} \prod_j [Q_j]^{e_{i,j}}. \tag{2.5}$$

It is a matter of convenience, that the relations among quantities hold also among the corresponding quantity values expressed in the same unit system, that is $\{Q_i\} = k \prod_j \{Q_j\}^{e_{i,j}}$. To this purpose, in (2.5) it must be

$$[Q_i] = \prod_j [Q_j]^{e_{i,j}}. \tag{2.6}$$

That is, all derived units must be defined in such a way that the corresponding relations are particular cases of (2.6). Such a unit system, where the relations among units are formally identical to those among quantities, with no coefficient different from unity, is called a *coherent system.*

In order to analyze the structure of the unit system, it is interesting to consider the mathematical equations corresponding to the definitions expressed in words.

The ideal structure of a coherent system of units regarding N quantities with M relations among them, as assumed above, is formed by the $R = N - M$ base units with as many reference quantities, homogeneous with the units, and M derived units.

[1] It may be useful to remark the different meaning of braces in the following with respect to the precedent use, where they indicate a set of elements.

The base units are defined by their relations with the respective reference quantities, which are, following (2.4),

$$\left[Q_{B_n}\right] = \frac{Q_{R_n}}{\left\{Q_{R_n}\right\}}, \tag{2.7}$$

with $n = 1, 2, ..., R$. Following the sense of the definition, as expressed by (2.7), the unit $[Q_{B_n}]$ is such that the value of the reference quantity results just $\{Q_{R_n}\}$. The choice of the reference quantities is subject to the constraint of being independent, that is, no known relation must exist among them, while the assigned values are arbitrary, corresponding to degrees of freedom of the system.

The derived units are defined by particular cases of the general equation

$$\left[Q_{D_h}\right] = \prod_{k=1}^{M} \left[Q_{D_k}\right]^{e_{h,k}} \prod_{m=1}^{R} \left[Q_{B_m}\right]^{e_{h,m}}, \tag{2.8}$$

with $k \neq h$.

The set of M Equations (2.8) can be solved with respect to the derived units eliminating them from the resulting expressions, obtaining

$$\left[Q_{D_h}\right] = \prod_{n=1}^{R} \left[Q_{B_n}\right]^{e_{h,n}}, \tag{2.9}$$

with $h = 1, 2, ..., M$. Equations (2.9) are formally identical to the equations on the corresponding quantities, which also express their dimensions.

2.4 EVOLUTION OF THE UNIT DEFINITION

Some information about the development of units along the main historical periods, until the institution of SI, has already been given in Chapter 1. Here, a more specific analysis of the evolution of the kind of reference quantities adopted to define the base units unit definition will follow, which can be useful for a deeper understanding of the present unit system and of his future perspective. The evolution has regarded essentially the base units, that is both the kind of reference quantity and its relation with the unit. Some stages of that evolution are still represented in the present definitions of the basic SI units.

When the first units were defined, the needs for measurements were limited to quantities involved in trade or simple technical fields, such as mass (weight), length, volume (measure). The definitions essentially stated that the reference quantity was a property of an artifact, the relation between reference quantity and unit was a simple identity and the unit realization consisted in building and maintaining the artifact.

With the subsequent development of metrology, some quantities related to natural bodies (length of earth meridian), natural events (time interval between astronomic configurations) or phenomena (state-change temperature) were adopted as reference quantities for the units. The relation with the unit was no longer of identity, but of ratio. On the other side, the unity value is not requested for a primary standard, the essential feature of the reference quantity being that its value is stable and reproducible, at least in principle, anywhere and at any time. Having released the identification between primary standard and unit reference, the unit must be realized through experimental processes aimed at assigning values to the primary standards, within proper realization uncertainties.

With the increasing involvement of metrology in experimental sciences, the macroscopic reference quantities were abandoned in favor of properties of microscopic entities (e.g., wavelengths or periods corresponding to specific atomic radiations or transitions).

A last kind of unit definition adopted fundamental constants as reference quantities (e.g., magnetic constant, velocity of light). In this case, the relation of the reference constant with the unit can also be established by the definition through an ideal experiment involving universally accepted scientific theories.

Of course, as already specified for any unit reference quantity, fixing the ratios of the reference constants with the units by definition implies to fix the values of the same reference constants, which are no longer to be determined. This is perfectly legitimate, as the values of the reference constants of the unit system can be established arbitrarily, provided the following conditions are met: (i) they are all and only those which are necessary and sufficient to define all the units of the system; (ii) they must be independent and if they are related through other constants the values of such constants must not be fixed by some other definition.

It is important to realize the advantages of adopting fundamental constants as references for the measurement units, both for metrology and for experimental sciences. Metrology profits from having as basic references some of those fundamental constants that are indicated by theory as the most reliable ones, universally invariant in any technological and experimental context. Experimental sciences not only can rely on some constants with fxed and exact values, with no need for periodical changes due to new determinations, but also on a significant number of other constants, related to the exact ones, whose uncertainties are even greatly reduced. Table 2.1 reports the main stages of the evolution in the reference quantities, with some examples and the advantages that stimulated the changes.

2.5 THE SI SYSTEM OF UNITS

A system of units is not just a scientific matter, but different aspects are also involved: legal, commercial, educational and traditional. This has made the development of the system subject to compromises which altered the ideal structure outlined in Section 2.3.

Starting from the beginning of the decimal metric system, at the time of the French Revolution, until the MKS system, and including the intermediate CGS system, the structure defined by (2.7) and (2.8) was maintained. A first derogation from the ideal structure came with the introduction of a fourth base unit, besides the three mechanical

TABLE 2.1 Main stages of the evolution in the references of the base units

Kind of reference	Examples	Advantages
Artifact as unique reference	Prototype kilogram, meter as a rule	Unity value, no realization uncertainty
Macroscopic natural invariant	Earth meridian, ephemeris time, water triple point	Universal access, invariance
Microscopic natural invariant	Krypton radiation, cesium hyperfine transition	Universal access, invariance, high accuracy
Fundamental constant	Magnetic constant, velocity of light	Universal access, invariance, high accuracy, scientific implications

ones. In fact, after a first proposal by Giorgi in 1901, the *ampere* was adopted in 1948 with a definition involving the force between two conductors, with defined geometry, in which the electric current is maintained (see the ampere definition below). Because the force is expressed in newton and the geometry involves the meter, the base unit of current is not independent of the other units, as it would be expected. In particular, that definition does not comply with (2.7), but the general equation is to be modified as

$$[Q_{\mathrm{B}i}] = \prod_h [Q_{i,h}]^{e_{i,h}} \frac{Q_{\mathrm{R}i}^{e_{\mathrm{R}i}}}{\{Q_{\mathrm{R}i}\}^{e_{\mathrm{R}i}}}, \qquad (2.10)$$

where $Q_{i,h}$ may be any base or derived quantity, except for $Q_{\mathrm{B}i}$ and $Q_{\mathrm{R}i}$.

Therefore, the unit of the reference quantity $Q_{\mathrm{R}i}$ is no longer the same unit of $Q_{\mathrm{B}i}$ being defined.

The derogation from the ideal model of a base unit definition was the result of a compromise between the reasons of tradition, for which the set of base units should have been maintained, and the convenience in adopting some particular reference quantities.

The unit system including the new ampere definition was officially approved in 1954, with the addition of two further base units for temperature (*kelvin*) and luminous intensity (*candela*), and in 1960 the *Conférence Générale des Poids et Mesures* (CGPM) gave to the unit system the name *Système International d'unités* (SI). Afterwards, in 1971, the SI base units reached its present number of seven with the introduction of the *mole*, a base unit for chemical quantities.

Since the SI institution, two base units were redefined: the second in 1968, and the meter in 1983. The new definition of the meter has been another particular case of (2.10), as it will be shown in the following.

At the present stage of development of the SI, the base units and the respective SI base quantities are those listed in Table 2.2, where also the reference of each unit is indicated.

TABLE 2.2 Present SI base units with the respective quantities and references

Base unit	Symbol	Quantity	Unit reference
Meter	m	Length	Velocity of light c_0
Kilogram	kg	Mass	Prototype mass $m(\mathcal{K})$
Second	s	Time	Radiation of ^{133}Cs
Ampere	A	Electric current	Magnetic constant μ_0
Kelvin	K	Temperature	Triple point of H_2O
Mole	mol	Amount of substance	Molar mass of ^{12}C
Candela	cd	Luminous intensity	Luminous efficiency K_m

The details of every base unit are reported below, with the official wording of definition and the corresponding mathematical expression, also useful to put in evidence the links of dependence on other units where relevant.

- **Unit of length**
  The meter is the length of the path travelled by light in vacuum during a time interval of 1/299 792 458 of a second.
  That is

$$\mathrm{m} = \frac{\mathrm{s}}{299\,792\,458}\,c_0 = \mathrm{s}\,\frac{c_0}{\{c_0\}}, \tag{2.11}$$

  where $c_0/\{c_0\}$ is the unit of velocity. The last term of (2.11) confirms that the equation of units is formally the same as the equation of the corresponding quantities. Furthermore, it is a particular case of (2.10) and in fact the definition is not independent, but implies a link with the definition of the second.
  The definition has the effect of imposing $c_0 = 299\,792\,458$ m/s exactly.
- **Unit of mass**
  The kilogram is the unit of mass; it is equal to the mass of the international prototype of the kilogram.
  That is, simply,

$$\mathrm{kg} = m(\mathcal{K}), \tag{2.12}$$

  where $m(\mathcal{K})$ is the mass of the international prototype. Equation (2.12) complies with (2.7), where it is $\{QRn\} = \{m(\mathcal{K})\} = 1$.
- **Unit of time**
  The second is the duration of 9 192 631 770 periods of the radiation corresponding to the transition between the two hyperfine levels of the ground state of the cesium 133 atom.

That is

$$\mathrm{s} = 9\,192\,631\,770 \frac{1}{\Delta\nu_{\mathrm{Cs}}} = \frac{1/\Delta\nu_{\mathrm{Cs}}}{\{1/\Delta\nu_{\mathrm{Cs}}\}}, \tag{2.13}$$

where $\Delta\nu_{\mathrm{Cs}}$ is the frequency corresponding to the ground-state hyperfine splitting of the ^{133}Cs atom, to which the definition assigns the value of 9 192 631 770 Hz. Equation (2.13) is a particular case of (2.7), if the period of the radiation is considered as the reference quantity of the unit.

- **Unit of electric current**
  The ampere is that constant current which, if maintained in two straight parallel conductors of infinite length, of negligible circular cross-section, and placed 1 meter apart in vacuum, would produce between these conductors a force equal to 2×10^{-7} *newton per meter of length.*
  That is, from the force equation

$$F = \frac{\mu_0\,\mathrm{A}^2}{2\,\pi} = 2 \times 10^{-7}\,\mathrm{N}, \tag{2.14}$$

where μ_0 is the magnetic constant,

$$\mathrm{A} = \mathrm{N}^{\frac{1}{2}} \left(\frac{\mu_0}{4\pi \times 10^{-7}} \right)^{-\frac{1}{2}}. \tag{2.15}$$

Thus the ampere is not an independent unit, as it implies the force unit N (see Table 2.3), and (2.15) is a particular case of (2.10), being $\mu_0/\{\mu_0\}$ (with $\{\mu_0\} = 4\,\pi \times 10^{-7}$) the unit of magnetic permeability.

As an effect of the definition, from (2.15) it results $\mu_0 = 4\pi \times 10^{-7}\,\mathrm{N\,A^{-2}}$ exactly.

- **Unit of thermodynamic temperature**
  The kelvin, unit of thermodynamic temperature, is the fraction 1/273.16 of the thermodynamic temperature of the triple point of water.
  That is

$$\mathrm{K} = \frac{T_{\mathrm{tpw}}}{273.16}, \tag{2.16}$$

where T_{tpw} is the thermodynamic temperature of the triple point of water, whose value therefore results $T_{\mathrm{tpw}} = 273.16$ K exactly. So this definition fully complies with (2.7).

- **Unit of amount of substance**
  1. *The mole is the amount of substance of a system which contains as many elementary entities as there are atoms in 0.012 kilograms of carbon 12; its symbol is "mol".*

TABLE 2.3 SI derived units with special name. For each unit the symbol, the quantity and the expression in terms of the base units, according to (2.9), are indicated

Derived unit	Symbol	Quantity	Base unit expression
Hertz	Hz	Frequency	$m^0\ kg^0\ s^{-1}\ A^0\ K^0\ mol^0\ cd^0$
Newton	N	Force	$m^1\ kg^1\ s^{-2}\ A^0\ K^0\ mol^0\ cd^0$
Pascal	Pa	Pressure	$m^{-1}\ kg^1\ s^{-2}\ A^0\ K^0\ mol^0\ cd^0$
Joule	J	Energy	$m^2\ kg^1\ s^{-2}\ A^0\ K^0\ mol^0\ cd^0$
Watt	W	Power	$m^2\ kg^1\ s^{-3}\ A^0\ K^0\ mol^0\ cd^0$
Coulomb	C	Electric charge	$m^0\ kg^0\ s^1\ A^1\ K^0\ mol^0\ cd^0$
Volt	V	Electromotive force	$m^2\ kg^1\ s^{-3}\ A^{-1}\ K^0\ mol^0\ cd^0$
Farad	F	Electric capacitance	$m^{-2}\ kg^{-1}\ s^4\ A^2\ K^0\ mol^0\ cd^0$
Ohm	Ω	Electric resistance	$m^2\ kg^1\ s^{-3}\ A^{-2}\ K^0\ mol^0\ cd^0$
Siemens	S	Electric conductance	$m^{-2}\ kg^{-1}\ s^3\ A^2\ K^0\ mol^0\ cd^0$
Weber	Wb	Magnetic flux	$m^2\ kg^1\ s^{-2}\ A^{-1}\ K^0\ mol^0\ cd^0$
Tesla	T	Magnetic flux density	$m^0\ kg^1\ s^{-2}\ A^{-1}\ K^0\ mol^0\ cd^0$
Henry	H	Inductance	$m^2\ kg^1\ s^{-2}\ A^{-2}\ K^0\ mol^0\ cd^0$
Degree C.	°C	Celsius temperature	$m^0\ kg^0\ s^0\ A^0\ K^1\ mol^0\ cd^0$
Lumen	lm	Luminous flux	$m^0\ kg^0\ s^0\ A^0\ K^0\ mol^0\ cd^1$
Lux	lx	Illuminance	$m^{-2}\ kg^0\ s^0\ A^0\ K^0\ mol^0\ cd^1$
Becquerel	Bq	Radionuclide activity	$m^0\ kg^0\ s^{-1}\ A^0\ K^0\ mol^0\ cd^0$
Gray	Gy	Absorbed dose	$m^2\ kg^0\ s^{-2}\ A^0\ K^0\ mol^0\ cd^0$
Sievert	Sv	Dose equivalent	$m^2\ kg^0\ s^{-2}\ A^0\ K^0\ mol^0\ cd^0$

2. *When the mole is used, the elementary entities must be specified and may be atoms, molecules, ions, electrons, other particles, or specified groups of such particles.*

The atoms of ^{12}C are intended to be unbound, at rest and in their ground state. The definition can be expressed by the equation

$$\text{mol}\,(X) = \frac{0.012\ \text{kg}}{m(^{12}\text{C})}\, n_e(X), \tag{2.17}$$

where $n_e(X)$ is the amount of substance of one elementary entity X, $m(^{12}C)$ is the mass of the ^{12}C atom (intended as unbound, at rest and in its ground state) and the dimensionless ratio $0.012\ \text{kg}/m(^{12}C)$ stays for the count of ^{12}C atoms contained in a system of a 0.012 kg mass. The same ratio corresponds to the numerical value of the Avogadro constant N_A, that is $0.012\ \text{kg}/m(^{12}C) = \{N_A\}$, formerly referred to as Avogadro number.

A quantity strictly related to the amount of substance, and linked to the traditional concept of *gram-atom* or *gram-molecule*, is the *molar mass* $M(X)$ of a given entity X, intended as $M(X) = m(X)/n_e(X)$ whose unit is the kilogram per mole.[2]

[2]The molar mass of a given entity is also intended as the mass of one mole of the same entity, that is $M(X) = N_A\, m(X)$.

Readjusting (2.17) to comply with the general Equation (2.10) and considering that for any X it is $n_e(\mathrm{X}) = n_e(^{12}\mathrm{C})$, it results

$$\mathrm{mol(X)} = \mathrm{kg}\,\frac{n_e(\mathrm{X})/m(^{12}\mathrm{C})}{1/0.012} = \mathrm{kg}\,\frac{1/M(^{12}\mathrm{C})}{\{1/M(^{12}\mathrm{C})\}}. \tag{2.18}$$

Thus, the definition of the mole imposes to be $M(^{12}\mathrm{C}) = 0.012\ \mathrm{kg/mol}$ exactly.

- **Unit of luminous intensity**
  The candela is the luminous intensity, in a given direction, of a source that emits monochromatic radiation of frequency 540×10^{12} hertz and that has a radiant intensity in that direction of 1/683 watt per steradian.
  That is

$$\mathrm{cd} = \frac{\mathrm{W/sr}}{683}\,K_\mathrm{m} = \mathrm{W\,sr}^{-1}\,\frac{K_\mathrm{m}}{683}, \tag{2.19}$$

where K_m is the luminous efficacy corresponding to the radiation frequency of 540 THz. The definition is not independent, as it involves the derived units watt and steradian (see below and Table 2.3), and the last expression in (2.19) complies with (2.10), fixing the value $K_\mathrm{m} = 683$ lm/W exacly.[3]

The above analysis of the base unit definitions confirms that such units are no longer those independent corner stones of the unit system that might be expected. Nevertheless, the corresponding SI base quantities maintain an important role in the respective field from a practical point of view. Furthermore, the base units are the only ones including the reference quantities of the system in their definitions, yet not always explicitly.

A synoptic view of the derivation of each SI base units from the reference quantities and, where relevant, from other base units is given in Figure 2.3.

From the base units, using relations reflecting scientific theories, a large number of derived units have been defined. Many of them have been given special names, even if the units of many other quantities (e.g., velocity or torque) must be expressed as combinations of those (e.g., meter per second or newton meter). The names and symbols of the SI derived units with special names are reported in Table 2.3 together with the respective derived quantities and the expressions in terms of the base units, according to (2.9).

To the derived units with special names reported in the same table, two dimensionless units are to be added: the *radian* (symbol rad and dimension m/m), unit of plane angle as ratio between an arc and its radius, and the *steradian* (symbol sr and

[3]In the cited Table 2.3, the expression of lumen in terms of SI base units includes only the candela, as it is usual to omit the radian and steradian because they are dimensionless units and are not included in the set of the SI base units. However, the unit of luminous flux is expressed as lm = cd sr, as well as in the definition of candela the unit of radiant intensity is specified as watt per steradian.

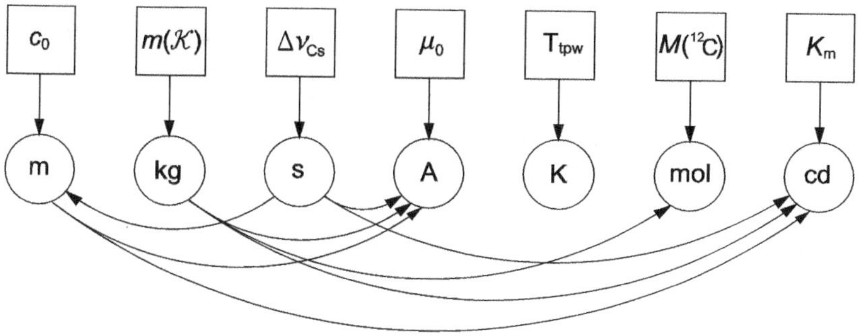

FIGURE 2.3 Links of dependence of the SI base units deriving from their definitions. Each unit depends on the reference quantity and the other units from which it receives an arrow.

dimension m^2/m^2), unit of solid angle as ratio between the area of a spherical cap and the square of its radius.

Of course, as observed above, the list of units limited to those with special name is largely incomplete. In particular, all the quantities whose units involve the mole (e.g., the molar mass) are excluded, as one can see from the last column of Table 2.3.

2.6 PERSPECTIVES OF FUTURE SI EVOLUTION

To understand the very substantial structure of the SI, as well as any its possible future evolution, it is useful to bring the mathematical expression of every unit definition to its ultimate form by expressing every unit directly in terms of the reference quantities of the system. In fact, substituting $[Q_{B,n}]$ in (2.9) with its expression given by (2.7), one can write

$$[Q_i] = \prod_{n=1}^{R} \left(\frac{Q_{Rn}}{\{Q_{Rn}\}} \right)^{e_{i,n}}, \tag{2.20}$$

where Q_i may be any base or derived quantity, whose unit is expressed by a monomial combination of the units $Q_{Rn}/\{Q_{Rn}\}$ of the system reference quantities Q_{Rn}.

Figure 2.4 represents graphically the direct relations among the present SI base units and the set of reference quantities of the system, according to Equation (2.20).

The same general equation can also be written as

$$[Q_i] = \prod_{n=1}^{R} \{Q_{Rn}\}^{-e_{i,n}} \prod_{n=1}^{R} Q_{R_n}^{e_{i,n}}, \tag{2.21}$$

where the first product is a numerical factor and the second product is a monomial combination of quantities with fixed values, which as a whole could be considered as the very reference quantity of the unit $[Q_i]$ homogeneous with it.

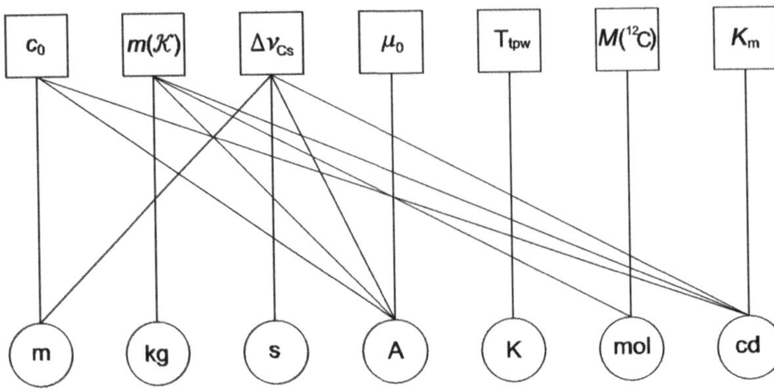

FIGURE 2.4 Direct relations of the SI base units with the references of the system.

Equation (2.21) could be the model for a general definition suitable both to base and derived units, thus making the distinction unnecessary.

As it looks evident from the preceding sections, a unit system is subject to changes, even if they are slowed by tradition and constraints due to the conventional nature of the unit definition, which requires a consensus to be reached on a potentially global basis. Redefinitions already occurred for the SI, as it was observed, the most recent regarding the length unit, which changed its reference from the wavelength of the krypton 86 radiation to the velocity of light through the second.

The change of the reference quantity of a unit in favor of a more convenient one is the normal reason for a redefinition. The convenience of a reference quantity may have more than one aspect: a new and better standard available for that unit, a more general definition that endures even if technology progress offers different and improved standards, advantages related to having a fixed value for the reference quantity. In the case of the light velocity c_0, the last two aspects were relevant. In fact, the length unit has always been realized through an interferometric system and the laser sources can change, following the technical development, without involving the unit definition. In addition, an exact value of c_0 turned out to be an advantage for science. In particular, being the magnetic constant already exactly fixed by the ampere definition, also the electric constant $\epsilon_0 = 1/\mu_0\, c_0^2$ and then the characteristic impedance of vacuum $Z_0 = \sqrt{\mu_0/\epsilon_0}$ became exact.

In case of redefinition of a unit, a conventional rule is that the value of a measurand must be the same before and after the change of definition (*continuity* requirement). Such a rule is almost automatically complied with if the value of the new reference quantity is determined in terms of the abandoned unit and the same value is fixed by the new definition.

The redefinition of units, and then the change of the reference quantities, is the action through which the unit system is evolved. Recalling Table 2.2, one can note that the most advanced stage of evolution corresponds to the adoption of fundamental constants as unit system references. Taking this as an indication of trend, a future

SI more completely based on fundamental constants could be imagined. In fact, proposals in that direction have been forwarded [1, 2], followed by recommendations of the *Comité International des Poids et Mesures* (CIPM), promoting the necessary research activity [3], and a detailed resolution of the CGPM which outlines the new SI [4].

The revision consists in the introduction of four fundamental constants as reference quantities for the redefinitions of kilogram, ampere, kelvin and mole. The proposed new reference constants are: the Planck constant h, the elementary (electron) charge e, the Boltzmann constant k and the Avogadro constant N_A. The values to be assigned to the constants would be those recommended by CODATA as a result of the most recent least square adjustment, at the time of redefinition [5].

As examples of unit redefinitions, in case the four proposed fundamental constants were adopted as SI references, those of the ampere (a base unit) and the joule (a derived unit) are considered. Ignoring the definition wordings and related discussion, their mathematical expressions would be respectively, according to (2.21),

$$\mathrm{A} = \{e\}^{-1} \left\{\Delta\nu_{\mathrm{Cs}}\right\}^{-1} e\, \Delta\nu_{\mathrm{Cs}} = 6.789\,687\,4... \times 10^{8}\, e\, \Delta\nu_{\mathrm{Cs}} \tag{2.22}$$

and

$$\mathrm{J} = \{h\}^{-1} \left\{\Delta\nu_{\mathrm{Cs}}\right\}^{-1} h\, \Delta\nu_{\mathrm{Cs}} = 1.641\,739\,26... \times 10^{23}\, h\, \Delta\nu_{\mathrm{Cs}}, \tag{2.23}$$

where the numerical values would be truncated according to the accuracies of the respective recommended values at the time of redefinitions.

In (2.22) the reference quantity $e\, \Delta\nu_{\mathrm{Cs}}$ is an electric current, homogeneous with the ampere, and in (2.23) $h\, \Delta\nu_{\mathrm{Cs}}$ is an energy, homogeneous with the joule. Thus, the structure of such definitions corresponds to that of the simplest SI base unit definitions reported above.

Again in the hypothesis of adopting the above fundamental constants as SI reference quantities, a great benefit in terms of reduced uncertainties would result for several other important constants, such as the electron and proton masses, the Josephson constant and the von Klitzing constant (on which most of the present electrical metrology is based), the Faraday constant, and the molar gas constant. As a drawback, the magnetic constant should be released, because it would no longer be fixed by the ampere definition, but it should be determined or derived from other constants, yet with a small uncertainty.

From a general point of view, referring the set of physical base units to basic invariants having theoretical implications seems most adequate for a unit system conceived also with a concern about its use in experimental sciences.

As an exception, with little coherence with such a concern, the cited redefinition proposals do not consider the case of $\Delta\nu_{\mathrm{Cs}}$, evidently because of the extremely high accuracy of atomic clocks based on particular parameters of ions or atoms. In reality, that constant, which is involved in four base units and in many derived units, has no fundamental role in theory and then his exact value does not improve the accuracy of other constants.

A possible redefinition of the time unit entirely referred to fundamental constants could be based on the Compton frequency of the electron $\nu_C = m_e c^2/h$ in place of $\Delta\nu_{Cs}$, where m_e is the electron mass. If the value of m_e were fixed by an easily understandable redefinition of the kilogram as a multiple of that mass [6], the above redefinition of the second would maintain the fixed value of h, with the consequent advantages for electrical units.

In spite of a much larger realization uncertainty, yet inconsequential for time metrology (see Section 2.7.4), such a definition of the second, when adopted together with the other proposed redefinitions, would have some advantages: a greatly improved accuracy of the mass of other elementary particles, such as the proton and neutron, of the fine structure constant and then of the electric constant, the magnetic constant and the characteristic impedance of vacuum.

Not all the units can be referred to a fundamental constant, if the quality of fundamental depends on the role of the constant in scientific theories. The first SI base units are certainly suitable to be defined by reference to fundamental constants, while for example, the candle, being a quantity of perception, is referred to a parameter related to the sensitivity of human eye, which could hardly be considered a fundamental constant, yet most important for photometric measurements.

In conclusion, a future possible system based on fundamental constants could have a basic core referred to a set of constants of high theoretical relevance, covering the field of physical quantities, which could be extended potentially to all measurable quantities by addition of other reference quantities not necessarily playing a fundamental theoretical role.

2.7 REALIZATION OF UNITS AND PRIMARY STANDARDS

For the definition of a unit to be effective in the measurement process, a quantity of the same kind must be made available as a reference with a known value. When the value of such a quantity is coherently derived from the unit definition, the procedure is considered a realization of the unit and the quantity becomes a primary standard of the same unit.

The unit realization is an essential step of the logical path underlying any measurement process performed within a unit system, as shown in Section 2.2 and Figure 2.2. Of course, the primary standard is not directly involved in measurements, but any standard acting as the reference in a measurement process is linked to a primary standard of the relevant unit through a chain of intermediate standards, hierarchically ordered, whose values are assigned and periodically verified by calibration procedures.

The realization process has different degrees of complexity depending on the kind of unit definition. The simplest case is that of a unit defined by a prototype, as the mass unit at present is, because the prototype itself is also the primary standard. Only a little less simple are the realizations of base units whose definition complies with the ideal model (2.7), as also a ratio is concerned.

A base unit with a definition complying with (2.10) requires a more complex realization procedure. However, even where an ideal experiment is described in the definition, any other experimental setup and procedure can be adopted for practical convenience, provided the equivalence to the ideal one can be proved through consolidated scientific theories.

In principle, once the very reference quantity of the unit is identified, according to its definition, any experiment establishing a relation between a measurement standard and that reference quantity can be considered as a realization of the unit.

As a general approach, considering the ultimate expression (2.21) of any unit as a function of the reference quantities of the system and recalling (2.4), if Q_{PS} is the quantity to be used as a primary standard, it can be written

$$Q_{\mathrm{PS}} = \{Q_{\mathrm{PS}}\} \prod_{n=1}^{R} \{Q_{\mathrm{R}n}\}^{-e_n} \prod_{n=1}^{R} Q_{\mathrm{R}n}^{e_n}. \tag{2.24}$$

As the very objective of the unit realization is the value $\{Q_{\mathrm{PS}}\}$ to be assigned to the primary standard, one can readjust (2.24) to obtain

$$\{Q_{\mathrm{PS}}\} = \frac{Q_{\mathrm{PS}}}{\prod_{n=1}^{R} Q_{\mathrm{R}n}^{e_n}} \prod_{n=1}^{R} \{Q_{\mathrm{R}n}\}^{e_n} = \rho_{\mathrm{E}} \prod_{n=1}^{R} \{Q_{\mathrm{R}n}\}^{e_n} . \tag{2.25}$$

In Equation (2.25), ρ_{E} is evaluated as the ratio between Q_{PS} (or its expression derived from a mathematical model of the realization experiment) and the reference quantity of the related unit $[Q_{\mathrm{PS}}]$, while the values $\{Q_{\mathrm{R}n}\}$ are exactly known. Thus, the realization of a unit can be reduced to the experimental determination of the ratio ρ_{E} between homogeneous quantities, in principle independent of any measurement unit.

A realization process as that outlined above is suitable for both base and derived units, whenever the measurement of ρ_{E} is experimentally convenient. Otherwise, it is possible to obtain the value of the primary standard from primary standards of other quantities, thus relying on their respective realizations and on the coherence of the system. In this case, the suitability of an experiment to perform as a unit realization can be verified on its mathematical model. In fact, let $Q_{\mathrm{PS}} = \prod_i Q_i^{e_i}$ be the mathematical model, then recalling (2.4) it must be

$$Q_{\mathrm{PS}} = \prod_i \{Q_i\}^{e_i} \prod_i [Q_i]^{e_i} = \prod_i \{Q_i\}^{e_i} \cdot [Q_{\mathrm{PS}}] . \tag{2.26}$$

If all the units in the expression $\prod_i [Q_i]^{e_i}$ cancel except for the one of Q_{PS} and the numerical values $\{Q_i\}$ are determined by traceable measurements, then Equation (2.26) establishes the relation between primary standard and measurement unit required by the realization process.

Whatever realization process is implemented, the related experimental system must be represented by a mathematical model which necessarily implies a simplification of the real experimental system. Therefore, the residual between the model and the real system contributes more or less to the realization uncertainty of the unit, that is to the uncertainty associated to the value of the primary standard.

To evaluate the components of the realization uncertainty related to model approximation may be very difficult, so that they are rarely accounted for in uncertainty budgets. In fact, values of primary standards obtained with different procedures sometime result incompatible.

For this reason, it is important to carry out independent realizations of the same unit and to compare the related primary standards, so that the result of each realization is validated by its compatibility with the others, or a possible neglected systematic effect is revealed by a discrepancy.

In practice, the primary standards are available at the national metrological laboratories and the international laboratory, the *Bureau International des Poids et Mesures* (BIPM) [7]. Primary standards of different laboratories are intercompared through international exercises where travelling standards are circulated and measured by reference to the different primary standards. Then from the measurement results a reference value is obtained and the residuals of every standard are made known.

From the primary standards, secondary and lower rank standards are derived with the purpose of expanding as far as possible the reference to the SI units. This activity, referred to as *dissemination of units*, is carried out through a network of metrological laboratories specifically devoted to calibration of reference standards or instruments. The result of the unit dissemination is the *traceability* of potentially all the measurements results expressed in SI units to the respective unit definition. The traceability implies the existence of a documented chain of calibrations of all of the standards, hierarchically ordered, through which a measurement result is related to the relevant unit.

The calibration of a measuring instrument has been outlined at Section 2.2 as a means to implement the function of a measurement process, leaving aside the analysis of the operations performed inside the instrument. The result of an instrument calibration is the set of deviations of the reading values from the corresponding values of the applied standards. Such deviations can be simply compared with the specified uncertainties associated with the different ranges of values, to verify their compliance. Otherwise, the deviations can be registered in the form of a table, or diagram, which allows the reading values to be converted into measurement values. In digital instruments, the deviations can be stored in a memory and used to automatically correct the output values, so that the displayed values, together with the associated uncertainties, can be directly considered as measurement results.

Going backward along the traceability path, the calibrations concern measurement standards rather than instruments. The calibration of a measurement standard consists in the comparison with another standard whose value is traceable to the unit through a primary standard. The output of the comparison system is a deviation, which is used to assign a value to the standard under calibration or to update a value coming from a previous calibration.

In general, a standard calibration is a measuring process where the standard quantity submitted to calibration takes the place of the measurand and the reference quantity is usually a standard of higher hierarchical rank. Therefore, the uncertainty associated with the value assigned to the calibrated standard is higher than the

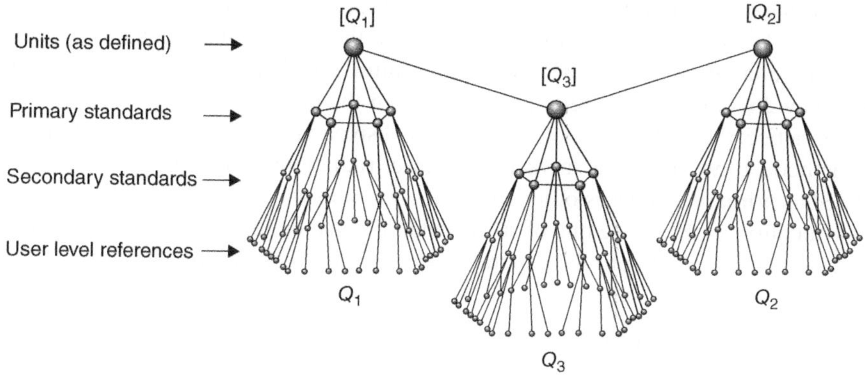

FIGURE 2.5 Extended traceability in a coherent system. An indirect measurement of $Q_3 = f(Q_1, Q_2)$, being the same function f of direct measurements of Q_1 and Q_2, is compatible with a direct measurement of Q_3 only if the coherence of the system guarantees that for the respective units it is $[Q_3] = f([Q_1], [Q_2])$. In other words, the indirect measurement of Q_3 can be considered as traceable to $[Q_3]$ through the links among the units established by their definitions in a coherent system.

one of the reference, whose uncertainty should be possibly negligible or very smaller.

The time of validity of a calibration is limited and may range from months to years, depending on the performance of the measurement standard or instrument.

In the conceptual diagram of Figure 2.5, the measurement traceability in the SI is addressed. Within each metrological system of a given quantity, the compatibility of measurements performed at the user level is guaranteed on a global basis by the traceability to even different primary standards, provided their equivalence has been verified by means of intercomparisons. Again within the same metrological system, the realization uncertainty of the relevant unit is irrelevant for the compatbility of direct measurements [6].

On the contrary, the unit realization uncertainty is an important component in the uncertainty budget of indirect measurements, whose traceability to the relevant unit relies on the coherence of the unit system, guaranteed by the definitions of units and their realizations.

In the following subsections of Section 2.7, the realization of SI base units and of other units of utmost importance in metrological activity is outlined and a general information about the measurement standards used in the first steps of the dissemination activity is given.

2.7.1 Meter Realization and Length Standards

Starting from the meter definition expressed in mathematical terms by (2.11) and substituting the second with its mathematical expression given by (2.13), one obtains

$$\mathrm{m} = \frac{\{\Delta\nu_{\mathrm{Cs}}\}}{\{c_0\}} \frac{c_0}{\Delta\nu_{\mathrm{Cs}}}, \tag{2.27}$$

which is the meter definition in the most general form (2.21). In (2.27), the numerical values $\{c_0\}$ and $\{\Delta\nu_{Cs}\}$ are exactly stated by the meter and second definitions, while the quantity $c_0/\Delta\nu_{Cs}$, which is the vacuum wavelength corresponding to the frequency $\Delta\nu_{Cs}$, is the very reference quantity of the meter.

In practice, the realizations of the meter are displacement-related, by interferometric methods, to wavelengths in vacuum of laser radiations. Thus, the quantities considered as primary standards of length are the vacuum wavelengths of well-identified laser radiations. Let λ_{PS} be one of such standard wavelength and λ_{Cs} the wavelength $c_0/\Delta\nu_{Cs}$, then according to the general model for unit realizations (2.25) it is

$$\{\lambda_{PS}\} = \frac{\lambda_{PS}}{\lambda_{Cs}}\{\lambda_{Cs}\} = \frac{\Delta\nu_{Cs}}{\nu_{PS}}\{\lambda_{Cs}\}, \tag{2.28}$$

where the ratio $\rho_E = \lambda_{PS}/\lambda_{Cs}$, to be determined by experiment, is much better measured as the reciprocal of the frequency ratio $\Delta\nu_{Cs}/\nu_{PS}$, being $\nu_{PS} = c_0/\lambda_{PS}$.

Interferometric systems of primary level include a sub-system based on frequency comb techniques, to measure the above frequency ratio for any laser radiation used. Otherwise, one can choose a laser radiation among those recommended by CIPM for the *mise en pratique* of the meter definition [8]. The relative uncertainty level of these unit realizations is, at present, better than 10^{-11} in vacuum, reduced to 10^{-8} in air.

Secondary length standards are interferometric systems calibrated by comparison with a primary system in terms of frequency of the laser source. Lower level standards of several shapes are used to match the different geometries of the objects to be measured. Among these, the gauge blocks, or Johansson gauge, are the most diffused standards, which are parallelepiped with a pair of opposite parallel faces as the standard length. They are usually grouped in sets of standards for various lengths, the most used being sets from 1 mm to 100 mm, while sets of longer blocks range from 100 mm to 1000 mm.

The gauge blocks of the highest grade are directly calibrated by interferometric method with a resolution of less than 1 nm and have associated relative uncertainties ranging from the level of 2×10^{-5}, for 1 mm, to 5×10^{-8} for 1 m. Block sets of lower grades are calibrated by comparison and provide a basis for the dissemination of length to the measuring instruments, such as coordinate measuring machines and calipers of various kinds.

A dissemination line, parallel to the one based on gauge blocks, uses optical line scales as length reference included in linear encoders, which can provide displacements with a resolution of a few nanometers and relative accuracy levels approaching those of the gauge blocks.

2.7.2 Kilogram Realization and Mass Standards: Present Situation

Different from other units, the kilogram realization is unique and the primary standard is coincident with the international prototype, made of 90% platinum and 10% iridium, conserved at the BIPM. That is, according to (2.12) and (2.25),

$$\{m_{PS}\} = \{m(\mathcal{K})\} = 1. \tag{2.29}$$

The advantages of a unity value and of a null realization uncertainty is balanced by some disadvantages: the primary standard is neither available unless at BIPM, nor reproducible, while its time invariance is put in question for possible gas absorption or desorption and surface alterations.

In order to compensate for the first disadvantage, a number of copies of the international prototype have been compared with the primary standard and distributed to the countries members of the Metre Convention, so that they can perform as primary standards of the unit on a national basis.

These copies have a typical relative uncertainty of 3 to 5 parts in 10^9. By comparison with them, secondary kilogram standards, usually made of stainless steel, are calibrated with a relative uncertainty of 6 to 8 parts in 10^9. From these, reference multiples and submultiples of the kilogram are obtained in a range normally from 1 mg to 500 kg. This reference scale is used to calibrate stainless steel working standards, which are used for dissemination.

The copies of the international prototype have been compared with it several times and a drift has been put in evidence of a few parts in 10^8 in a century. As the copies are similar to the prototype, one could suspect that their drift and that of the prototype itself might be even larger. Hence the proposals of a redefinition of the kilogram by reference to physical invariants.

2.7.3 Kilogram Realization: Future Perspective

Referring to the set of reference constants indicated in the resolution of the 24th CGPM [4] for the revised SI, the new definition of the kilogram would result, in substantial terms,

$$\mathrm{kg} = \frac{\{c_0\}^2}{\{h\}\,\{\Delta\nu_{\mathrm{Cs}}\}}\,\frac{h\,\Delta\nu_{\mathrm{Cs}}}{c_0^2}. \tag{2.30}$$

Then, for a unit realization following (2.25), the measurement of the ratio ρ_{E} between a material standard mass and that immaterial reference quantity would seem a quite problematic experiment.

In reality, an experiment has been pursued for a long time which can relate a material mass to the Planck constant through kinematic quantities. The experiment, referred to as *moving-coil balance* or *watt balance* [9, 10], was initially conceived as a realization of the ampere and volt, but has been also finalized to the determination of that fundamental constant.

The name of watt balance comes from the performed comparison between electrical and mechanical power, which can be considered as a realization of the electrical watt (see Section 2.7.5). The experiment is based on two different conjugated interactions between a coil and a magnetic flux of induction B.

One interaction concerns the force F developed by a current I flowing in the coil. The other interaction regards the electromotive force E induced in the same coil as an effect of a velocity v in the direction of the force (hence the name of moving-coil balance). Both effects depend on the same geometrical configuration of coil and magnetic flux, that can be accounted for with a factor γ_B in both the force equation $F = \gamma_B I$ and the velocity equation $E = \gamma_B v$. Eliminating γ_B from both these equations yields the power equation

$$F v = m_{PS} g v = E I, \tag{2.31}$$

where the first equation refers to the comparison performed by the balance between F and the force acting on the standard mass m_{PS} by the effect of the gravitational acceleration g.

The relation between a material mass and the physical constants whose combination constitutes the unit reference quantity can be established measuring E and I in (2.31) by reference to electrical standards based on quantum phenomena, namely the Josephson effect [11] and the quantum Hall effect [12].

The standards based on quantum effects are dealt with in some more details in Section 2.7.6. Here it is sufficient to anticipate that the Josephson voltage is proportional to $\nu_J \, h/e$, ν_J being the frequency of a microwave involved in the standard, and the quantum Hall resistance is proportional to h/e^2. The quantum Hall resistance is concerned as a reference in the measurement of a shunt resistance in series with the coil, while the respective voltage drop is measured by reference to the Josephson standard to determine the value of I, which results proportional to $e \, \nu_J$.

Therefore, for (2.31), where also E is measured by reference to the Josephson standard, we can write

$$m_{PS} = \kappa \, \frac{h \, \nu_J^2}{g \, v} = \kappa \, \frac{\{h\} \, \{\nu_J\}^2}{\{g\} \, \{v\}} \, \text{kg}, \tag{2.32}$$

where κ is a numerical factor accounting for all the above mentioned proportionality coefficients, h would have an exactly fixed value and the other quantities can be measured during the experiment. The last equation in (2.32) is obtained following the procedure indicated in (2.26) to verify that in the expression all units cancel except for the kilogram, thus confirming the experiment as a realization of the unit and m as a primary standard of mass.

One could come to the same conclusion through the generalized model for unit realizations (2.25). However, in this case, apart from the conceptual interest for a realization procedure independent of any measurement unit, the recourse to traceable measurements of some quantities is acceptable and more convenient in practice.

At present, the uncertainties of experiments like those outlined above can attain uncertainty levels of some parts in 10^8, which could be adequate to dissemination requirements through secondary metrological laboratories.

An alternative realization of the kilogram redefined as in (2.30) is offered by an experiment finalized to the determination of the Avogadro constant (see section 2.3.8). The experiment consists of two main steps: one determines the number of ^{28}Si atoms in a macroscopic sphere obtained from a single silicon crystal; the other relates the mass of the ^{28}Si atom to the Planck constant through other fundamental constants.

The relation between the atomic scale dimensions and those of a macroscopic body is made possible by the extreme regularity of the crystal lattice and the high accuracy measurement of the lattice spacing performed with a combined X-ray and optical interferometric technique. Let m_{PS} be the mass of the silicon sphere, assumed as primary standard, V_{PS} its volume and $V(^{28}\mathrm{Si})$ the volume taken by a single atom. Then, it is

$$m_{PS} = \frac{V_{PS}}{V(^{28}\mathrm{Si})}\, m(^{28}\mathrm{Si}), \tag{2.33}$$

where $V(^{28}\mathrm{Si})$ is determined through the measurement of the edge of the cubic elementary crystal lattice, at present performed with a relative uncertainty of a few parts in 10^9.

The mass of the silicon atom $m(^{28}\mathrm{Si})$ can be indirectly related to h through the electron mass m_e, which is included in the expression of the Rydberg constant R_∞, one of the best known fundamental constants (relative uncertainty of parts in 10^{12}, at present), together with h, c_0 and the fine structure constant α. Namely, being $R_\infty = \alpha^2 m_e\, c_0/\, 2h$, one can write

$$m(^{28}\mathrm{Si}) = \frac{m(^{28}\mathrm{Si})}{m_e}\, m_e = \frac{A_r(^{28}\mathrm{Si})}{A_r(\mathrm{e})}\, \frac{2\, R_\infty}{\alpha^2\, c_0}\, h, \tag{2.34}$$

where the ratio between relative atomic masses $A_r(^{28}\mathrm{Si})/A_r(\mathrm{e})$ is very well measured with mass spectrometry techniques.

Thus, from (2.33) and (2.34) an expression of m_{PS} is obtained as a function of exact or well-known constants, except for the volume of the silicon sphere, whose measurement determines the unit realization uncertainty.

Other proposals of kilogram redefinition refer more intuitively to the real mass of a particle, such as the electron mass m_e, which is also a fundamental constant [6, 13, 14]. In that case, the new definition would simply correspond to the equation $\mathrm{kg} = \{m_e\}^{-1} m_e$, thus fixing $\{m_e\}$.

One can easily verify, from (2.33) and (2.34), that the last experiment can realize the kilogram also by reference to an exact value of m_e. Moreover, also the experiment of the moving coil or watt balance could realize the kilogram. In fact, introducing in (2.32) the Compton frequency of the electron (cf. 2.6 and see section 2.7.4) as a time reference for the measurement of ν_J, g and v, one can verify that the mass m_{PS} results expressed as a multiple of m_e.

2.7.4 Realization of the Second and Time Standards

The reference quantity of the unit of time, as appearing in the mathematical expression of its definition (2.13), is the period of a particular radiation of ^{133}Cs atom. As a radiation is usually characterized rather by its frequency than by its period, all the standards used in time metrology are characterized by their frequency, whose well known SI unit is the hertz. In particular, a realization of the second is also considered as a realization of the hertz.

A realization of the second consists of an experiment where a radiation corresponding to the hyperfine splitting of ^{133}Cs in its ground state is made physically available with a minimum associated uncertainty. The period $1/\Delta\nu_{Cs}$ of such a radiation, coincident with the reference quantity of the time unit, can be considered as a primary standard of time. Then, in (2.25) the experimental ratio ρ_E is equal to 1 and the equation degenerates in the simple identity $\{t_{PS}\} = 1/\{\Delta\nu_{Cs}\}$.

The most advanced experimental system to realize the second is the cesium fountain oscillator, where the name is due to the particular movement impressed to the cesium atoms, similar to that of water in a fountain. In fact, the ^{133}Cs atoms are grouped together and cooled near a temperature of 0 K by a series of infrared lasers. Then, by a pair of vertically opposed lasers, slightly untuned to produce moving stationary waves, the atoms are slowly tossed upward and left to fall down by switching off the lasers. In an intermediate region of their trajectory, the atoms pass through a cavity where they interact with a microwave whose frequency is tuned so as to produce a maximum of transitions to the altered hyperfine level. In this condition, the microwave frequency is just $\Delta\nu_{Cs}$ with an associated relative uncertainty of 10^{-15} or even slightly smaller.

Beside the cesium fountain standard, several other atomic clocks are used or developed for time and frequency metrology. The most performing, in terms of stability and reproducibility, are those based on radiations in the optical band, some of which with associated uncertainty comparable to that of the primary standards or even smaller. The same radiations recommended by the CIPM for the practical realization of the meter [8] are also recommended as secondary standards of time. The ratio between any optical frequency and $\Delta\nu_{Cs}$ is established with the maximum accuracy through the frequency comb technique, as it is for the primary length standards.

The dissemination of the second takes advantage of the property of time and frequency standards of being compared without physical transfer of instruments or systems. In fact, the time signal transmissions have practically completely replaced the transfer of standard clocks. At primary level, comparisons of different unit realizations are performed through two-way satellite communications. For distances up to some hundreds kilometers, optical fiber communication is also suitable. At secondary and lower levels, clocks can be synchronized through the Global Positioning System (GPS).

As it is also the case for length measurements, the reference quantity of the second is extremely smaller than the unit. This allows most time intervals to be generated by simple count of numbers of periods of the reference signal. In particular, permanent time scales, such as the International Atomic Time (TAI) and the Coordinated Universal Time (UTC), are maintained with the contribute of several national metrological laboratories.

Future long-term perspectives, not yet officially supported, regard the replacement of the reference frequency adopted by the second definition with the frequency of an atom or ion optical radiation with better performance.

Considering that the second is involved in almost all the present SI derived units (cf. Table 2.3) and above all that the proposed new definitions of the base units, except for the mole, depends on $\Delta\nu_{Cs}$, one may find peculiar the persistent proposals of time reference quantities of no relevance in scientific theories.

Indeed, should the electron Compton frequency $\nu_C = m_e c^2/h$ be adopted as the reference quantity of the second (cf. 2.6), together with m_e as the reference quantity of the kilogram (cf. 2.7.3), the realization uncertainty of the second would be some orders of magnitude larger than with a definition based on an atomic standard, and this may seem hardly acceptable. Therefore, the traditional use of assuming as the reference quantity of the unit the best available standard will be probably maintained.

However, one might consider that, as a matter of fact, the unit realization uncertainty does not affect the uncertainty associated with every step of the traceability chain [6], so in time and frequency metrology the best standards could continue to be used, but with values subject to possible readjustment after new improved realizations referred to ν_C. Only in indirect measurements the unit realization uncertainty is involved, but the one of time unit would always be easily negligible compared to those of the other involved units.

Furthermore, in addition to the already cited advantages for the experimental sciences due to reduced uncertainties of some other constants and to a more rational structure of the whole unit system, with a unit reference quantity based on a fundamental constant such as m_e the time unit definition should not be changed to follow the development of new and better atomic clocks.

2.7.5 Electrical Unit Realizations and Standards: Present Situation

The practical realizations of electrical units are rather independent of the wording of their definitions, including the realization of the base electrical unit, the ampere, which cannot follow the ideal experiment described in its present definition. Therefore, the realizations of the most important electrical units will be dealt with following the actual line of derivation of their primary standards.

2.7.5.1 Watt Realization After the introduction of the present definition of the ampere, the base electrical unit was realized by means of electrodynamic balances where the force produced by interaction between fixed and moving coils was measured opposing the gravitational force on a standard mass. The accuracy of this experimental systems was limited by the difficulty in establishing an equivalence with the ideal experiment of the definition, essentially for the complex geometrical characterization of the coils. A decisive improvement has come from an additional phase of the experiment, which avoids any geometrical measurement and calculation on the coils, replacing them by the measurement of the electromotive force induced in the moving coil by a constant-velocity displacement. Of course, we are dealing with the experimental system already outlined in Section 2.7.3 as moving-coil or watt balance. The result of the experiment is not the value of the current flowing in the coil, but the electric power corresponding to the product of this current by the induced electromotive force.

The second equation in (2.31) allows the product $E\,I$ to be measured in terms of three mechanical quantities. Such a product is only a virtual electric power, as it does not correspond to any energy flow. However, comparing E and I respectively with a voltage and a current at a given port of an electrical circuit, one could measure the actual electric power flowing through that port in terms of mechanical units. Thus, the experiment can be considered as a realization of the electric watt and the pair of quantities E and I, taken together, as a primary standard of electric power.

It is worthy of note that a recurrent comparison of the electrical to mechanical power, just as performed by the watt balance, was warmly recommended by Giorgi as essential for maintaining the coherence of the unit system.

The moving-coil balance has become a key experiment in primary metrology and different systems have been conceived, some of them still being developed. At present, the least relative uncertainty of the experiment is of a few parts in 10^8.

There are no maintained standards of electrical power, but a power standard at the secondary level is indirectly obtained combining voltage and resistance standards. Only for alternate regime, there are standard power generators (calibrators) and measuring instruments (standard wattmeters) operating at relative uncertainty levels up to about 10^{-4}.

2.7.5.2 *Ohm Realization* The reference quantity of the unit of electrical resistance $Q_R(\Omega)$, in the ultimate sense expressed by (2.21), can be obtained from the expression given for the ohm in Table 2.3. Substituting in that expression the base units with the respective reference quantities, one obtains $Q_R(\Omega) = \mu_0\, c_0$. Considering that $c_0 = 1/\sqrt{\mu_0\,\epsilon_0}$ and the characteristic impedance of vacuum is $Z_0 = \sqrt{\mu_o/\epsilon_0}$, one can also write $Q_R(\Omega) = Z_0$ and finally

$$\Omega = \{Z_0\}^{-1}\, Z_0, \tag{2.35}$$

where $\{Z_0\} = \{c_0\}\,\{\mu_0\}$ is exactly $299\,792\,458 \times 4\,\pi \times 10^{-7}$ and approximately 376.73.

Therefore, according to (2.25), if R_{PS} is a standard resistor to be qualified as a primary standard, it must be $\{R_{PS}\} = \rho_E\,\{Z_0\}$, where the ratio $\rho_E = R_{PS}/Z_0$ is to be determined.

The impedance Z_0 is indirectly accessible to measurement through an experiment referred to as *calculable capacitor* or *cross capacitor* and based on the Lampard theorem [15]. The theory can be applied to a variety of cylindrical conductive surfaces, but its metrological applications make use of circular-section cylindrical bars symmetrically disposed close one another with parallel axes. In its simplest form, the system has four bars acting as electrodes of two capacitors obtained with different connections of the bars. Each of the two capacitors is formed by a pair of opposite bars, the others being connected to the shield. The electric flux in the two configurations has cross directions, hence the name of cross capacitor.

When the two capacitances are adjusted to be equal, their common value is

$$C_C = \epsilon_0 \frac{\ln 2}{\pi} l, \quad (2.36)$$

where l is the axial length, delimited by shielding bars inserted at the opposite ends of the structure. Besides an intermediate step to the ohm realization, the calculable capacitor is considered as the realization of the farad, the unit of electrical capacitance.

The capacitance is measured with the highest accuracy through its low-frequency admittance $1/Z_C = 2\pi\nu\, C_C$, ν being a frequency of the order of the kilohertz. The measuring systems are comparison bridges based on fully coaxial circuitry, high-accuracy transformer ratio and capacitance standards configured as two-port or four-port devices.

The relation between Z_0 and Z_C is obtained from the above relations substituting ϵ_0 with its expression derived from (2.36). The result is

$$Z_0 = \frac{1}{c_0\, \epsilon_0} = \frac{\ln 2}{\pi} \frac{l}{c_0\, C_C} = 2 \ln 2 \frac{l}{c_0/\nu} Z_C. \quad (2.37)$$

With (2.37), the experimental ratio ρ_E reduces to the impedance ratio R_{PS}/Z_C and the measurement of the dimensionless quantity $l/(c_0/\nu)$. That is, the particular expression of (2.25) results in

$$\{R_{PS}\} = \frac{R_{PS}}{Z_0} \{Z_0\} = \frac{1}{2 \ln 2} \frac{R_{PS}}{Z_C} \frac{c_0/\nu}{l} \{Z_0\}. \quad (2.38)$$

The length c_0/ν, corresponding to a wavelength, could be expressed in terms of the wavelength $\lambda_I = c_0/\nu_I$ of the laser radiation of frequency ν_I used by the interferometer as $l/(c_0/\nu) = (l/\lambda_I)(\nu/\nu_I)$. Thus, theoretically, the required measurement could be split into the interferometric measurements of a length ratio and the measurement of a frequency ratio, without involvement of any units. Practically, ν and l may also be measured by reference to primary standards of length and frequency, with negligible loss of accuracy.

In reality, the measurement of the impedance ratio R_{PS}/Z_C is a complex procedure, involving capacitance ratios and, in particular, the capacitance–resistance comparison, carried out by means of the so-called *quadrature bridge*. At the end of the complex procedure, one may identify as the primary standard of resistance R_{PS} any of the two resistors, usually of 10 kΩ, directly involved in the quadrature bridge. However, the realization process is considered complete when a resistance value is assigned to a standard resistor in direct-current regime. This last step is performed through a special resistor with calculable frequency characteristic. The whole ohm realization procedure can be performed with a final relative uncertainty of a few parts in 10^8.

The different groups of reference resistors maintained at a national laboratory usually range from 1 Ω to the gigaohm and beyond, at reduced accuracy levels.

Only a few national laboratories perform the calculable capacitor experiment. More recently, the national metrological laboratories, beside the traditional resistance standards, are also equipped with a quantum Hall effect apparatus, already mentioned in Section 2.7.3. The resistance values produced by such a quantum effect (see Section 2.7.6) are apparently independent of its physical realization and this allows diffferent laboratories to be provided with highly consistent resistance references.[4] On the other side, assuming as consolidated theories both the relation $R_K = h/e^2$ and $\alpha = e^2/(2\,\epsilon_0\,h\,c_0)$, α being the fine structure constant, the most used quantum Hall resistance, $R_H(2) = R_K/2$, results in $\{R_H(2)\} = (R_H/Z_0)\,\{Z_0\} = \{Z_0\}\,/4\,\alpha$. Thus, the quantum Hall experiment could be considered a realization of the ohm alternative to the procedure based on the calculable capacitor, α being determined with a relative uncertainty of a few parts in 10^{10}.

The ohm dissemination is carried out with wire resistors, typically of 10 kΩ, and, at lower accuracy level, with multifunction digital calibrators and digital multimeters.

2.7.5.3 *Volt Realization* Traditionally, the volt was realized by means of a balance in which an electrostatic force acting on the moving electrode of a capacitor is measured in terms of the gravitational force on a known mass. The electrical force depends on the square of the voltage between the electrodes, whose value in volt is the objective of the experiment, and on their geometrical characteristics, which can be calculated as a function of dimensional parameters or determined as derivative of the capacitance with respect to a displacement of the moving electrode.

As it was for the current balance, the watt balance has been preferred for the volt realization. In fact, the realization of the watt in terms of electrical quantities determines the product $E\,I$ as a function of a mass and kinematic quantities, as expressed by (2.31). The realization also of the ohm allows us to separate either the voltage or the current, thus yielding realizations of the respective SI units.

In the experiment of the moving coil balance, the current flowing in the coil is measured through the voltage $U = R\,I$ produced on a shunt of resistance R, traceable to a primary resistance standard. If the ratio $\rho_U = E/U$ is also determined, then for (2.31) it is $E\,I = E^2/(R\,\rho_U) = m\,g\,v$, from which

$$U_{PS} = E = \sqrt{\rho_U\,R\,m\,g\,v} = \sqrt{\rho_U\{R\}\{m\}\{g\}\{v\}}\ \mathrm{V}. \tag{2.39}$$

In (2.39), the electromotive force E induced in the moving coil during the experiment is considered as the primary standard of voltage and the quantities under square root are to be measured during the experiment, while the last term of the equation applies the criterion of (2.26). The relative uncertainty associated with the volt realization is nearly at the same level as the one for the watt realization, that is, parts in 10^8.

In practice, the electromotive force E is compared with the voltage of a Josephson standard $U_J = n\,\nu_J/K_J$, where $K_J = 2e/h$ is the Josephson constant, which

[4]In 1990, the conventional value $R_{K-90} = 25,812.807\ \Omega$ was recommended by the CIPM for the von Klitzing constant, so that the quantum Hall resistance $R_H = R_K/2$ can be used worldwide as a common resistance reference.

then is determined by the experiment. The same standard, already mentioned in Section 2.6.3 and more specifically outlined in Section 2.7.6, can also be used to measure the ratio ρ_U with very high accuracy and allows the same voltage to be reproduced indefinitely, even in different laboratories.[5]

Good reference voltage standards were, in the past, electrochemical standard cells, then completely replaced by standards based on temperature-stabilized Zener diodes. However, by far the best voltage standards are now based on arrays of large numbers of Josephson junctions, to obtain voltages of 1 V, 10 V, or of programmable values. These standards are usually maintained by national metrological laboratories, while secondary laboratories and calibration centers are equipped with transportable digital calibrators, even multifunction, which make available standard voltage scales in a wide range of values, both in continuous (d.c.) and alternate (a.c.) regime. Such scales mainly correspond to those of digital measuring instruments to be calibrated or verified.

2.7.5.4 Ampere Realization The original realization of the ampere by means of current balances, as practical implementation of the ideal experiment of the definition, has been discussed above with regard to the watt realization, together with the improvements brought by the introduction of the moving-coil technique.

As it was for the volt realization, the realization of the ohm makes it possible to separate also the current from the product $E\,I$ determined by the moving-coil experiment. Again from (2.31) one can obtain $E\,I = \rho_U\,R\,I^2 = m\,g\,v$, from which

$$I_{PS} = I = \sqrt{\frac{m\,g\,v}{\rho_U\,R}} = \sqrt{\frac{\{m\}\ \{g\}\ \{v\}}{\rho_U\ \{R\}}}\ \text{A}. \tag{2.40}$$

The last equation in (2.40) corresponds to the procedure indicated by (2.26) to verify that all the units of the quantities appearing in the expression cancel excepted the ampere, and then the experiment can be considered a realization of that unit, I being a primary standard.

The accuracy of the ampere realization is at the same level as that of the volt realization (parts in 10^8), carried out with practically the same experiment.

The current I flowing in the moving coil during the experiment and considered a primary standard of current, is not easily reproducible, due to the complexity of the realization experiment, and then is not suitable to be used as a reference for the ampere dissemination. In practice, until now, the first steps of the ampere dissemination are carried out indirectly by reference to voltage and resistance standards traceable to the respective units. In particular, dedicated standard resistors (*shunt resistors*) are maintained as reference for calibrations. Only for calibration of current measuring

[5]Similar to what was done for the quantum Hall resistance standard, also for the Josephson standard in 1990 was recommended a conventional value of the constant K_{J-90}, to obtain a high degree of agreement among voltage measurements world wide. However, as it was already pointed out for resistance, in case of indirect measurements involving other quantities, the value in SI unit with the associated realization uncertainty is to be considered.

instruments, both in d.c. and a.c. regime, variable current sources (calibrators) are used.

2.7.6 Electrical Units Realization and Standards: Future Perspective

The officially supported proposals of unit redefinitions, fixing in particular the values of the Plank constant h and of the electron charge e, would greatly simplify and improve the primary metrology of electrical quantities. This is essentially due to the use of quantum standards of voltage and resistance, very strictly and neatly dependent on those fundamental constants.

2.7.6.1 *Volt Realization* In the proposed revision of the SI, the volt would be defined, in mathematical terms, as

$$\mathrm{V} = \frac{\{e\}}{\{h\}\left\{\Delta\nu_{\mathrm{Cs}}\right\}} \frac{h\,\Delta\nu_{\mathrm{Cs}}}{e}, \tag{2.41}$$

$h\,\Delta\nu_{\mathrm{Cs}}/e$ being the reference quantity of the unit.[6]

A voltage standard based on the Josephson effect is perfectly suitable as a volt realization with the new definition. This effect consists in the production of discrete voltage levels between weakly coupled superconductors irradiated with a microwave. The voltage levels are strictly dependent only on the frequency ν_{J} of the microwave and the Josephson constant $K_{\mathrm{J}} = 2\,e/h$. Namely, if n is the order of the voltage step, the corresponding voltage level is

$$U_{\mathrm{J}} = n\frac{\nu_{\mathrm{J}}}{K_{\mathrm{J}}} = n\frac{\nu_{\mathrm{J}}\,h}{2\,e}. \tag{2.42}$$

Considering a Josephson voltage as a primary voltage standard, its numerical value results, following (2.25), in

$$\left\{U_{\mathrm{PS}}\right\} = \left\{U_{\mathrm{J}}\right\} = \frac{U_{\mathrm{J}}}{h\,\Delta\nu_{\mathrm{Cs}}/e} \frac{\{h\}\left\{\Delta\nu_{\mathrm{Cs}}\right\}}{\{e\}} = n\frac{\nu_{\mathrm{J}}}{2\,\Delta\nu_{\mathrm{Cs}}} \frac{\{h\}\left\{\Delta\nu_{\mathrm{Cs}}\right\}}{\{e\}}, \tag{2.43}$$

where the only experimental datum is a frequency ratio, easily determined.

In fact, intercomparisons of standards constituted by arrays of thousands of Josephson junctions, carried out adopting the same conventional value for K_{J}, have shown average differences even below 1 nV (corresponding to relative uncertainties less than 10^{-9}, at 1 V level, and 10^{-10} at 10 V level), which demonstrates the level of accuracy of the unit realization in the prospected situation.

[6]The expression of the volt in (2.41), as also that of the ohm in (2.44) and of the ampere in (2.47), are obtained ignoring the quantum effects, in spite of their coincidence, but are derived from the expressions reported in Table 2.3 for ohm and volt as functions of the base units and the equations fixing the exact values of h, e and $\Delta\nu_{\mathrm{Cs}}$.

2.7.6.2 *Ohm Realization* In the same hypothesis of exact values for h and e, the new ohm definition would correspond to the equation

$$\Omega = \frac{\{e\}^2}{\{h\}}\frac{h}{e^2}, \tag{2.44}$$

where h/e^2 would be the reference quantity of the unit. Thus, a resistance standard based on quantum Hall effect fits very well in the new definition.

In order to outline such a quantum effect, let us refer to the well known Hall resistance, that is the ratio between a transversal voltage produced by a current, flowing in a conducting bar subject to a perpendicular magnetic flux, and the same current. Under special conditions,[7] the diagram of resistance versus magnetic induction shows plateaux corresponding to submultiples of a constant value, the von Klitzing constant R_K, which apparently is equal to h/e^2 in any experimental context.

In fact, the resistance values of those plateaux are extremely reproducible and independent of geometrical or material features. Namely they are

$$R_H = \frac{1}{i}R_K = \frac{1}{i}\frac{h}{e^2}, \tag{2.45}$$

where i is an integer, practically limited to a few units (Landau levels).

Assuming the experiment producing the quantum Hall effect as a realization of the ohm and the related resistance as a primary standard, for (2.25) its value will result in

$$\{R_{PS}\} = \{R_H\} = \frac{R_H}{h/e^2}\frac{\{h\}}{\{e\}^2} = \frac{1}{i}\frac{\{h\}}{\{e\}^2}, \tag{2.46}$$

with no experimental quantity to be determined beside the identification of the Landau level i.

The relative uncertainty of quantum Hall standards, evaluated in terms of average difference between standards adopting the same value of the von Klitzing constant, is of a few parts in 10^9 at present, with a trend of further reduction. Therefore, at the same level can be considered the realization uncertainty of the ohm in its prospected redefinition.

2.7.6.3 *Ampere Realization* The new ampere definition, in terms of number of elementary charges flowing in one second, would be

$$A = \frac{1}{\{e\}\left\{\Delta\nu_{Cs}\right\}} e\,\Delta\nu_{Cs}, \tag{2.47}$$

[7] The conditions are *(i)* the conducting bar reduced to a two-dimension separation surface between semiconductors in a heterostructure (two-dimensional electron gas); *(ii)* intense magnetic induction (around 10 T) produced by a superconducting coil; *(iii)* temperature close to 0 K.

$e\,\Delta\nu_{Cs}$ being the reference quantity of the unit. Evidently, such a reference quantity is a current equivalent to the flow of $\{\Delta\nu_{Cs}\}$ electron charges per second.

A current proportional to that reference current can be obtained with a combination of the experiments of quantum Hall and Josephson effects. In fact, suppose to balance the voltage U_J of a Josephson standard with the voltage produced in a quantum Hall standard by a current I. Then it will be $U_J = R_H\, I$, from which, recalling (2.42) and (2.45),

$$I = \frac{U_J}{R_H} = \frac{1}{2}\, i\, n\, e\, \nu_J. \tag{2.48}$$

Such a composite experiment could be a realization of the ampere and the current I a primary standard of current, whose value for (2.25) results in

$$\{I_{PS}\} = \frac{I}{e\,\Delta\nu_{Cs}}\ \{e\}\ \{\Delta\nu_{Cs}\} = \frac{1}{2}\, i\, n\, \frac{\nu_J}{\Delta\nu_{Cs}}\ \{e\}\ \{\Delta\nu_{Cs}\}, \tag{2.49}$$

where only the identification of integer numbers and the experimental determination of a frequency ratio are required.

The relative uncertainty associated with the ampere realization outlined above would be at the same level indicated for the ohm and the volt (parts in 10^9).

A more recent effect has been conceived, referred to as *single electron tunneling* (SET), where the transfer of single electrons through very thin barriers is obtained under the control of a high frequency alternate voltage, so that one electron per cycle is allowed to flow in an external circuit. Of course, an experiment based on SET effect would directly realize the definition expressed by (2.47). However, the current values so far attained with a reliability suitable for a metrological standard are still insufficient and different experimental approaches are being investigated.

2.7.7 Kelvin Realization and Temperature Standards: Present Situation

From the definition of the kelvin and from (2.16) it is clear that the reference quantity of the unit is the temperature of the triple point of water. Then, the realization of the kelvin consists in the implementation of a physical system where the three states of water are present together and the reference quantity is defined as the equilibrium temperature of vapour, liquid, and solid water, with the water having a defined isotopic composition.

As the realization experiment yields just the reference quantity of the kelvin, in (2.25) it is $\rho_E = 1$ and the equation degenerates in the simple form

$$\{T_{PS}\} = 1\ \{T_{tpw}\} = 273.16. \tag{2.50}$$

The uncertainty associated with the realization of the kelvin with the water triple point in sealed cell can approach 0.03 K (about 10^{-7} in relative terms).

It must be noted, because of its wide diffusion in practice, that another temperature scale is used whose unit is the Celsius degree (symbol °C) equal in magnitude to the kelvin. The Celsius temperature t is defined as $t = T - T_0$, being $T_0 = 273.15$ K, the ice point of water, slightly different from the triple point.

Measurement of other thermodynamic temperatures requires primary thermometers based on physical systems whose temperature may be derived from measurements of other quantities, such as gas pressure, electric resistance, radiation intensity.

Primary thermometry is not quite a practical means for dissemination of kelvin. To overcome the difficulty, the International Temperature Scale provides a worldwide accepted recipe for realizing a temperature scale in a practical way. Its version of 1990 (ITS-90) consists in a number of fixed-temperature points (triple points or freezing points) and interpolating measuring systems. The temperature scale extends from 0.65 K to the highest temperature practically measurable, divided into three main ranges: from 0.65 K to 13.8033 K (triple-point equilibrium of hydrogen), where interpolations are made with different gas thermometers; from 13.8033 to 961.78 K (freezing point of silver), where platinum resistance thermometers are used over all the temperature range; above 961.78 K, where temperature is measured by means of optical pyrometry techniques.

In the lowest range, from 0.9 mK to 1 K, the ITS-90 scale is integrated, since 2000, by the Provisional Low Temperature Scale (PLTS-2000), which defines the relation between temperature and melting pressure of ^{3}He.

The uncertainty at which the dissemination of the kelvin is carried out by the national laboratories has a minimum around the triple point of water, with a relative value of parts in 10^7. The absolute uncertainty increases toward the lowest temperatures, up to the level of 0.1 mK, and toward the highest temperatures, up to about 30 mK.

2.7.8 Kelvin Realization and Temperature Standards: Future Perspective

With a fixed value of the Boltzmann constant k, together with the fixed value of h, the kelvin could be defined as the change of thermodynamic temperature corresponding to the energy change of one joule, whose new reference quantity would be $h\,\Delta\nu_{\mathrm{Cs}}$. Thus, the new reference quantity of the kelvin would be $h\,\Delta\nu_{\mathrm{Cs}}/k$ and the new definition would correspond to the equation

$$\mathrm{K} = \frac{\{k\}}{\{h\}\,\{\Delta\nu_{\mathrm{Cs}}\}}\,\frac{h\,\Delta\nu_{\mathrm{Cs}}}{k}. \qquad (2.51)$$

As already pointed out for other redefinitions, the same experiments performed to determine the value of the relevant constant can realize the redefined unit. At present, the most promising method for determining k is the so-called acoustic gas thermometry (AGT). Other actively pursued methods are based on gas refractive index (RIGT), gas dielectric constant (DCGT), Johnson noise (JNT) and Doppler effect broadening (DBT) involving gas molecule velocity.

In the AGT method, the relation between the thermodynamic temperature T and the velocity of sound u (extrapolated to zero pressure) is determined with a spheroidal resonator. The theoretical relation is

$$u^2 = \frac{\gamma R T}{M} = \frac{\gamma k T}{m}, \tag{2.52}$$

where γ is the ratio of the ideal-gas heat capacities, $R = N_A k$ the molar gas constant, $M = N_A m$ the molar mass of the specific gas, m being the mass of the gas molecule. The last term in (2.52) is obtained dividing both R and M by the Avogadro constant.

Before the unit redefinition, k is determined, according to (2.52) as a function of T measured in terms of the triple point of water. After the unit redefinition, the same experiment could become a realization of the new kelvin definition and the temperature of any fixed point, considered as a primary standard, could be determined by the same (2.52) as

$$T_{PS} = \frac{m u^2}{\gamma k} = \frac{\{m\}\ \{u\}^2}{\{\gamma\}\ \{k\}}\ \mathrm{K}, \tag{2.53}$$

where the last equation verifies, according to (2.26), that T_{PS} can be a primary reference of thermodynamic temperature if the quantities involved in (2.53) are measured in SI units.

2.7.9 Mole Realization: Present Situation

The realization of the mole regards a great variety of entities, going from subatomic particles to molecules, so that dealing with primary procedures or primary methods is more appropriate than with primary standards.

The most usual case of mole realization consists in assigning the value to the amount of substance of a pure sample containing only particles of the same entity X, the amount of substance being the number of entities contained in the sample divided by the number of entities contained in one mole. In general, this is obtained by simply measuring the mass m_S of the sample and calculating the corresponding amount of substance as

$$n_S = \frac{m_S}{M(\mathrm{X})}, \tag{2.54}$$

where the value of the molar mass $M(\mathrm{X})$ can be easily found with great accuracy for a large number of simple entities [16] and calculated for compound entities.

With this procedure the mole can be realized with a relative uncertainty of even less than 10^{-6}, the main uncertainty component being the limited purity of the sample.

If the sample is a pure gas, contained in a volume V, at a pressure p and a temperature T, the amount of substance of the sample can also be determined as

$$n_S = \frac{p\,V}{R\,T}, \tag{2.55}$$

where R is the above introduced molar gas constant (cf. 2.7.8). In most cases and up to normal pressure, Equation (2.55) holds with a sufficient accuracy. More refined results, at higher pressure, are obtained introducing correction terms as functions of the same measured quantities.

When the sample is formed by entities liberated at an electrode during an electrolysis process, where a constant electric current I is flowing for a time t, the amount of substance of the sample can be determined according to the equation

$$n_S = z\,\frac{I\,t}{F}, \tag{2.56}$$

F being the Faraday constant and z the number of elementary charges for every entity liberated at the electrode.

2.7.10 Mole Realization: Future Perspective

One could note some singularity in the quantity amount of substance and in its unit. In fact, in the quantity "amount of substance", referred to a sample as a system of nominally identical entities, apparently one can not identify any property beside its numerousness and, when referred to each entity, beside its existence as a specified countable item. However, since a proper dimension was attributed to the amount of substance, also the Avogadro number, the number of entities contained in one mole, became the numerical value of the Avogadro constant, so that $N_A = \{N_A\}\ \mathrm{mol}^{-1}$. Therefore, from (2.17) and related comments, $n_e(\mathrm{X}) = (1/\{N_A\})\ \mathrm{mol(X)} = 1/N_A$ for any kind of entity.

If the Avogadro constant were one of the reference constants of the SI, as officially proposed, the mole definition expressed by (2.17) would become

$$\mathrm{mol} = \left\{N_A\right\}\frac{1}{N_A} = \frac{1/N_A}{\left\{1/N_A\right\}}, \tag{2.57}$$

complying with (2.7) with $1/N_A$ as the reference quantity of the unit. In the new definition, for a requirement of continuity, the numerical value $\left\{N_A\right\}$ would be stated as corresponding exactly to the best experimental determination of the Avogadro constant.

The absence in (2.57) of any reference to the kilogram and to $m(^{12}\mathrm{C})$, would make the mole independent of the mass unit. As a consequence, the molar mass $M(^{12}\mathrm{C})$ would no longer be 0.012 kg/mol exactly, but its value should be determined by experiment as $M(^{12}\mathrm{C}) = m(^{12}\mathrm{C})\,N_A$. This consequence is important because the molar-mass constant $M_u = M(^{12}C)/12 = 0.001$ kg/mol, is the basis for the calculation of the molar mass of any entity as $M(X) = A_r(X)\, M_u$, where the relative atomic mass $A_r(X) = m(X)/m_u)$, $m_u = m(^{12}C)/12$ being the atomc mass unit, is determined with mass spectrometry techniques.

Regarding the realization of the mole, the present situation outlined above would substantially hold also in the future situation. In fact, in the main primary procedure, expressed by (2.54), the additional uncertainty component introduced by having released the presently exact reference for $M(X)$ would be mostly negligible and of rather conceptual than practical relevance. As for the procedures expressed by (2.55) and (2.56), both the molar gas constant $R = k\, N_A$ and the Faraday constant $F = e\, N_A$ would become exact, as related to proposed fixed reference constants of the SI.

2.7.11 Candela Realization and Photometric Standards

The most recent procedure to realize the unit of luminous intensity consists in measuring the radiant intensity I_e of laser sources by means of a cryogenic radiometer, which compares the radiation power to an electric power with a substitution method. The optical power incident on an absorbing cavity is considered equal to the electrical power required to heat the cavity to the same temperature. The absorbing cavity is operated near liquid helium temperature to better comply with the black body conditions, thus improving the sensitivity and the accuracy of the radiometer.

Different laser sources are used with different frequencies, none of which coincident with the one indicated by the definition, which is not available as a laser frequency. Therefore, the luminous intensity I_v of each laser source is obtained multiplying the respective radiant intensity $I_e(\lambda)$ by the specific luminous efficacy $K(\lambda) = K_m\, V(\lambda)$, $V(\lambda)$ being the spectral luminous efficiency function, a bell–shaped function of wavelength having the unity as its maximum. That is, considering a given laser source as a primary standard of luminous intensity, it is

$$I_{vPS}(\lambda) = I_{ePS}(\lambda)\, K(\lambda), \tag{2.58}$$

where $I_{ePS}(\lambda)$ is determined by a radiometer, and the numerical coefficient $V(\lambda)$ is available from tabular data.

The realization uncertainty of the candela, in relative terms, is at the level of 10^{-4}. At nearly the same uncertainty level, standard photometers can be calibrated, by a substitution method, to be used as reference standards by the national metrological laboratories.

With the reference photometers, sets of incandescent standard lamps are calibrated, at a relative uncertainty level about 10^{-3}. These are used to calibrate instruments used in the dissemination activity, and also as travelling standards in intercomparisons among laboratories.

As a future perspective, it is worth mentioning the experimental introduction in the photometric metrology of the single-photon source, as a radiation generator, and of the photon number resolving detector, as a radiometer. As for the candela definition, there is no officially recognized proposal for its redefinition.

2.8 CONCLUSIONS

While the book is being published, a deep revision of SI is in progress, even if so far no formal change has been accomplished. The general trend is toward a system of units mainly based on fundamental constants. This implies changes in most of the unit definitions and also different experiments to realize them with the corresponding primary standards.

A consistent part of the present chapter is devoted to a theoretical arrangement of the subjects of general relevance and of course these sections will hold in a revised SI. Of these, the most basic ones concern the role of the measurement unit and of its definition in the measurement process, the structure of the unit system and the rules for its evolution in the future and for the realization of units.

The other parts of the chapter, regarding the detailed wording of units and the application of theory to each single unit definition and realization, had to be developed both for the present SI and for the one most authoritatively prospected, without excluding the possibility of some change before the upcoming decisions.

However, in view of the transitory context, sufficient hints have been given to the reader for understanding the sense of any future evolution of the unit system and also for adapting the theoretical notions to any new situation.

The ultimate decisions about SI definitions are up to the CGPM, but are prepared by the CIPM on the basis of studies and advices of the Comités Consultatifs, specific for each field of measurement. As for the proposed unit redefinitions, after the initial stimulus from the articulate proposals appeared in the specialized literature [1, 2] and taken into account by the Comité Consultatif des Unités, the due procedure has been followed until the Resolution of the CGPM without substantial changes, except for some difference in the wording of definitions. The final CGPM Resolution officially adopting the new definitions is expected in 2018.

FURTHER READING

Several of References offer a wide range of information beyond the specific citation purpose. However, indications for further reading are given in the following:

An official information about all the subjects, rules and definitions regarding the present unit system can be found in

BIPM. *The International System of Units,* BIPM, Sèvres Cedex (2006). Available at http://www.bipm.org/en/publications/si-brochure/

For an in-depth analysis of physical and technological subjects concerning the measurement units, their realization and the related metrology, one can consult

Petley, B.W. *The Fundamental Physical Constants and the Frontier of Measurement,* Adam Hilger, Briston (1985).

Bava, E., Kühne, M., Rossi, A.M. *Proceedings of the International School of Physics "Enrico Fermi"*. Metrology and Physical constants, IOS Press, Amsterdam, (2013).

REFERENCES

1. Mills, I.M., Mohr, P.J., Quinn, T.J., Taylor, B.N., Williams, E.R. Redefinition of the kilogram: a decision whose time has come, *Metrologia*, **42**, 71–80 (2005).
2. Mills, I.M., Mohr, P.J., Quinn, T.J., Taylor, B.N, Williams, E.R. Redefinition of the kilogram, ampere, kelvin and mole: a proposed approach to implementing CIPM recommendation 1 (CI-2005), *Metrologia*, **43**, 227–246 (2006).
3. BIPM. *Procés-Verbaux des Séances du Comité International des Poids et Mesures, 94th meeting (2005)*, p. 233, BIPM, Sèvres Cedex, 2006.
4. BIPM, *Comptes Rendus de la 24e CGPM (2011)*, p. 532, BIPM, Sèvres Cedex, (2013).
5. Mohr, P.J., Taylor, B.N., Newell, D.B. CODATA recommended values of the fundamental physical constants: 2010, *Reviews of Modern Physics*, **84**, 1527–1605 (2012).
6. Cabiati, F., Bich, W. Thoughts on a changing SI, *Metrologia*, **46**, 457–466 (2009).
7. Available at http://www.bipm.org.
8. Available at http://www.bipm.org/en/publications/mep.html
9. Kibble, B.P., Robinson, I.A., Belliss, J.H. A realization of the SI watt by the NPL moving-coil balance, *Metrologia*, **27**, 173–192 (1990).
10. Williams, E.R., Steiner, R.L., Newell, D.B., Olsen, P.T. Accurate measurement of the Planck constant, *Physical Review Letters*, **81**, 2404–2407 (1998).
11. Josephson, B.D. Possible new effects in superconductive tunneling, *Physics Letters*, **1** (7), 251–253 (1962).
12. von Klitzing, K. The quantized Hall effect, *Review Modern Physics*, **58**, 519–531 (1986).
13. Becker, P., De Bièvre, P., Fujii, K., Gläser, M., Inglis, B., Luebbig, H., Mana, G. Considerations on future redefinitions of the kilogram, the mole and of other units, *Metrologia*, **44**, 1–14 (2007).
14. Leonard, B.P. Comments on recent proposals for redefining the mole and kilogram, *Metrologia*, **47**, L5–L8 (2010).
15. Thompson, A.M., Lampard, D.G. A new theorem in electrostatics and its application to calculable standards of capacitance, *Nature*, **177**, 888 (1956).
16. Cohen, E.R., Cvitaš, T., Frey, J.G., Holmström, B., Kuchitsu, K., Marquardt, R., Mills, I., Pavese, F., Quack, M., Stohner, J., Strauss, H.L., Takami, M., Thor, A.J. *Quantities, Units and Symbols in Physical Chemistry*, IUPAC Green Book, 3rd edition, IUPAC and RSC Publishing, Cambridge (2008), pp. 113–128.

EXERCISES

2.1 Referring to the definition of the meter, write the corresponding equation in the form of (2.7). Then identify the very reference quantity of the length unit and calculate its value.

2.2 Referring to the ampere definition, prove the force expression (2.14) leading to the equation underlying the wording of the definition.

2.3 Starting from the defining equation of the ampere (2.15), rewrite the same equation in the form of (2.7). Then identify the very reference quantity of the electric current unit and calculate its value.

2.4 In the hypothesis of redefining the kilogram with a fixed value of the Planck constant $h = 6.626\,069... \times 10^{-34}$ J s (last figures to be fixed exactly), show that the reference quantity expressed in (2.30) is a virtual mass and find its size. Then observe the implication of the fundamental equations of energy $E = m\,c_0^2$ and $E = h\,\nu$ in the expression of the above reference quantity and try to define the kilogram as the mass whose equivalent energy is that of a photon of a calculated frequency.

2.5 Find the reference quantity of the capacitance unit and verify that the experiment of the calculable capacitor (cf. 2.7.5) is a realization of the farad. Then show that its value can be determined without involvement of any other unit.

2.6 Energy is a very important quantity, because of its different forms, and so it is also the joule, yet not a base unit. Starting from the joule expression given in Table 2.3, which is immediately referred to mechanical energy, find the reference quantity of the unit. Then verify that the electrical energy also has the same unit reference.

In the hypothesis of the possible revised SI definitions, show what would be the joule reference quantity in both the mechanical and electrical cases.

2.7 Should also the second be referred to fundamental constants through the Compton frequency of the electron $\nu_C = m_e\,c_0^2/h$ with an exact value of the electron mass m_e, what would be the defining equations of the second and the kilogram? Calculate also the involved numerical values.

2.8 The introduction of the candela within the SI units, through the constant K_m, can be seen as the first example of extension of the core units based on fundamental constants to units of measurement related to non-physical quantities, such as the visual sensation. Starting from the candela definition, find the very reference quantity of the unit.

CHAPTER 3

DIGITAL SIGNAL PROCESSING IN MEASUREMENT

ALESSANDRO FERRERO[1] and CLAUDIO NARDUZZI[2]
[1]Department of Electrical Engineering, Milan Polytechnic, Milan, Italy
[2]Department of Information Engineering, University of Padua, Padua, Italy

3.1 INTRODUCTION

Measuring instruments are increasingly based on digital signal processing (DSP) techniques. This means any input quantity of interest, described in mathematical form as a continuous-time signal $s(t)$ (e.g., a time-varying voltage), is first of all *digitized* by the instrument. That is, a set of values, representing samples of the signal, are acquired at suitable time instants t_n, $n \in \mathbb{Z}$, with $\mathbb{Z}$ being the set of integer numbers. These samples are then processed by numerical algorithms to obtain the desired measurements.

In signal processing terms, a measurement is seen as the *estimate* of some relevant signal parameter. Since each measurement function may require a separate algorithm, by implementing multiple algorithms a single instrument can provide several different measurements in parallel, all drawn from the same initial set of samples.

For measurements obtained through DSP algorithms, accuracy is dependent both on the performance of data acquisition hardware and on the characteristics of signal processing algorithms. Whereas specifications are usually comprehensive enough for the former, the evaluation of uncertainty for a measurement algorithm is not always straightforward. The analysis of the main features and limitations is thus useful to correctly employ DSP-based measurement functions.

Digitising a signal implies both sampling, that is, the observation of the measurand at a discrete set of instants, and quantization, by which a numerical value taken from a suitable finite set is assigned to each sample. In a signal acquisition channel, sample-and-hold circuits take care of sampling and are followed by analog-to-digital

Modern Measurements: Fundamentals and Applications, First Edition.
Edited by Alessandro Ferrero, Dario Petri, Paolo Carbone, and Marcantonio Catelani.

converters for quantization. The total time required to complete the two steps is also the minimum time separation between two consecutive samples. Thus, it determines the best *time resolution* achievable by a digitizer.

Reflecting actual practice in a wide majority of measurement applications, *uniform sampling* will be assumed throughout the chapter. Therefore, sampling time instants are integer multiples of a constant time length T_s, called the *sampling interval*: $t_n = nT_s$. With a constant time interval, it becomes unnecessary for the instrument to record a couple of time-amplitude values for each sample point. Digitized data can simply be stored in an acquisition memory as a linear array of amplitude values, with an integer array index n associated with each sample in the sequence.

In DSP theory it is a common convention to altogether drop the indication of the sampling interval, referring to a sequence $s[n]$, where the equality $s[n] = s(nT_s)$ is implied. This should be more correctly thought of as a *normalization* of the time axis with respect to the sampling interval T_s, where time is expressed in terms of the variable t/T_s. Samples are thus referred to the integer values of normalized time. Of course, if such scaling is applied within a measurement algorithm, results related to time or frequency then need to be scaled back appropriately.

An obvious limitation in any DSP algorithm follows from the fact that no measuring instrument is able to acquire, store, and process whatever number of signal samples, with algorithms of any degree of complexity, in an arbitrarily predefined finite time. Available computing power determines the speed performance of an algorithm and, accordingly, the instrument capability to track the measurand as it evolves with time.

Measurements are considered *real-time* whenever results are provided at the same rate as samples are acquired. This may have a different meaning depending on algorithm features, as a new measurement could either be generated after each new sample is acquired, that is, at a rate $1/T_s$, or be obtained by processing a block of N samples. In the latter case, the measurement rate is $1/NT_s$ and measurements are produced after a delay equal to the algorithm processing time. Whenever a real-time requirement cannot be met or is not critical to operations, a *dead-time* may occur between consecutive measurements, as in any other kind of instrument.

In a block-processing measurement algorithm the length N of the processed sample sequence can often be seen as a parameter that allows to trade accuracy (which typically improves for larger N) for measurement rate (which increases as N gets smaller). Since the internal memory of a digital instrument, however large, has finite size N_{mem}, the *observation interval* length is limited as well. For uniform sampling, this sets the maximum length to $N_{mem}T_s$.

3.2 SAMPLING THEORY

One of the main issues in the application of DSP theory in measurement is whether *information* associated with the original signal $s(t)$ is correctly carried over to a digitized and possibly finite-length sequence $s[n]$. If this is not the case, alterations introduced in the process need to be analyzed, to determine what kinds of errors may affect measurements based on digitized data. This kind of study is made possible by sampling theory and the application of Fourier analysis.

It should be remarked that sampling theory is primarily concerned with the problem of *exact reconstruction* of a continuous-time signal from its samples. Reconstructing a whole signal from its samples is actually the most stringent requirement one may place. If this condition is satisfied, any kind of measurement information one may wish to extract from $s(t)$ can be retrieved, unaltered, from the sequence of samples $s[n]$. However, in some cases less restrictive conditions might still allow to obtain accurate measurements.

3.2.1 Sampling and Fourier Analysis

From a very intuitive viewpoint, since sampling alters the original signal $s(t)$, it can be expected that, somehow, the associated information is also changed. However, if a "sufficient" number of samples is taken, the effect of the change should be limited and the original signal still recoverable "almost exactly" from the sequence of sampled data $s[n]$.

It is also immediate to think that an accurate representation of the original signal shape requires samples to be "'dense" enough. This is illustrated in Figures 3.1 and 3.2 which show, respectively, a continuous-time periodic signal $s(t)$ and a sequence $s[n]$ where samples have been obtained by taking approximately 40 samples per period. The original signal $s(t)$ can be rather easily recognized in the sequence of sampled data represented in Figure 3.2.

If now a much lower *sampling rate* of only approximately 10 samples per period is considered, as shown in Figure 3.3, the same original signal $s(t)$ can no longer be recognized from the sequence of samples.

It should be noted that the sampling "density" shown in these figures actually exceeds requirements for exact reconstruction in both cases. This points to another important aspect, that is, the difference between the information that *may* allow to reconstruct a signal by an appropriate procedure and what is required, in this instance, to graphically reproduce a signal in a faithful way as would be required, say, of a digitising oscilloscope. In the latter case, samples may need to be processed to determine how the signal behaves "in between." This emphasizes the other important,

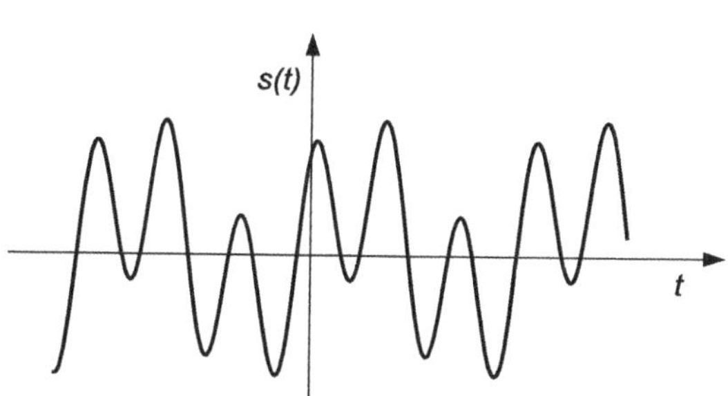

FIGURE 3.1 A periodic, continuous-time signal.

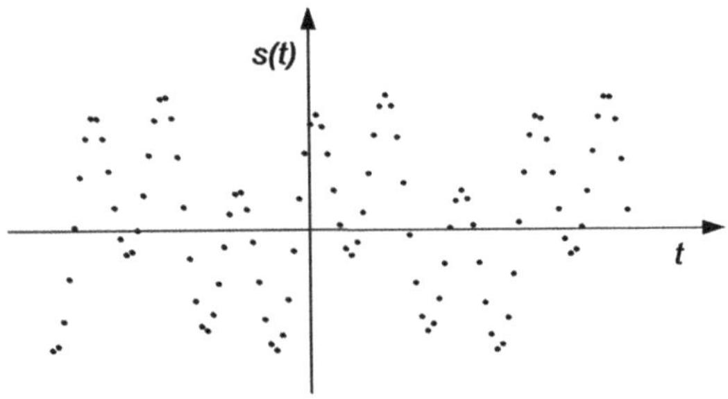

FIGURE 3.2 Sequence of samples of the signal in Figure 3.1, obtained with "dense" sampling (approximately 40 samples per period).

though less apparent, facet of sampling theory, that is, reconstruction by *interpolation* of samples.

To understand how information associated with a signal is transferred into the discrete-time domain after sampling, let us consider the continuous-time signal $s(t)$ and assume its Fourier transform can be defined:

$$S(f) = \int_{-\infty}^{+\infty} s(t)e^{-j2\pi ft}dt \quad \text{or} \quad S(\omega) = \int_{-\infty}^{+\infty} s(t)e^{-j\omega t}dt, \tag{3.1}$$

which maps $s(t)$ into the *frequency domain*. The variable ω is called the *angular frequency*, with $\omega = 2\pi f$. Either of the two definitions is in common use, but in the following the frequency f will be used as the frequency domain variable.

The inverse Fourier transform is

$$s(t) = \frac{1}{2\pi}\int_{-\infty}^{+\infty} S(\omega)e^{+j\omega t}d\omega = \int_{-\infty}^{+\infty} S(f)e^{+j2\pi ft}df. \tag{3.2}$$

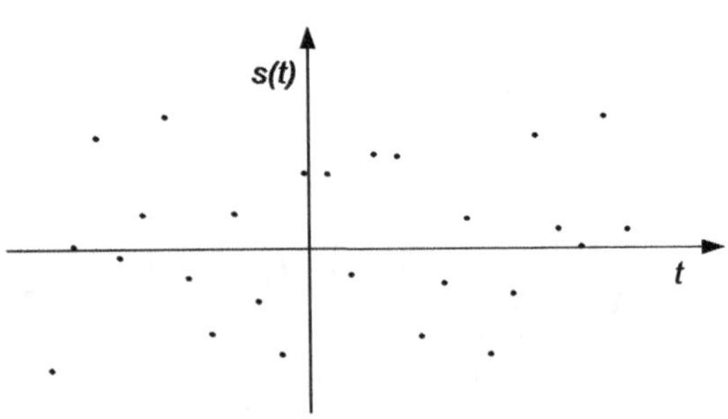

FIGURE 3.3 Sequence of samples of the signal in Figure 3.1, obtained with "thin" sampling (approximately 10 samples per period).

The Fourier transform of the discrete-time sequence $s(nT_s)$, $-\infty < n < +\infty$, obtained by sampling $s(t)$ with a constant sampling interval T_s, is defined as [1,2]

$$\tilde{S}(f) = \sum_{n=-\infty}^{+\infty} T_s \cdot s(nT_s)e^{-j2\pi fnT_s} \tag{3.3}$$

and can be related to the (continuous-time) Fourier transform $S(f)$ through *Poisson summation formula*

$$\sum_{n=-\infty}^{+\infty} T_s \cdot s(nT_s)e^{-j2\pi fnT_s} = \sum_{k=-\infty}^{+\infty} S\left(f - \frac{k}{T_s}\right) \tag{3.4}$$

whose left-hand side can also be interpreted as a first-order numerical approximation of the integral in Equation (3.1). It follows that the relationship between the Fourier transform of the continuous-time signal $s(t)$ and the discrete-time Fourier transform (DTFT) of its sample sequence is

$$\tilde{S}(f) = \frac{1}{T_s} \cdot \sum_{k=-\infty}^{+\infty} S\left(f - \frac{k}{T_s}\right). \tag{3.5}$$

This expression allows to understand the effect of uniform sampling, as it can be immediately recognized that $\tilde{S}(f)$ is a sum of shifted replicas of $S(f)$ and, as such, is periodic in f with period $F_s = \frac{1}{T_s}$. This shows that *any* sequence obtained by sampling a continuous-time signal has a periodic spectrum in the frequency domain, no matter what the spectrum of the original signal is (Figure 3.4). The quantity F_s, that is, the reciprocal of the sampling interval, is called the *sampling rate* or, commonly but less appropriately, the *sampling frequency.*

According to DSP theory conventions, the sampled signal is often expressed as an integer-index sequence $s[n]$. As already remarked, this is equivalent to introducing

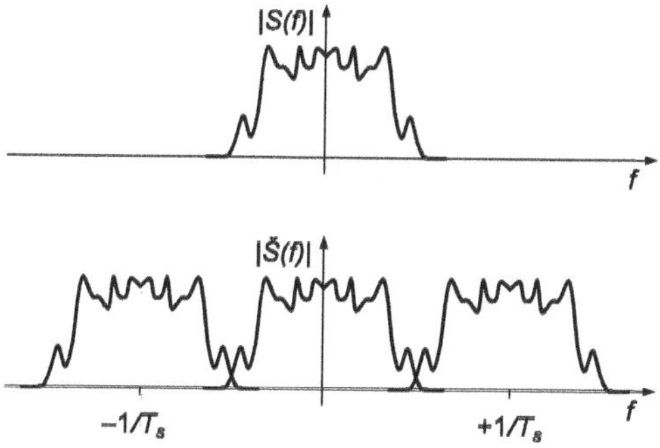

FIGURE 3.4 Fourier transform of a generic signal and its periodic repetition due to sampling.

a normalization of the time axis with respect to the sampling interval T_s. Applying basic Fourier transform properties with regard to time scaling, the corresponding discrete-time Fourier transform is then

$$\sum_{n=-\infty}^{+\infty} s[n]e^{-j2\pi f T_s n} = T_s \cdot \tilde{S}(fT_s) = \sum_{k=-\infty}^{+\infty} S(fT_s - k). \tag{3.6}$$

This equation shows why it is also common practice in DSP theory to refer to the *normalized frequency* $\nu = fT_s$ (or the normalized angular frequency: $\Omega = \omega T_s = 2\pi fT_s$). Using a normalized frequency variable, (3.6) becomes

$$\sum_{n=-\infty}^{+\infty} s[n]e^{-j2\pi \nu n} = T_s \cdot \tilde{S}(\nu) = \sum_{n=-\infty}^{+\infty} S(\nu - k). \tag{3.7}$$

Thus, after normalization, the Fourier transform is periodic in the normalized frequency ν with period equal to 1 or, in the normalized angular frequency Ω, with period 2π.

3.2.2 Band-Limited Signals

Given that sampling does modify the nature of a signal, it can be expected that measurement information associated with $s(t)$ is retrievable from the sequence $s[n]$ only when such modifications do not interfere destructively with the original information. This consideration is compatible with the condition that no overlap occurs between the functions $S(f)$ associated to consecutive periods in (3.5), so that each period of the Fourier transform can be considered as a separate replica of the Fourier transform of $s(t)$.

For this condition to hold, the signal $s(t)$ must be assumed to be *band-limited*, that is, its Fourier transform has to satisfy the condition

$$S(f) = 0 \quad \text{for} \quad |f| > F_0. \tag{3.8}$$

In this case, no signal component has frequency higher than F_0. The situation is graphically represented in Figure 3.5, where an example of Fourier transform of a band-limited signal is shown. The width of the frequency interval where $S(f) \neq 0$ is called the signal *bandwidth* and is equal to $2F_0$.

Looking at Figure 3.5 and Equation (3.8) it can be concluded that, if the condition $F_s \geq 2F_0$ is satisfied, each period of $\tilde{S}(f)$ is an undistorted replica of $S(f)$. Information about the continuous-time signal can then be retrieved, unmodified, from any single period of $\tilde{S}(f)$ and, in particular, for $|f| \leq F_0$ using an ideal low-pass filter with bandwidth $2F_0$. Therefore $s(t)$ can be exactly reconstructed from the sequence of sampled values $s[n]$.

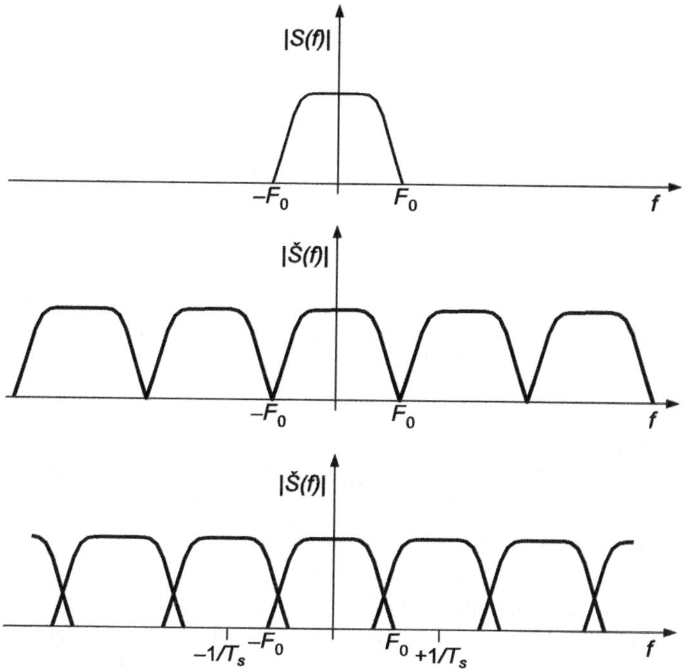

FIGURE 3.5 Fourier transform of a generic signal (above) and the effect of sampling: Nyquist-rate sampling (middle); incorrect sampling at lower-than-Nyquist rate (below).

Hence, the *lowest limit* for the sampling frequency is

$$F_{s(min)} = 2F_0, \tag{3.9}$$

which is called the *Nyquist rate*, F_N.

If, on the contrary, $F_s < 2F_0$, the shifted replicas of $S(f)$ in (3.5) are partly overlapped over frequency intervals of width $F_s - F_0$. High frequency components in the Fourier transform of the continuous-time signal appear to be folded back onto the lower frequency components modifying them. It should be remembered that the sum in (3.5) is referred to complex quantities. Therefore, while the magnitudes $|S(f - \frac{k}{T_s})|$ do overlap, as shown in the lower part of Figure 3.5, functions $S(f - \frac{k}{T_s})$ sum up as vectors and the actual result generally differs from the sum of magnitudes. Overlap is progressively increased as the sampling frequency gets lower, causing information about $s(t)$ to be modified irreversibly, so that the signal can no longer be retrieved from the sequence of sampled values. This effect is called *aliasing*.

In some instances, aliasing can become an extremely misleading phenomenon, because the sequence of sampled data might reproduce a signal with features apparently similar to those of the original sampled signal. For instance, Figure 3.6 shows the results of sampling two sinewaves, with frequencies 2.5 kHz and 97.5 kHz, at the

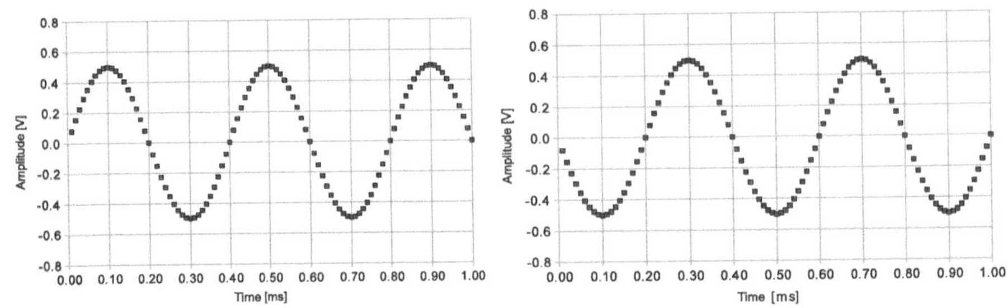

FIGURE 3.6 Aliasing effects can become particularly misleading when the waveform shape remains unaltered. Here, two sinewaves with frequencies 2.5 kHz (left) and 97.5 kHz (right) are both sampled at a rate $F_s = 100$ kHz.

same rate $F_s = 100$ kHz. Due to aliasing, the frequency-domain components of the sinewave at frequency 97.5 kHz have been aliased into the lower frequency 2.5 kHz. If no information but the waveform shape is known about the original signal, the sequence obtained from the incorrect sampling operation might appear correct, thus leading to a dramatically inaccurate measurement of the signal period and frequency.

To obtain correct measurements by DSP, aliasing should in general be absolutely avoided. In practice, aliasing effects can hardly be eliminated entirely, and need to be carefully analyzed to ensure their impact remains bounded to negligible levels. Possible countermeasures depend on the nature of the signals to be sampled. It is possible to either modify the signals themselves, in order to adapt their bandwidth to the available sampling rate, or adopt different sampling strategies. The first approach requires suitable hardware devices to be used as the input stage of a DSP-based instrument (such as analog anti-aliasing filters or frequency prescalers [2]), while the second approach requires to modify the way in which signals are sampled and acquired samples are stored into the data memory (for instance by adopting equivalent-time sampling strategies, or multi-converter devices [2]).

In summary, given a constant sampling interval T_s a continuous-time signal is correctly sampled if the following conditions are met:

- the Fourier transform of the continuous-time signal is upper-bounded at frequency F_0, as shown by (3.8);
- the sampling frequency F_s is at least equal to the Nyquist rate: $F_s \geq 2F_0$;
- the signal is sampled over the entire time axis, that is: $t_n = nT_s$ with $-\infty < n < +\infty$).

These conditions ensure that exact reconstruction from the sequence of sampled values is possible and aliasing is avoided.

The statements above are often referred to as the *Shannon sampling theorem*, since the works by Shannon emphasized their relevance to the communications engineering world [3]. Most of the theory discussed so far was actually developed in a mathematical context, over about a century, mainly by Newton and Gauss.

3.2.3 Interpolation

The conditions placed by the sampling theorem ensure that information associated with a continuous-time signal can be retrieved from its sampled values, since the value taken at any time t by a band-limited signal can be determined from the knowledge of its sampled values at times $t_n = nT_s$, $-\infty < n < +\infty$. Fourier theory provides a perfect equivalence between the time and the frequency domains whenever transforms do exist. Thus, although conditions are referred to the frequency domain, when sampling theorem conditions are met, the continuous-time signal is retrievable as well by a suitable time-domain procedure.

To prove this statement, consider a band-limited signal $s(t)$, sampled correctly with $F_s \geq F_N = 2F_0$. Under this condition, (3.5) leads to

$$S(f) = T_s \cdot \tilde{S}(f) \quad \text{for} \quad -\frac{1}{2T_s} \leq f \leq +\frac{1}{2T_s}, \tag{3.10}$$

as graphically shown in the middle graph of Figure 3.5.

Since $s(t)$ is band-limited by assumption, computation of the integral in the continuous-time inverse Fourier transform (3.2) can be similarly restricted to the interval $-\frac{1}{2T_s} \leq f \leq +\frac{1}{2T_s}$, yielding

$$s(t) = \int_{-\frac{1}{T_s}}^{+\frac{1}{T_s}} T_s \tilde{S}(f) e^{+j2\pi ft} df. \tag{3.11}$$

Replacing $\tilde{S}(f)$ in (3.11) by its expression (3.5), the equation becomes

$$s(t) = \int_{-\frac{1}{2T_s}}^{+\frac{1}{2T_s}} \left[\sum_{n=-\infty}^{+\infty} s(nT_s) e^{j2\pi f(t-nT_s)} df \right].$$

The order of integration and summation can now be interchanged to obtain

$$s(t) = \sum_{n=-\infty}^{+\infty} s(nT_s) \left[\int_{-\frac{1}{2T_s}}^{+\frac{1}{2T_s}} e^{-j2\pi fnT_s} \right] e^{j2\pi ft} df.$$

Finally, computation of the integral within square brackets yields

$$s(t) = \sum_{n=-\infty}^{+\infty} s(nT_s) \frac{\sin\left[\frac{\pi}{T_s}(t - nT_s)\right]}{\frac{\pi}{T_s}(t - nT_s)}. \tag{3.12}$$

This is generally called the Shannon *interpolation formula* and shows how the value taken by the continuous-time signal $s(t)$ can be calculated at any time t from known values at the sampling times nT_s (Figure 3.7).

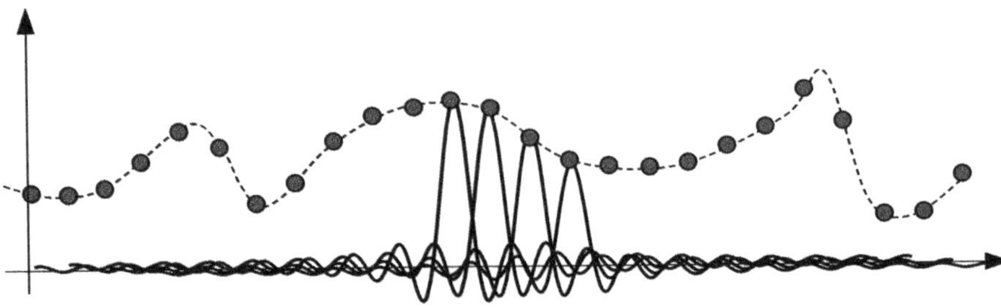

FIGURE 3.7 Interpolation of sampled values by the function $\frac{\sin\left[\frac{\pi}{T_s}(t-nT_s)\right]}{\frac{\pi}{T_s}(t-nT_s)}$.

In spite of its theoretical importance, the practical usefulness of equation (3.12) is limited. Although it shows how a band-limited continuous signal can be reconstructed from its samples, provided the sampling theorem conditions are met and no aliasing occurs, exact reconstruction requires the interpolation of an infinite number of samples. Therefore, the observation interval should be of infinite length as well.

This statement actually agrees with sampling theorem conditions, which assume no restrictions for the signal, except band-limitedness. Since a priori the signal might be aperiodic and have unlimited time duration, observation over an infinite-length interval is needed, in general, to retrieve the whole amount of information associated with the signal.

Of course, in practical situations no signal corresponds to this kind of mathematical abstraction. Phenomena that are generally dealt with in measurement applications are represented either by *finite-length signals*, that do not require to be observed beyond their duration, or by *periodic signals*, whose associated information can be retrieved by observing just one single period.

Since a signal cannot simultaneously be bandlimited and have finite duration, sampling a finite length signal necessarily introduces aliasing so that, from a strictly theoretical point of view, exact reconstruction is not possible. In practice, any signal can be forced to become band-limited by *filtering* with a low-pass (or a band-pass) *anti-aliasing filter*. In a finite-length signal this will introduce some decaying exponential terms, that extend the resulting duration to infinity. However, a suitably designed filter can generally ensure that these components fall below the analog-to-digital conversion amplitude resolution, so that the signal length can still be considered limited for all practical purposes.

3.3 MEASUREMENT ALGORITHMS FOR PERIODIC SIGNALS

The application of numerical algorithms in the case of periodic signals needs a specific analysis that requires, in the first place, to clarify the distinction between signals that are known to be periodic in the continuous-time domain and signals that retain this property after sampling.

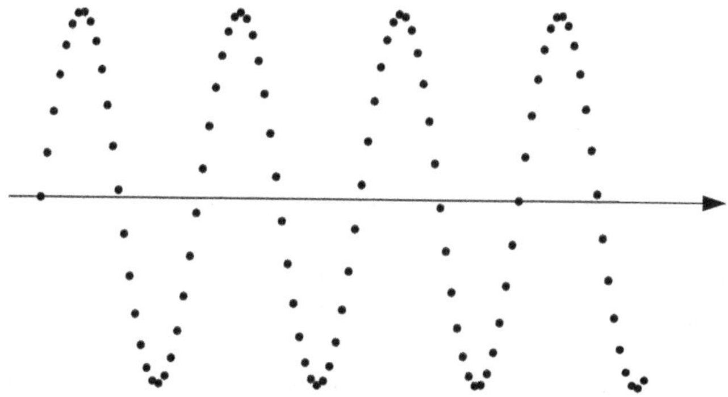

FIGURE 3.8 Sampled sinewave. The period of the sequence $s[n]$ is equal to 79 samples and corresponds to three periods of the continuous-time signal.

Let signal $s(t)$ be periodic in the continuous time domain, that is, $s(t) = s(t + mT)$ for $m \in \mathbb{Z}$. Sampling this signal with an interval T_s produces a sequence that might no longer be periodic. In fact, in the discrete-time domain a periodicity property only holds when the period can be given as an integer multiple of T_s. This is possible if the signal period T and the sampling interval are related by an expression of the form

$$\frac{T}{T_s} = \frac{N_P}{P}, \tag{3.13}$$

with the integers N_P and P mutually prime. This condition results in *coherent sampling* and is based on the assumption that, somehow, the sampling rate of the data acquisition system is synchronous with the signal period.

In this case it follows immediately that, since (3.13) can also be written as: $N_P T_s = PT$, the sampled sequence $s[n] = s(nT_s)$ is indeed periodic. Its period has a length of N_P samples, which corresponds to values repeating after a time PT. Thus, even when $s[n]$ is itself periodic, its period N_P may actually correspond to a multiple of the original period in the continuous-time domain (Figure 3.8).

3.3.1 Sampling Periodic Signals

A bandlimited signal that is periodic in the continuous time domain is no different from any other bandlimited signal. Hence, the general sampling theorem conditions can be applied. However, a periodic signal can also be described by means of its Fourier coefficients and, if it is assumed to be band-limited, its harmonic components will be at most of some order K, so that $s(t)$ can always be written in the form

$$s(t) = A_0 + \sum_{k=1}^{K} A_k \cdot \cos\left(2\pi k \frac{1}{T} t + \phi_k\right), \tag{3.14}$$

where

$$A_k = \frac{1}{T}\int_{t_0}^{t_0+T} s(t)e^{-j2\pi\frac{k}{T}t}dt. \tag{3.15}$$

Expression (3.14) has $2K + 1$ degrees of freedom, represented by parameters A_k and ϕ_k. Intuitively, it can be expected that a similar number of samples should suffice to reconstruct the signal exactly. In fact, this turns out to be the case.

The sampling theorem for periodic signals states that a real signal $s(t)$ having period T and whose harmonic components are at most of order K, can be exactly reconstructed from the knowledge of $2K + 1$ values sampled in one period at uniform intervals

$$T_s = \frac{T}{2K + 1}. \tag{3.16}$$

The corresponding reconstruction formula is

$$s(t) = \sum_{k=0}^{2K} s(nT_s) \cdot \frac{\sin \pi \left(\frac{t}{T} - n\right)}{\sin \pi \left[\frac{t-nT_s}{(2K+1)T_s}\right]}, \tag{3.17}$$

where the interpolation function is a periodic repetition of the $sinc(\cdot)$ function that appears in the general form (3.12).

It is interesting to note that, in this particular case, the required sampling frequency is strictly greater than twice the maximum signal frequency. Equation (3.17) also shows that periodic signals can be characterized from the knowledge of a very small number of samples: for a pure sinewave, just three suitably acquired samples are enough (Figure 3.9).

This is a very interesting aspect of sampling theory: whereas the general theorem requires $F_s > \frac{T}{2K}$ but does not provide any guarantee that a finite-length sequence

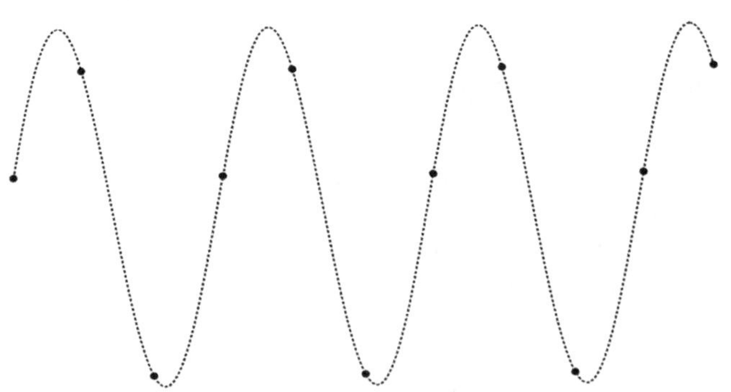

FIGURE 3.9 Three samples per period suffice to exactly reconstruct a sinusoid.

of samples could be acceptable, this specialized formulation ensures that a *finite* number of samples suffices to reconstruct a periodic signal. However, this remarkable efficiency comes at a price: in fact, it is easy to see that the sampling interval required by (3.16) is a special case of (3.13) where $N_P = 2K + 1$ and $P = 1$. Thus, to preserve periodicity also in the discrete-time domain, sampling would have to be perfectly synchronized with the signal period.

Achieving synchronous or, more generally, coherent sampling conditions requires that the signal period is known a priori and remains stable for the whole observation interval. Alternatively, data acquisition circuitry could be designed to track the signal period by synchronising the sampling frequency to it, for example, by phase-lock techniques. In the latter case, the sampling interval is forced to depend on the period of the measurand which becomes, to all purposes, the time reference for the acquisition system. Thus, the measuring instrument would no longer have a calibrated *time base* of its own.

Advantages may still overcome this limitation in a number of applications, such as in sampling RMS voltmeters and power-line harmonic analyzers, where absolute time or frequency measurements are unnecessary. Also, when stimulus/response measurements are carried out, like in network analysis, coherent sampling can be easily realized by using a digital-synthesis waveform generator that is set up to share the master clock with the measuring unit.

In any other case, a sequence of samples from a periodic, continuous-time signal will most likely loose the periodicity property. Consequently, even if the sampling rate is greater than the Nyquist rate, the information associated with the signal itself may not be retrieved correctly. It is therefore important to understand what kind of errors are introduced, quantify them and investigate possible methods for their reduction.

3.3.2 Estimation of the RMS Value

For electrical quantities like voltage and current, the root-mean-square (RMS) value of a periodic waveform is equal to the value of a corresponding continuous voltage or, respectively, current that dissipates on a resistance the same amount of active power. Its mathematical definition, given a periodic continuous-time signal $s(t)$, is

$$S_{RMS} = \sqrt{\frac{1}{T}\int_{t_0}^{t_0+T} s^2(t)dt}\,, \tag{3.18}$$

where T is the signal period.

Given a sequence $s[n]$, it seems reasonable to consider the closest numerical approximation to (3.18), that means taking a finite number N of samples and computing

$$\hat{S}_{RMS} = \sqrt{\frac{1}{N}\sum_{n=n_0}^{n_0+N-1} s^2[n]}. \tag{3.19}$$

When sampling conditions specific to periodic signals hold and $N = N_P$, this equation is in fact exact and accurate. Under more general assumptions, however, analysis of (3.19) is necessary to understand what is required to ensure that $\hat{S}_{RMS}$ correctly estimates the RMS value S_{RMS} of $s(t)$.

Since the input bandwidth of any instrument is limited, the hypothesis that, for a periodic signal $s(t)$, harmonic components of order $k > K$ have negligible amplitudes is always satisfied in practice for some value of K. Therefore, the signal has the general form (3.14) and, given the Fourier coefficients A_k, the RMS value of the signal can be written also as

$$S_{RMS} = \sqrt{A_0^2 + \sum_{k=1}^{K} \frac{A_k^2}{2}}. \tag{3.20}$$

This can be compared with the expression for sequence $s^2[n]$, that takes the form

$$\begin{aligned} s^2[n] = {} & \left[A_0^2 + \sum_{k=1}^{K} \frac{A_k^2}{2}\right] + \sum_{k=1}^{K} \frac{A_k^2}{2} \cos\left(2\pi \frac{2k}{T} nT_S + 2\phi_k\right) \\ & + \sum_{j,k=1, j \neq k}^{K} \frac{A_j A_k}{2} \left\{ \cos\left[2\pi \frac{j-k}{T} nT_S + \phi_j - \phi_k\right] \right. \\ & \left. + \cos\left[2\pi \frac{j+k}{T} nT_S + \phi_j + \phi_k\right] \right\}. \end{aligned} \tag{3.21}$$

An example is provided in Figure 3.10, where the squared values of the sequence of samples shown in Figure 3.2 is presented. The first term within brackets on the

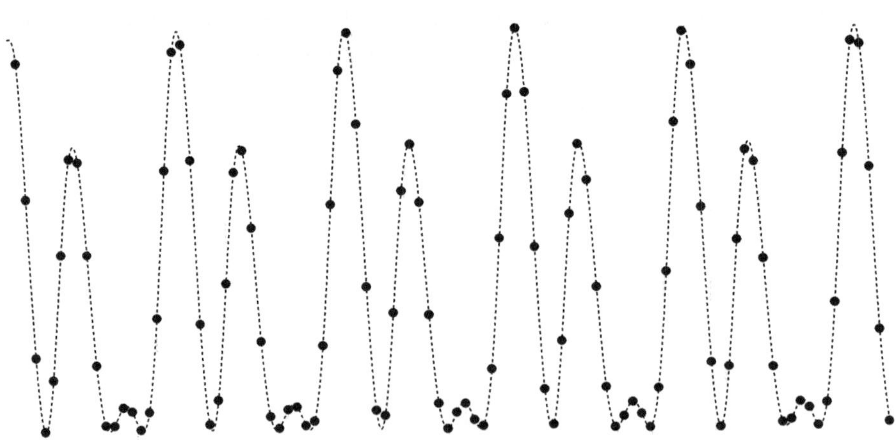

FIGURE 3.10 Sequence of squared values of samples in Figure 3.2.

right-hand side of the equation (that is, the constant term) is exactly the RMS value of the continuous-time signal $s(t)$. To obtain an accurate estimate, the remaining terms should vanish after averaging N values of $s^2[n]$. Thus, the accuracy of $\hat{S}_{RMS}$ depends on a suitable choice of N.

It is useful to recall the identity

$$\frac{1}{N}\sum_{n=0}^{N-1}\cos(An+B) = \frac{\sin\frac{A}{2}N}{N\cdot\sin\frac{A}{2}}\cdot\cos\left[\frac{A}{2}(N-1)+B\right]. \tag{3.22}$$

Using this relationship to analyze the effect of averaging over N values, it is easy to see that the same general form is obtained for any of the undesired terms in (3.21)

$$\frac{\sin\pi\frac{l}{T}NT_s}{N\cdot\sin\pi\frac{l}{T}T_s}\cos\left[\pi\frac{l}{T}(N-1)T_s+\phi_l\right]\quad\text{with:}\quad\begin{array}{c} l=2k,\ j-k,\ j+k\\ \phi_l=2\phi_k,\ \phi_j-\phi_k,\ \phi_j+\phi_k\end{array}, \tag{3.23}$$

where l is an integer index, with $1\le l\le 2K$. It can be noticed that (3.23) vanishes if N is allowed to tend to infinity, which suggests that the presence of undesired terms in (3.21) is related to the finiteness of the sample sequence.

The possibility to cancel those terms depends on the values of N and the ratio NT_S/T. Disregarding the (possible but unlikely) chance that $NT_s = PT$ for some value of P, two different conditions can be considered.

- **Asynchronous sampling**: The sequence length N bears no relation to the period of the measured signal. Sampling is *asynchronous* with the signal period and the sequence is aperiodic, so it can be assumed that the ratio NT_S/T yields an irrational number. Hence the length of the sample sequence will not coincide with any integer number of signal periods.

  The worst-case condition occurs when $NT_S \simeq (M+\frac{1}{2})T$, that is, the observation period extends over an integer number of periods, plus one half-period. The contribution from undesired terms is then maximized, but becomes negligible when $M \gg \frac{1}{2}$ and, consequently, for large values of N.
- **Quasi-coherent sampling**: It is assumed that, as a preliminary step, the measurement algorithm can determine the approximate signal period, or integer multiple of the signal period. That is, after a sequence of generic length N has been acquired, the algorithm first finds a value $N_P \le N$ such that $(N_P-1)T_S \le PT \le (N_P+1)T_P$. The calculation (3.19) is then restricted to the subset of N_P consecutive samples. In this case PT is known with a resolution equal to the sampling interval T_S and the algorithm can be analyzed assuming *quasi-coherent* sampling, which means that the following inequality holds:

$$\frac{P}{N_P+1}\le\frac{T_S}{T}\le\frac{P}{N_P-1}. \tag{3.24}$$

Then, it follows that

$$\left| \frac{\sin \pi l \frac{N_P T_S}{T}}{N_P \cdot \sin \pi l \frac{T_S}{T}} \right| \leq \frac{1}{N_P}, \qquad \forall l. \tag{3.25}$$

Thus, the contributions of all undesired terms in (3.21) can be reduced to a negligible level by considering a suitably large, but finite number of samples.

3.4 DIGITAL FILTERS

A digital filter is a numerical algorithm that implements a transformation on a sequence of samples. Specifically, linear time-invariant digital filters are considered here. Rather than discussing the theory, for which the reader is addressed to [1, 2, 4], the usefulness of digital filters in measurement applications will be shown by discussing a couple of examples relevant to measurement algorithms.

An algorithm for RMS value estimation based on (3.19) can be decomposed into a sequence of three operations, namely, squaring, averaging and the calculation of a square root. The RMS formula seemingly points to an implementation by block calculation on a set of N samples, and this is actually necessary for the quasi-coherent case, where preliminary estimation of N_P is required. However, in the most general asynchronous sampling case an algorithm implementation providing an updated estimate $\hat{S}_{RMS}$ after each new sample is possible.

The averaging operation can in fact be interpreted as a filter, that sequentially processes the sequence $s^2[n]$ and provides an average value after each new sample is received. To see this, it suffices to rewrite the averaging expression in a different form

$$\bar{s}[n_0 + N - 1] = \frac{1}{N} \sum_{n=n_0}^{n_0+N-1} s^2[n] = \sum_{m=0}^{N-1} s^2[(n_0 + N - 1) - m] \cdot h[m]. \tag{3.26}$$

This equation shows that the average value at time $t_n = (n_0 + N - 1)T_s$ is computed using the current squared sample value $s^2[n_0 + N - 1]$ as well as the samples taken at the $N - 1$ preceding sample times. Of course, the expression can be written in the same form for any value of the time index. Furthermore, the right-hand side term presents averaging in the form of a discrete-time convolution by the finite-length sequence $h[m]$, which is defined as

$$h[m] = \begin{cases} \frac{1}{N} & \text{for } 0 \leq m < N \\ 0 & \text{for } m < 0 \text{ and } m \geq N \end{cases} \tag{3.27}$$

Clearly, $h[m]$ can be interpreted as a discrete-time impulse response. This particular filter is called a *moving average* (MA) filter. Using the DTFT with normalized

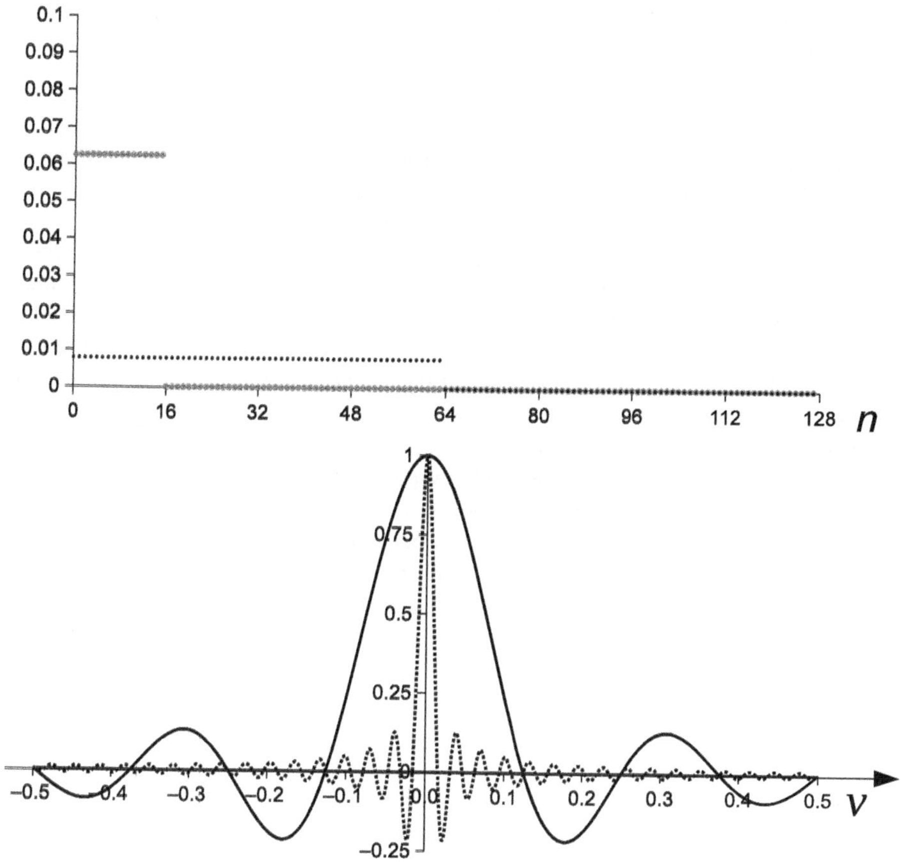

FIGURE 3.11 Impulse response (top) and frequency response (bottom) of two MA filters of length, respectively, $N = 16$ and $N = 64$. The frequency axis is normalized, with $\nu = 0.5$ corresponding to half the Nyquist rate. For $N = 16$ impulse response is shown by grey dots, frequency response by continuous line. For $N = 64$, black dots and dashed line.

variables given in (3.7), its frequency response can be written as

$$\tilde{H}(\nu) = \sum_{n=-\infty}^{+\infty} h[m]e^{-j2\pi\nu m} = \sum_{n=0}^{N-1} \frac{1}{N} e^{-j2\pi\nu m} = \frac{\sin(\pi\nu N)}{N \cdot \sin(\pi\nu)} \cdot e^{-j2\pi\nu\left(\frac{N-1}{2}\right)}. \quad (3.28)$$

The plot of $\tilde{H}(\nu)$ presented in Figure 3.11 shows that, obviously, the moving average filter has a low-pass behavior.

For the RMS estimation algorithm to operate sequentially as each new sample arrives, calculations involved at each step are

- compute the square of the new sample;

- calculate the convolution sum (3.26). This requires N multiplications and $N - 1$ sums;
- finally, compute the square root of the moving average filter output.

If the total time required by a measuring instrument to carry out these operations is less than T_s, the instrument has real-time capability.

It has already been noted that, under the assumption of asynchronous sampling, accuracy in the estimation of $\hat{S}_{RMS}$ can be improved by letting N get larger. Figure 3.11 shows that the effect of increasing N is the reduction of the filter bandwidth. This agrees with the observation that, in the expression of $s^2[n]$ given in (3.21), only the constant term contains the useful information. On the other hand, calculation of the convolution sum tends to become cumbersome, as an ever larger number of terms is involved.

Since low-pass filtering is the desired effect, it is possible to replace the moving average filter by some different form of low-pass filter. For instance, another filter described by a finite-length impulse response $h[m]$, generally called a *finite impulse response* (FIR) filter, could be designed with better low-pass properties.

Narrow-band FIR filters tend to have long impulse responses, requiring the DSP algorithm to keep a large number of samples in memory and calling for adequate processing power. An alternative formulation of a digital filter response is based on finite differences and has the general form

$$y[n] + a_1 y[n-1] + \cdots + a_k y[n-k] = b_0 x[n] + b_1 x[n-1] + \cdots + b_m x[n-m], \tag{3.29}$$

where $x[n]$ is the input sequence and $y[n]$ is the corresponding output. This recursive expression requires that a number of previous values of both the input and output sequences are stored in memory and may give rise to an infinite-length impulse response, although this is not always the case. When it happens, the filter is termed *infinite impulse response* (IIR).

IIR filters are often more efficient in the realization of narrow-band frequency responses, although they may miss other useful properties. For instance, the moving average filter is guaranteed to have linear phase, unlike a low-pass IIR filter.

For the purpose of filtering out variable components and retaining only the continuous component in (3.21), a recursive IIR filter design can be more efficient. Software tools for designing digital filters with given specifications (i.e., for determining the values of either the coefficients a_k, b_m or the impulse response samples $h[m]$) can be found in several digital signal processing, analysis and simulation environments.

It should be remembered that the behavior of digital filters is not dissimilar, in general principles, from that of their analog counterparts, including the fact that filter response time is, roughly, inversely proportional to its bandwidth. Hence, in the design of a low-pass filter for an RMS estimation algorithm, bandwidth should be specified as a tradeoff between the desired attenuation of unwanted variable components in $s^2[n]$ and the ability to follow fluctuations of the RMS value.

Another significant example of digital filter application is interpolation. In some cases, the need of reconstructing the behavior of a sampled signal on a denser time grid may arise, so that the sampling period appears to be shorter than the actual one. This may be required to obtain better definition in displaying a waveform, or to simplify comparison between waveforms acquired at different sampling rates.

The simplest case considered here is interpolation by an integer factor M, that means waveform values have to be calculated at new sampling points, spaced in time by the interval $T'_s = \frac{T_s}{M}$. In principle, the interpolation equation can be obtained by *resampling* (3.12)

$$s(mT'_s) = \sum_{n=-\infty}^{+\infty} s(nT_s) \frac{\sin\left[\frac{\pi}{T_s}(mT'_s - nT_s)\right]}{\frac{\pi}{T_s}(mT'_s - nT_s)}. \tag{3.30}$$

If the sample sequence $s[n]$ is redefined in the new discrete-time domain as

$$s_{int}[m] = \begin{cases} s(nT_s) & \text{for} \quad m = lM, \quad\quad l \in \mathbb{Z} \\ 0 & \text{for} \quad m = lM + i, \quad l \in \mathbb{Z}, \quad i = 1, \ldots, M-1 \end{cases} \tag{3.31}$$

the equation can be interpreted as a discrete convolution between $s_{int}[m]$ and a sinc-type interpolating filter. Of course, this filter has infinite duration and is not suitable for DSP implementation.

For this aim, a filter having finite and possibly short duration is preferred. Furthermore, to avoid distortion, an interpolating filter is also required to preserve the linear-phase characteristic of the sinc filter. Several implementations are possible, the simplest one being *linear interpolation*. This can be described in filter form by the impulse response

$$h[m] = \begin{cases} \frac{m}{M} & \text{for} \quad 1 \le m \le M \\ \frac{2M-m}{M} & \text{for} \quad M+1 \le m \le 2M-1 \\ 0 & \text{otherwise} \end{cases} \tag{3.32}$$

whose total length is $2M - 1$ samples. It can be seen that the impulse response is symmetric around its middle sample ($m = M$), thus preserving phase linearity. The impulse response for $M = 4$ and the magnitude of the resulting filter frequency response are shown in Figure 3.12. It can be noticed that the interpolator is, as expected, a low-pass filter and its normalized (−3 dB) bandwidth is approximately 0.16. By comparison, the bandwidth of the ideal interpolator (3.30) is equal to the normalized Nyquist rate (= 1) divided by the interpolation factor and, in this example, would be 0.25. This evidences that the simplified time-domain beahviour of a linear interpolating filter is paid for in terms of a less-than-ideal frequency response, with rather large attenuation at the higher frequencies of interest. This explains why linear

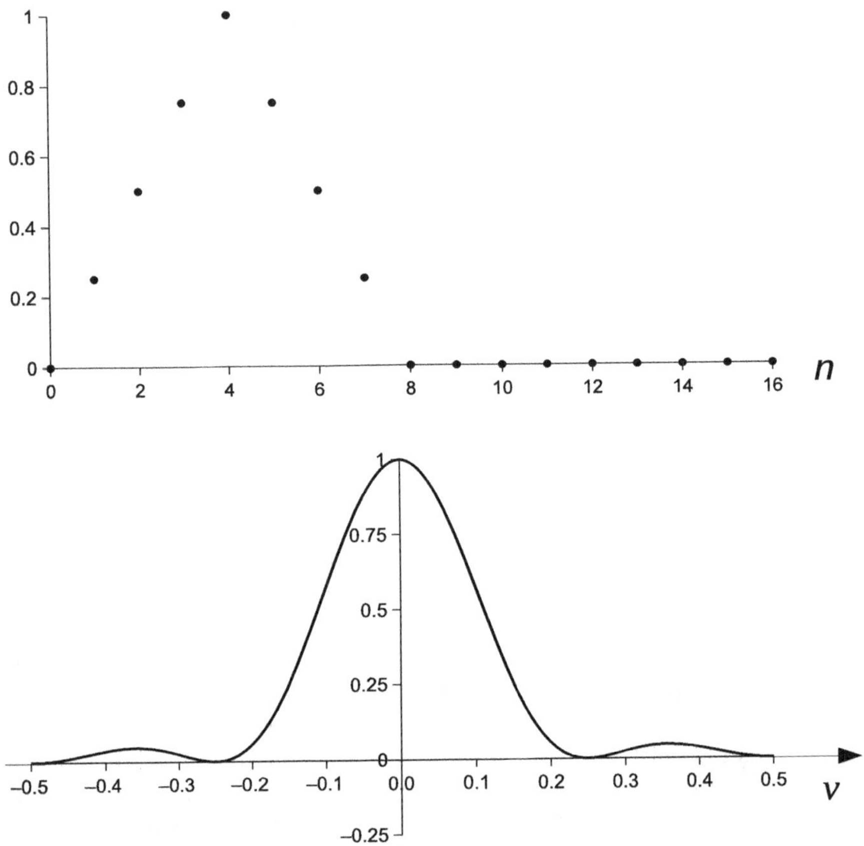

FIGURE 3.12 Impulse response (top) and magnitude of the frequency response (bottom) of a linear interpolating filter with $M = 4$.

interpolation is usually employed only when significant oversampling has taken place in the signal acquisition phase.

3.5 MEASURING MULTI-FREQUENCY SIGNALS

Multifrequency signals are a generalization of the class of periodic signals, whose Fourier transform assumes non-zero values only at a discrete set of frequencies F_k, $1 \leq k \leq K$. Given the Fourier transform $S(f)$ of one such signal, its time domain expression takes the form

$$s(t) = |S(0)| + \sum_{h=1}^{H} 2\,|S(F_h)| \cos\left(2\pi F_h t + \arg\left[S(F_h)\right]\right), \tag{3.33}$$

where $\arg[S(f)] = \arctan\{\mathrm{Im}[S(f)]/\mathrm{Re}[S(f)]\}$ is the phase component of $S(f)$.

The only difference between (3.14) and (3.33) is that frequencies F_k are no longer assumed to be integer multiples of a fundamental frequency $F_1 = 1/T$, hence a multifrequency signal can be aperiodic. Considering that, in the general asynchronous sampling case, any assumption on the periodicity of the underlying continuous-time signal would be disregarded in any case, this generalization allows a broader class of signals to be dealt with, without increasing the level of complexity.

A multifrequency signal is completely characterized when the $2K + 1$ non-zero elements of its Fourier transform are known. If these quantities are accurately measured, the whole signal can be reconstructed and any other quantity related to it can be measured as well. Thus, the analysis of this specific problem has general relevance.

DSP-based measurement algorithms for multifrequency signals usually process blocks composed of a finite number of samples. Since, in general, the analyzed signal is non-periodic, considering a finite-length sample sequence may cause some loss of information, whose consequences need to be analyzed.

3.5.1 Finite-Length Sequences

In the following, it is assumed that measurement information is to be obtained from a sequence $s[n]$ having a finite length N. To consider the case in its most general form, it will be assumed that the coherent sampling condition (3.13) does not hold. Maximum length may be determined by system design constraints, but is not related to any particular signal feature.

Relationship (3.5) between the Fourier transforms of a continuous-time signal and that of the sequence obtained by sampling it was based on the mathematical connection between representations in the time domain and in the frequency domain. A different approach to the derivation of that result, which provides some useful insight, can be followed if mathematical *distributions* are considered instead of functions.

In the continuous time domain, sampling can be described as the multiplication of $s(t)$ by a *sampling distribution*, defined as a combination of impulses, represented by Dirac delta distributions $\delta(t)$. Impulses repeating at equally spaced intervals T_s, as shown in Figure 3.13, produce a regular *pulse train*

$$\sum_{n=-\infty}^{+\infty} \delta(t - nT_s), \tag{3.34}$$

which could be seen as an idealization of the sequence of pulses that control sampling circuits in a data acquisition device.

The sampled signal can then be written as the product (in the distribution domain) of $s(t)$ by the pulse train (3.34) and results in a distribution of regularly spaced impulses, whose amplitudes are weighted by the values taken by $s(t)$ at each integer multiple of the sampling interval T_s

$$\tilde{s}(t) = s(t) \cdot \left[\sum_{n=-\infty}^{+\infty} \delta(t - nT_s) \right] = \sum_{n=-\infty}^{+\infty} s(nT_s) \cdot \delta(t - nT_s). \tag{3.35}$$

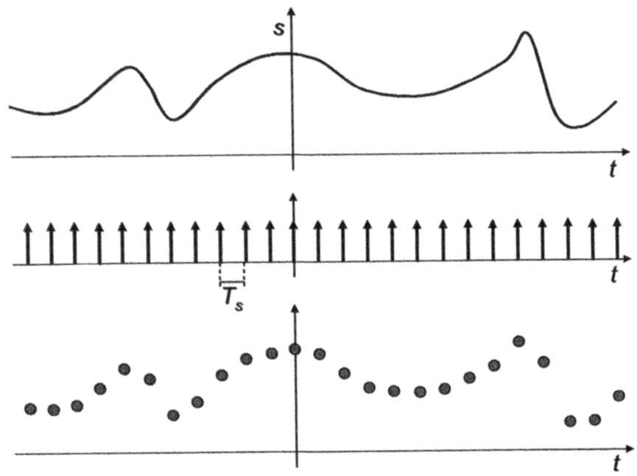

FIGURE 3.13 Sampling representation as multiplication by a pulse train.

Equation (3.35) is just a different way to formally express the fact that values of $s(t)$ are considered only at time instants $t = nT_s$. However, it also suggests a different path for determining the Fourier transform of the sampled signal, since time-domain multiplication between two signals corresponds, in the frequency domain, to the convolution of their respective Fourier transforms.

The formal definition of a Fourier transform for a pulse train relies again on Dirac distributions, and takes the form

$$\frac{1}{T_s} \cdot \left[\sum_{k=-\infty}^{+\infty} \delta\left(f - \frac{k}{T_s}\right) \right].$$

It can be noted that this transform is also a regular pulse train, whose period in f is equal to the sampling rate $\frac{1}{T_s}$ (Figure 3.14).

Using the convolution properties of the Dirac distribution, the (continuous-time) Fourier transform of the sampled signal $\tilde{s}(t)$ can be written as

$$\tilde{S}(f) = S(f) \cdot \frac{1}{T_s} \cdot \left[\sum_{k=-\infty}^{+\infty} \delta\left(f - \frac{k}{T_s}\right) \right] = \frac{1}{T_s} \cdot \sum_{n=-\infty}^{+\infty} S\left(f - \frac{k}{T_s}\right), \tag{3.36}$$

which yields the same expression obtained in (3.5).

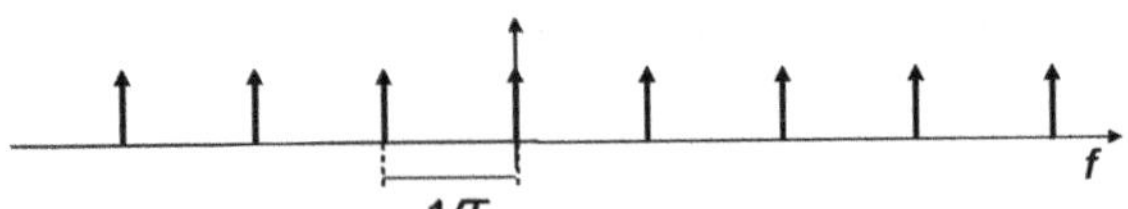

FIGURE 3.14 Frequency domain representation of an infinite duration pulse train.

When a finite-length sequence is assumed, it is still possible to represent the sampling operation as the multiplication by a pulse train. Of course, the latter will now be the sum of a *finite* number N of Dirac impulses

$$\tilde{w}_{(n_0,N)}(t) = \sum_{n=n_0}^{n_0+N-1} \delta(t - nT_s), \tag{3.37}$$

where the index n_0 indicates the starting point of the pulse train.

Direct computation of the corresponding Fourier transform yields

$$\tilde{W}_{(n_0,N)}(f) = \frac{\sin(\pi f T_s N)}{\sin(\pi f T_s)} \cdot e^{-j2\pi f T_s\left(n_0+\frac{N-1}{2}\right)}, \tag{3.38}$$

where the complex exponential term indicates that the finite pulse train is implicitly centred in time at $t = \left(n_0 + \frac{N-1}{2}\right) T_s$, that is, at the mid point, regardless of N being odd or even.

It can be immediately recognized that $\tilde{W}_{(n_0,N)}(f)$ is a continuous function of f and, like (3.36), is periodic with period equal to the sampling rate $\frac{1}{T_s}$. The shape of $|W_{(n_0,N)}(f)|$ is graphically presented in Figure 3.15 for the case $N = 32$. Notable features are

- a main lobe with amplitude N is present for each period, its centre frequency being an integer multiple of $\frac{1}{T_s}$. The main lobe width is $\frac{2}{T_s N}$;
- zero crossings are found at frequencies:

$$\frac{k}{NT_s} \quad \text{for} \quad k \in \mathbb{Z} \quad \text{and} \quad k \neq pN, \;\; p \in \mathbb{Z},$$

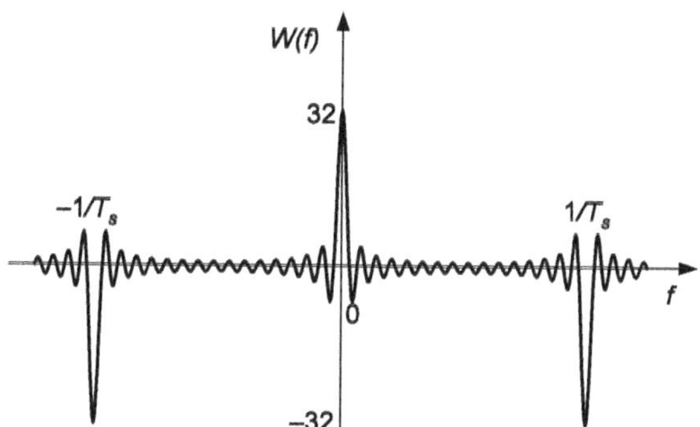

FIGURE 3.15 Fourier transform of a finite pulse train with length $N = 32$.

- side lobes have progressively decreasing amplitudes as their distance from the main lobe increases, but their width is constant and equal to $\frac{1}{T_s N}$.

It is instructive to follow a slightly different path in the derivation of (3.38), by thinking of the finite pulse train as the product of an infinite duration regular pulse train and a rectangle function $rect_{(n_0,N)}(t)$ defined as[1]

$$\mathrm{rect}_{(n_0,N)}(t) = \begin{cases} 1 \text{ for } \left(n_0 - \frac{1}{2}\right) T_s \leq t < \left(n_0 - \frac{1}{2} + N\right) T_s \\ 0 \text{ for } t < \left(n_0 - \frac{1}{2}\right) T_s \text{ and } t \geq \left(n_0 - \frac{1}{2} + N\right) T_s \end{cases} \tag{3.39}$$

The rectangle function explicitly accounts for the *truncation* of the pulse train to a finite length. Its Fourier transform is

$$W_{(n_0,N)}(f) = NT_s \frac{\sin \pi f T_s N}{\pi f T_s N} \cdot e^{-j2\pi f\left(n_0 + \frac{N-1}{2}\right)}. \tag{3.40}$$

Then, the Fourier transform of the finite pulse train can be expressed as the convolution of the frequency-domain regular pulse train (3.36) by $W_{(n_0,N)}(f)$, which yields

$$\tilde{W}_{(n_0,N)} = \frac{1}{T_s} \cdot \sum_{k=-\infty}^{+\infty} W_{(n_0,N)} \left(f - \frac{k}{T_s}\right). \tag{3.41}$$

This equation is just (3.38) written in a different form, but it evidences more clearly the effect of a finite observation interval, which is associated with the rectangle function (3.39).

Given the finite number of signal samples available in practice, calculation of the discrete-time Fourier transform is, by necessity, restricted to a finite sequence of N samples and actually becomes a finite sum of products $\tilde{S}_{(n_0,N)}(f)$, where the subscripts n_0 and N again indicate the starting point and the finite length of $s[n]$, respectively

$$\tilde{S}_{(n_0,N)}(f) = \sum_{n=n_0}^{n_0+N-1} s(nT_s) e^{-j2\pi f n T_s}. \tag{3.42}$$

Expressing this transform as a frequency-domain convolution yields the general form

$$\tilde{S}_{(n_0,N)}(f) = S(f) * \tilde{W}_{(n_0,N)}(f). \tag{3.43}$$

[1]The left-shift by half sampling interval in (3.39) allows to place the rectangle in a symmetric position with respect to the pulses.

This expression evidences that, even for a suitably bandlimited signal, working from a finite-length sample sequence has the effect of altering the information contained in the original signal, to the extent that exact reconstruction is no longer possible. In fact, in $\tilde{S}(f)$ periodic replicas of $S(f)$ are left unaltered. On the contrary, $\tilde{W}_{(n_0,N)}(f)$, while still periodic, is not a pulse train and the convolution with $S(f)$ does alter the whole shape of the resulting Fourier transform $\tilde{S}_{(n_0,N)}(f)$.

3.5.2 Discrete Fourier Transform

It has to be noticed that (3.42) does not yet provide a practical way to analyze a signal in the frequency domain. In fact, frequency is still defined as a continuous variable in $\tilde{S}_{(n_0,N)}(f)$, whereas any signal processing algorithm can only operate on discrete domains. The introduction of a discretized frequency domain is the final step that leads from a purely mathematical problem analysis tool to a DSP-based algorithm that can process digitized samples and provide actual measurements.

When measuring a multifrequency signal, the interest in frequency-domain algorithms is in fact motivated by two main reasons:

- in the frequency domain, a band-limited multifrequency signal can be characterized by a small number of parameters;
- very efficient numerical algorithms for computing the finite sum of products (3.42) on a finite set of frequency values exist [5].

The discrete Fourier transform (DFT) of a signal $s(t)$ is defined in terms of the finite-length, discrete sequence $s[n]$, as [5]

$$S_{DFT}[k] = \frac{1}{N} \sum_{n=n_0}^{n_0+N-1} s[n] e^{-j2\pi \frac{kn}{N}}. \tag{3.44}$$

The sequence of *complex Fourier coefficients* $S_{DFT}[k]$ defined by this expression is *periodic* in k with period N. Of course, calculations in the DFT algorithm are restricted to one period, that is, $k = 0, 1, \ldots, N-1$.

It can be noted that (3.44), which shows the most common form of writing the DFT, refers to the Fourier coefficients by means of an integer index k. The corresponding frequency variable should be correctly expressed either as $\frac{k}{NT_s}$ or, in a properly normalized form, as $\nu_k = \frac{k}{N}$, and the sequence period accordingly given as either $1/T_s$ or 1. For the sake of convenience in the algorithm implementation, the integer index k is often preferred. This can still be seen as a different form of frequency normalization, namely: $k = \nu_k \cdot N = (\frac{k}{NT_s} T_s) \cdot N$.

The fact that a frequency domain quantity is provided only at discrete, integer-step increments hints at some underlying periodicity assumption for the original signal.

This proves to be the case with the DFT, as the inverse transformation, given by

$$s_{IDFT}[n] = \sum_{k=0}^{N-1} S_{DFT}[k]e^{+j2\pi\frac{kn}{N}}, \tag{3.45}$$

is periodic in the index n, unlike $s[n]$, and can be non-zero for any value of n.

It should be remembered that periodicity in the time domain is an *implied consequence* of the use of a DFT-based algorithm: even though nothing is known about the signal beyond the finite-length observation interval, it is thought of as being extended to an infinite observation interval by periodic repetition.

This conclusion is intuitively explained in the plots of Figure 3.16, where a sampled sinewave is compared with the sampled waveform that would have been obtained by periodically repeating a sequence of just 70 samples, taken from the same sinewave. It can be immediately recognized that a distortion appears in the periodic repetition of the acquired sinewave segment, as evidenced by an ellipse in the figure.

In the following, analysis of the frequency-domain measurements provided by the DFT algorithm will of course be restricted to one period of $S_{DFT}[k]$, that is, to the

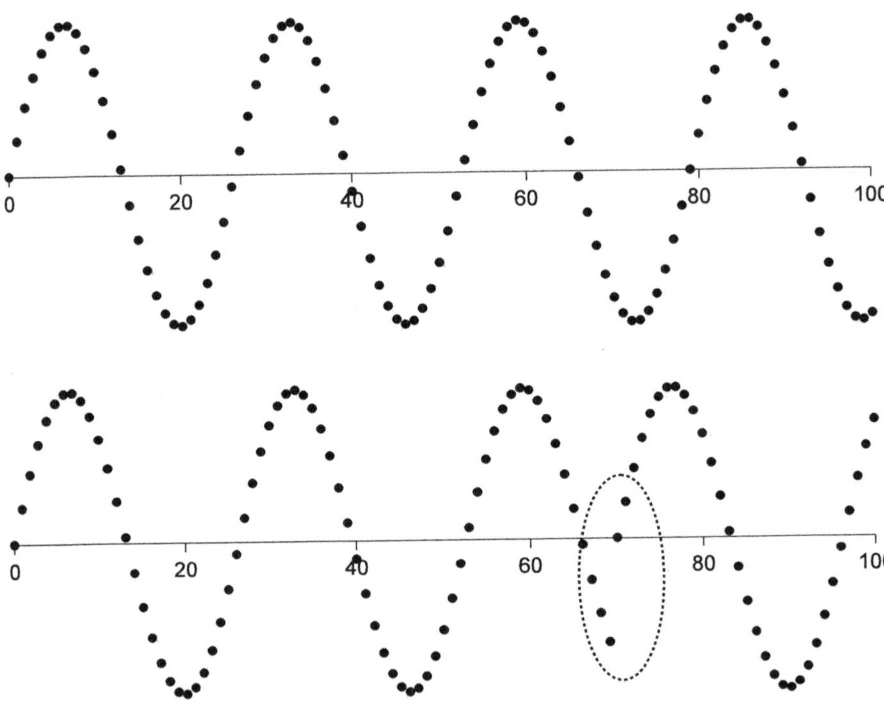

FIGURE 3.16 Implied periodicity introduced by the use of the DFT algorithm. A discontinuity (evidenced by the ellipse in the bottom trace) occurs when a finite sequence is periodically repeated. In this case the period is equal to 70 samples.

range of frequencies such that $|f| \leq \frac{1}{2T_s}$. It is important to remember that the sequence is only defined at frequencies that are integer multiples of $\frac{1}{NT_s}$. This quantity, which is the reciprocal of the observation interval, will be called the *frequency granularity* of the measurement algorithm and shall be indicated as F. Comparison between (3.42) and (3.44) shows that the following equality holds:

$$S_{DFT}[k] = \frac{1}{N}\tilde{S}_{(n_0,N)}(kF). \tag{3.46}$$

3.5.3 Uniform Window

As already noted, acquisition of a finite sequence of samples can be described as the multiplication of a signal $s(t)$ by a finite pulse train, whose Fourier transform is a periodic repetition of $W_{(n_0,N)}(f)$, the Fourier transform of the rectangle function that restricts the observation interval to a finite length NT_s. Multiplication by a rectangle leaves all samples unaltered within this time interval. In other words, all non-zero samples are given the same *weight* in the DFT algorithm. This explains the use of the term *uniform window* to indicate the purpose of the rectangle function.

The effect of a finite-length observation interval on the analysis of a multifrequency signal can be evidenced by writing its discrete Fourier transform. Using the convolution approach of (3.43) and recalling (3.46) one has

$$\begin{aligned} S_{DFT}[k] = \frac{1}{N}\Bigg[& |S(0)| + \sum_{h=1}^{H} |S(F_h)|\, e^{j\arg[S(F_h)]} \cdot \tilde{W}_{(n_0,N)}(kF - F_h) \\ & + \sum_{h=1}^{H} |S(-F_h)|\, e^{j\arg[S(-F_h)]} \cdot \tilde{W}_{(n_0,N)}(kF + F_h)\Bigg]. \end{aligned} \tag{3.47}$$

Since superposition holds, it suffices at first to consider only one term of the multifrequency signal at the generic frequency F_h, indicated as

$$s_h(t) = 2\,|S(F_h)| \cos\left(2\pi F_h t + \arg\left[S(F_h)\right]\right).$$

Its contribution to (3.47) is expressed by

$$\begin{aligned} & |S(F_h)|\, e^{j\arg[S(F_h)]} \cdot \frac{1}{N}\tilde{W}_{(n_0,N)}(kF - F_h) \\ & + |S(-F_h)|\, e^{j\arg[S(-F_h)]} \cdot \frac{1}{N}\tilde{W}_{(n_0,N)}(kF + F_h), \end{aligned} \tag{3.48}$$

where each term is proportional to one of the two complex Fourier coefficients $S(F_h)$ and $S(-F_h)$ that represent the component $s_h(t)$ of the multifrequency signal.

To determine whether an accurate measurement can actually be obtained, the effect of the multiplying factor $\tilde{W}_{(n_0,N)}(f)$ has to be considered. It should be recalled that this function has zero crossings at

$$\frac{k}{NT_s} \quad \text{for} \quad k \in \mathbb{Z} \quad \text{and} \quad k \neq pN, \; p \in \mathbb{Z},$$

therefore, if an index $\bar{k}$ such that $F_h = \bar{k}F$ existed, the contribution of $s_h(t)$ to $S_{DFT}[k]$ would be limited to only two non-zero terms, whose values would be exactly $S(F_h)$ and $S(-F_h)$. However, having assumed that F_h, in principle, will not satisfy any coherent sampling relationship, no value of the index k allows to express F_h as an integer multiple of the frequency granularity F. Thus, no element in the sequence $S_{DFT}[k]$ can provide a direct estimate of a component in a multifrequency signal and all terms can be non-zero. This phenomenon is called *spectral leakage* and, in this case, makes a sinusoidal component practically undiscernible from a signal with multiple frequency components, unless some *a priori* knowledge is at hand to settle the ambiguity.

3.5.4 Spectral Leakage

To understand the relationship between $\tilde{S}_{(n_0,N)}(f)$ and $S(f)$ and, in general terms, how uniform windowing affects an estimate, spectral leakage needs to be investigated more closely, by analysing the behavior of $W_{(n_0,N)}(f)$. The study will focus first on *short-range* leakage, that is, on how spectral leakage affects the estimation of a single component $s_h(t)$, assuming no interference may come from any other sinusoidal component of the multifrequency signal.

In view of the symmetries that follow from the assumption that $s(t)$ is a real-valued signal, only the first term in (3.48) will be analyzed, that means possible interference from the image component is disregarded as well. The term under analysis can be written more explicitly as

$$|S(F_h)| \, e^{j\arg[S(F_h)]} \frac{1}{N} \frac{\sin \pi(k - F_h T_s N)}{\sin \frac{\pi}{N}(k - F_h T_s N)} e^{-j\frac{2\pi}{N}(k - F_h T_s N)\left(n_0 + \frac{N-1}{2}\right)}, \tag{3.49}$$

where $\tilde{W}_{(n_0,N)}(f)$ has been replaced by its expression (3.40).

The consequences of frequency discretization are shown in Figure 3.17, where dots correspond to integer values of νN, that is, to the values of the index k, where DFT coefficients are calculated. The thin dashed line shows the corresponding values of the DTFT.

Whereas in the continuous-time domain a single component $s_h(t)$ of a multifrequency signal is described by two Fourier coefficients only, DFT calculation generally yields non-zero coefficients for all values of the frequency index k. The largest ones are found in the neighbourhood of the actual frequency F_h (and of its image).

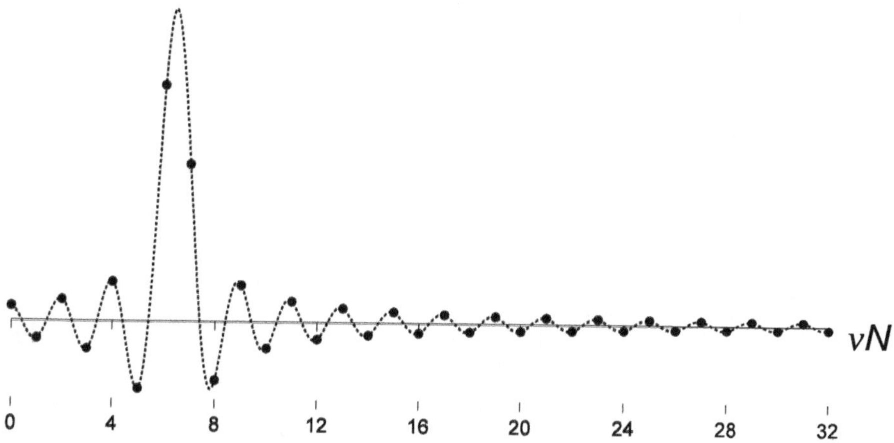

FIGURE 3.17 DFT coefficients of a single sinusoidal component, under the assumption of asynchronous sampling and a uniform window having length $N = 32$.

Considering the shape of the function $\tilde{W}_{(n_0,N)}(f)$, the best estimate of $S(F_h)$ generally corresponds to a peak value of $|S_{DFT}[k]|$. The associated index $\hat{k}$ gives the frequency estimate $\hat{F}_h = \hat{k}F$, that is the closest to F_h. The difference $F_h T_s N - \hat{k}$ expresses the distance from the actual frequency as a "fractional part" of the frequency index

$$F_h T_s N - \hat{k} = \frac{F_h - \hat{k}F}{F} = \delta \qquad \text{with:} \quad |\delta| \leq \frac{1}{2}.$$

Consequently, frequency estimation uncertainty is bounded by $\pm\frac{\delta}{2}F$.

Whenever $\delta \neq 0$, the estimated amplitude is always smaller than the actual one. In fact, for index $\hat{k}$, one has:

$$\left| |S(F_h)|\, e^{j\arg[S(F_h)]} \cdot \frac{1}{N}\tilde{W}_{(n_0,N)}(\hat{k}F - F_h) \right| = |S(F_h)| \cdot \frac{1}{N} \left| \frac{\sin \pi\delta}{\sin \frac{\pi}{N}\delta} \right|.$$

This attenuation effect depends on the shape of $\tilde{W}_{(n_0,N)}(f)$ and is described by a quantity called the *scalloping loss* (Figure 3.18). It is easy to realize that attenuation is maximum for $|\delta| = \frac{1}{2}$, that corresponds to the *worst-case scalloping loss* (WCSL), defined as

$$WCSL = \frac{|\tilde{W}_{(n_0,N)}|(\frac{F}{2})}{\tilde{W}_{(n_0,N)}(0)}. \tag{3.50}$$

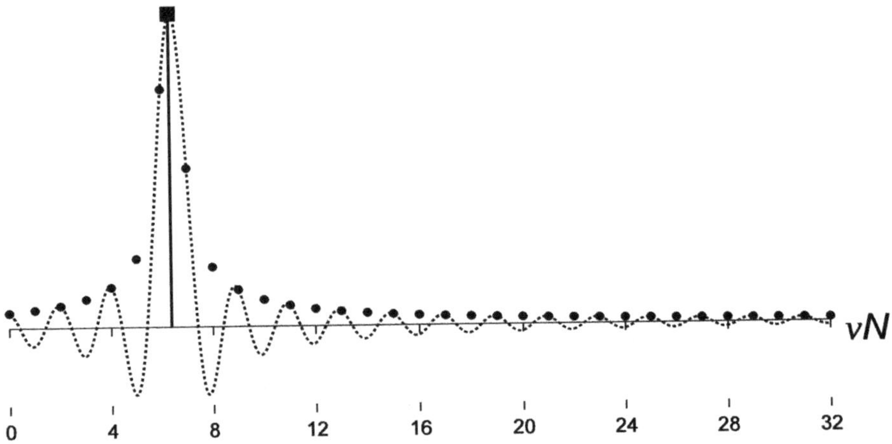

FIGURE 3.18 Magnitude of DFT coefficients of an asynchronously-sampled single sinusoidal component (dots), compared with the amplitude of the continuous-time sinusoid Fourier coefficient (square).

For a uniform window, assuming a sufficiently large value of N, WCSL is approximately $\frac{2}{\pi}$ and causes a reduction by $\frac{\hat{A}_0 - A_0}{A_0} \leq WCSL - 1 \simeq 35\%$. This is a very significant deviation, which calls for some form of correction of the raw estimate.

It is also important to consider the possibility of *spectral interference* among different components, which is associated with *long-range* leakage and the behavior of the side lobes of $\tilde{W}_{(n_0,N)}(f)$. This can be understood by considering that, for all practical purposes, the following relationship holds:

$$\frac{\tilde{W}_{(n_0,N)}(f)}{N} \leq \frac{1}{\pi f T_s N}. \tag{3.51}$$

The inequality evidences that side lobe amplitudes are inversely proportional to frequency. Thus, they decay rather slowly, so that spectral interference is possible when different signal components are found at comparativley close frequencies.

Recalling the general expression (3.47) for the DFT of a multifrequency signal, it has to be also remembered that all sums are referred to complex quantities and, when spectral interference occurs, the contributions from different components combine, affecting both the amplitude and phase of $S_{DFT}[k]$. This can produce inaccurate estimates, and may even cause the smaller components to be masked and hidden by the interference effects of the larger ones. An example is shown in Figure 3.19.

3.5.5 Leakage Reduction by the Use of Windows

Spectral leakage has been shown to be a consequence of the finite length of the sample sequence $s[n]$. A straightforward application of the DFT algorithm leads to uniform

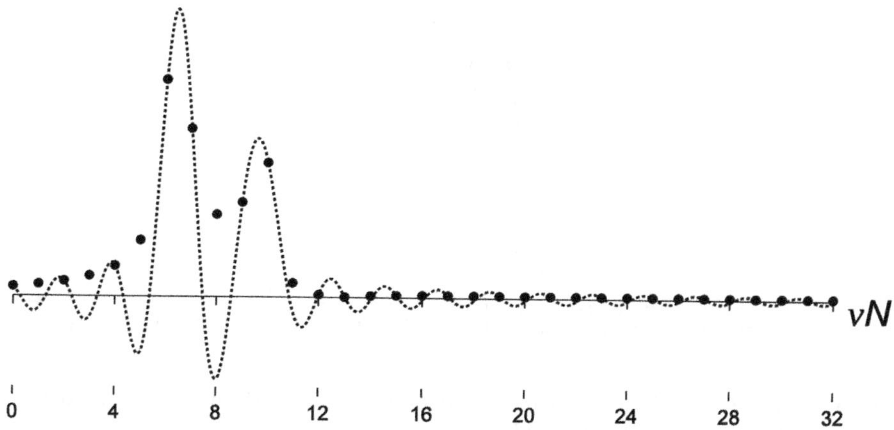

FIGURE 3.19 An example of interference between two sinusoidal components at different frequencies: both scalloping loss and long-range leakage effects are present.

weights being assigned to all samples. Of course, one may think of introducing different weights for each sample, replacing $\text{rect}_{(n_0,N)}(t)$ by other functions that might, hopefully, improve accuracy in the estimation of a multifrequency signal parameters.

The purpose of a *window* is the reduction of both spectral leakage effects, that is, scalloping loss and spectral interference. In principle, this requires the Fourier transform of the candidate window to be as flat as possible in the central part of the mainlobe, and to have either very small or fast decaying side lobes. Equivalently, in the time domain the purpose of a window is the reduction of the edge effects caused by restricting the sequence $s[n]$ to a finite number of samples.

In most cases window functions share a mathematical expression based on trigonometric functions, having the general form [2]

$$w_{(0,N)}[n] = \sum_{m=0}^{M} a_m(-1)^m \cos\left(2\pi m \frac{n}{N}\right) \qquad n = 0, \dots, N-1. \tag{3.52}$$

The *raised cosine*, or *Hanning* window is, possibly, among the best known examples

$$w_{Hann}[n] = \frac{1}{2} - \frac{1}{2}\cos\left(2\pi \frac{n}{N}\right) \qquad n = 0, \dots, N-1. \tag{3.53}$$

Its discrete-time Fourier transform is compared with that of the uniform window in Figure 3.20. It can be seen that side lobes decay faster, while the reduction in scalloping loss is basically obtained by widening the main lobe, which is twice as wide as for a uniform window. Worst-case amplitude attenuation is better in this case, resulting in $WCSL - 1 \simeq 15\%$.

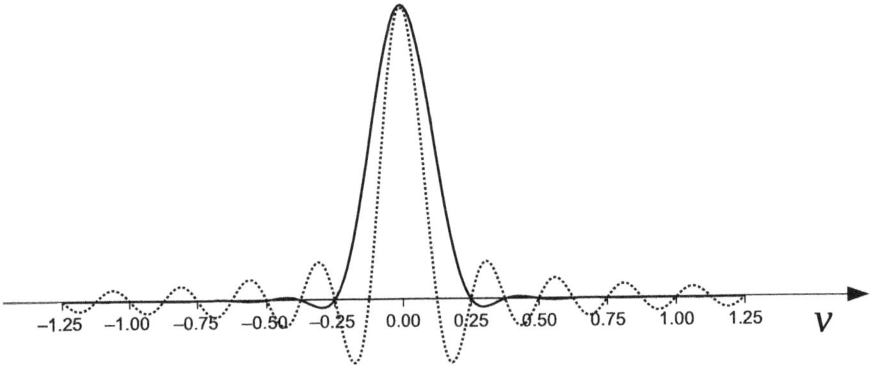

FIGURE 3.20 Comparison between the Fourier transforms of a Hanning window (continuous line) and a uniform window (dashed line).

When seen in the time domain, the application of a window to acquired signal samples has the purpose of reducing any edge effect. This is shown, as an example, in Figure 3.21, where weighting a finite-length observation of a sinewave by a Hanning window is shown. It can be noticed that the amplitudes of the start and end points of the unmodified waveform are significantly different, hence non-negligible truncation effects would be present. These are strongly attenuated by windowing, which forces the signal to a small or zero amplitude at the edges of the observation interval. On the other hand, in doing so the signal is somewhat distorted and this shows up in the frequency domain.

The magnitudes of DFT coefficients for the windowed signal are shown in Figure 3.22. Due to the wider mainlobe of the Hanning window, a sinusoidal component is associated to at least three non-zero DFT coefficients. Comparison of Figure 3.22 with Figure 3.18 also shows that long-range leakage effects are much lower for a Hanning window.

Different window designs can provide a whole range of tradeoffs between conflicting requirements, allowing to deal with a wide variety of signal measurement needs. In particular, *flat-top* windows feature a small WCSL, that minimizes inaccuracies in amplitude measurement, at the price of a reduced capability to resolve signal components at very near frequencies [2].

DFT-based measurement algorithms allow to accurately estimate the components of a multifrequency signal from the knowledge of its DFT coefficients, provided

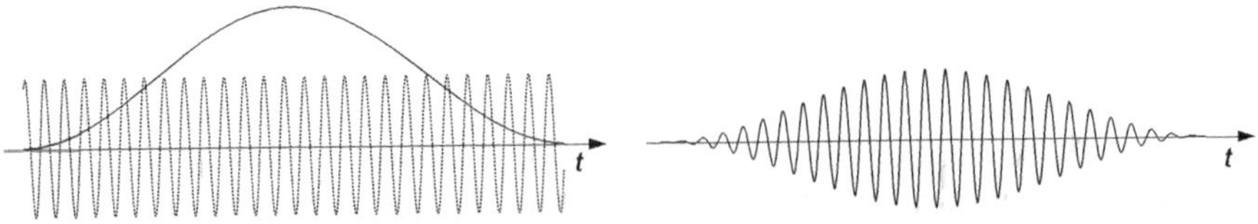

FIGURE 3.21 Application of a Hanning window (continuous line, left) to a finite-length observation of a sinewave. The windowed waveform (right) presents no edge effects.

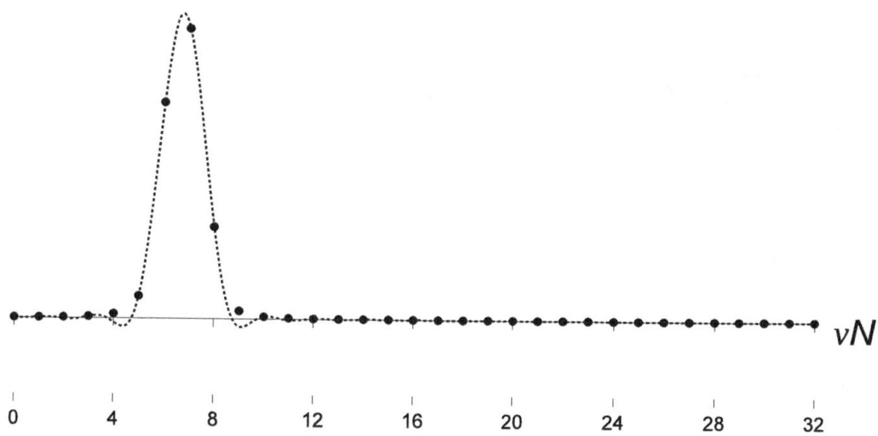

FIGURE 3.22 Magnitude of DFT coefficients of a windowed sinusoidal component (dots). A Hanning window has been applied, its shape being shown by a dashed line.

the finite-length sample sequence is suitably windowed. This approach is commonly employed in signal analysis and is finding widespread acceptance for harmonic analysis of electrical systems and power quality measurements.

3.6 STATISTICAL MEASUREMENT ALGORITHMS

When a sequence $s[n]$ is composed of a sufficiently large number of samples, algorithms based on statistical methods can be useful to estimate some signal parameters of interest.

For example, a periodic square wave includes steep edges, as well as longer intervals where the signal amplitude is, ideally, constant. Although analysis by DFT-based algorithms is possible, the approach would be exceedingly complex when simple parameters such as the *top* and *base* levels have to be measured. On the other hand, when a circuit generates a steep edge some oscillatory overshoot (or undershoot) is often introduced and, of course, some noise can be present. Thus, the top and base levels of an actual square wave may not be directly related to such easily identified parameters as the maximum and minimum values, as would suit the ideal case. Acknowledging this discrepancy, the top and base levels of a square wave-like signal are defined as the *prevailing* high and low amplitudes, respectively.

Given N samples, measurements of this kind can be obtained by means of an *amplitude histogram*. The measurement algorithm scans the acquired sequence and simply counts how many samples are associated to each quantization level, which is taken to correspond to one bin in the amplitude histogram. The shape of a square wave is such that the resulting histogram approximates a bimodal probability density function and shows two clearly detectable peaks, as shown in Figure 3.23.

A reasonably accurate estimate of the two levels can then be obtained by selecting the two separate subsets composed of the histogram bins in the neighbourhood of

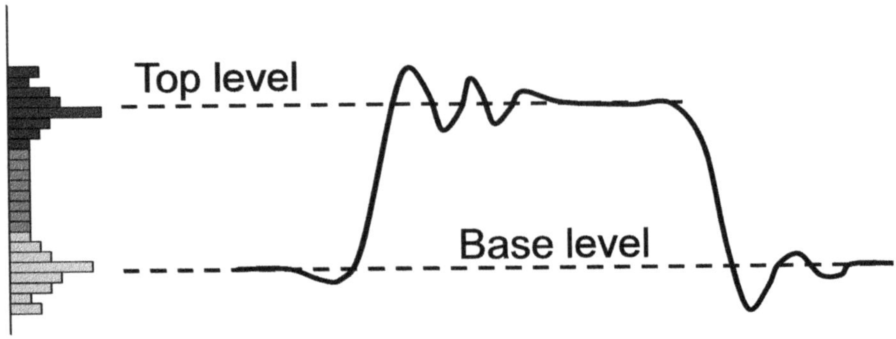

FIGURE 3.23 Estimation of the *top* and *base* levels of a square wave by means of an amplitude histogram.

one of the two peaks, and calculating the respective averages. This provides correct results even in the presence of noise or slight amplitude fluctuations.

Several other measurements can be obtained by statistical approaches. It can be noted, for instance, that a statistical interpretation can be given for the RMS measurement algorithm of a periodic signal in the asynchronous sampling case. In fact, sampling instants $t_n = nT_s$ randomly fall in any position within a period. If the measured signal is stationary and remains constant long enough, acquired samples will tend to cover the equivalent of a whole period with uniform probability. It follows that the instantaneous phase of any of the harmonic components can be seen as a random variable having uniform probability density function between 0 and 2π.

From this viewpoint, an amplitude histogram of the measured signal would be regarded as an estimate of the probabilty density function of the underlying stationary random process. More importantly, the RMS estimation formula (3.19) can be interpreted as the estimator of the standard deviation for this stationary process, evidencing the very tight link between the two quantities.

A critical issue for the accuracy and repeatability of the results yielded by statistical measurements is the number of samples on which estimates are based. The confidence level of a measurement typically improves, and variability decreases, as the number of samples gets larger, consequently a given performance target may require the definition of a minimum size for the sample sequence. Particular care should be taken when algorithms actually operate on just small subsets of the whole acquired sample sequence.

3.7 CONCLUSIONS

This chapter has provided a brief survey of the very fundamentals of DSP theory. The main goal was to give the readers an overview of the most important conditions that have to be satisfied in order to preserve, in the acquired samples, the information carried by the sampled continuous-time signal. We are confident that, once the reader has become familiar with the concept explained in this chapter, he or she can employ DSP techniques without incurring in dangerous mistakes.

On the other hand, DSP is a much more complex theory than the one we sketched in these few pages. Therefore we address the interested readers to both the referenced books and those listed in the next section.

FURTHER READING

General texts on DSP are plentiful. A few have been mentioned in References 1, 4, and 6 while, in addition to the general theory, Reference 2 is expressly written with measurement applications in mind.

An overview of developments in sampling theory over the decades since Reference 3 can be found in:

Jerri, A.J. The shannon sampling theorem—its various extensions and applications: a tutorial review, *Proceedings of the IEEE*, 65 (11), 1565–1595 (1977).

Unser, M. Sampling – 50 years after Shannon, *Proceedings of the IEEE*, 88 (4), 569–587 (2000).

Frequency-domain analysis of signals has been the subject of extensive research in many disciplines. Several good texts on the topic, as well as research papers, are available. The few mentioned here are a small selection focussed on general issues, an online search will easily produce many more papers discussing also the application side.

Rife, D.C., Boorstyn, R.R. Single-tone parameter estimation from discerete-time observations, *IEEE Transactions on Information Theory*, 20 (5), 591–598 (1974).

Rife, D.C., Boorstyn, R.R. Multiple tone parameter estimation from discrete-time observations, *Bell System Technical Journal*, 55, 1389–1410 (1976).

Jain, V.H., Collins, W.L., Davis, D.C. High accuracy analog measurements via interpolated FFT, *IEEE Transactions on Instrumentation and Measurement*, 28 (1), 113–122 (1979).

Marple, Jr., S.L. *Digital Spectral Analysis with Applications*, Prentice-Hall, Englewood Cliffs, NJ (1987).

Offelli, C., Petri, D. The influence of windowing on the accuracy of multifrequency signal parameter estimation, *IEEE Transactions on Instrumentation and Measurement*, 41(2), 256–261 (1992).

Ferrero, A., Ottoboni, R. A low-cost frequency multiplier for synchronous sampling of periodic signals, *IEEE Transactions on Instrumentation and Measurement*, 41(2), 203–207 (1992).

Stoica, P., Moses, R. *Spectral Analysis of Signals*, Prentice-Hall, Upper Saddle River, NJ (2005).

REFERENCES

1. Oppenheim, A.V., Schafer, R.W., Buck, J.R. *Discrete-Time Signal Processing*, Prentice Hall, Upper Saddle River, NJ (1999).
2. Ferrero, A., D'Antona, G. *Digital Signal Processing for Measurement Systems – Theory and Applications*, Springer, New York (2006).
3. Shannon, C.E. Communication in the presence of noise, *Proceedings of the IRE*, **37** (1), 10–21 (1949).
4. Bellanger, M. *Digital Processing of Signals – Theory and Practice*, Wiley, Chichester, UK (2000).
5. Oran Brigham, E. *The Fast Fourier Transform and Its Applications*, Prentice Hall, Upper Saddle River, NJ (1988).
6. Lyons, R.G. *Understanding Digital Signal Processing*, Prentice Hall, Upper Saddle River, NJ (2011).

EXERCISES

3.1 Let $x(t)$ be a zero-mean (also called alternating) triangular wave, having period $T_0 = 100$ ms. Would a sampling interval $T_S = 10$ ms suffice to obtain a correct acquisition of this signal?

3.2 The signal

$$x(t) = 3sin(2\pi 324t) + 0.8cos(2\pi 648t + \pi/5) + 0.1cos(2\pi 972t + \pi/2)$$

is sampled at the rate of 1600 Hz. Indicate the frequencies f_1, f_2, and f_3 of the three components, in the range 0–800 Hz, for the corresponding sampled signal.

3.3 A digital algorithm measures the RMS value of a sinewave whose frequency is about 50 Hz employing the quasi-coherent sampling approach. For this purpose, the algorithm processes a set of 1000 samples acquired with a sampling interval of 400 μs. Provide an order of magnitude assessment of the uncertainty associated with the digital algorithm.

3.4 The RMS value of a waveform, described by the equation

$$x(t) = 10sin(2\pi 200t) + 3cos(2\pi 350t + \pi/5) + 1cos(2\pi 400t + \pi/2)$$

is measured by a digital algorithm employing the quasi-coherent sampling approach. The algorithm employs zero-crossing to determine signal periods. A set of 400 samples, acquired with a sampling interval of 350 μs, is available for processing and is plotted in the figure.

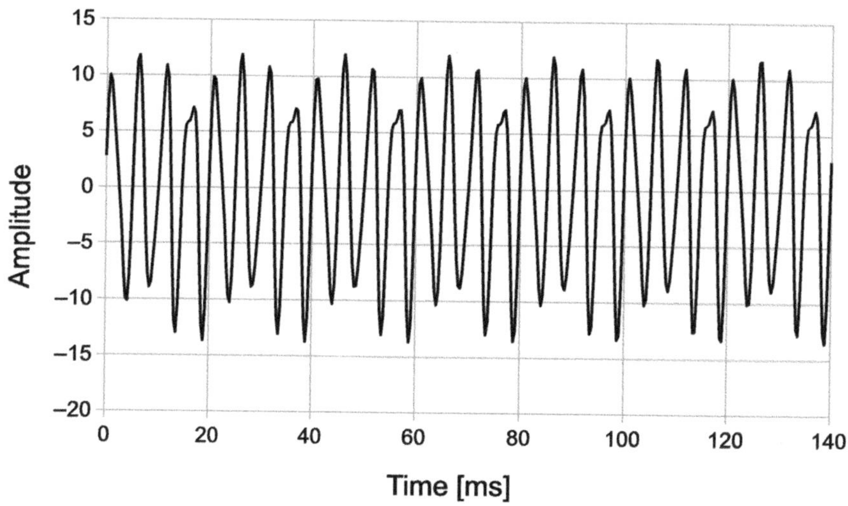

Discuss the uncertainty associated with the digital algorithm.

3.5 Consider a signal having three sinusoidal components at frequencies $f_1 = 18264$ Hz, $f_2 = 20750$ Hz, and $f_3 = 21688$ Hz. The signal is sampled at $F_S = 100$ kHz and a sequence of $N = 4000$ samples is acquired. The spectral composition of the signal is analyzed by means of a DFT algorithm, without application of any window. What frequency granularity is obtained? Does DFT coefficient magnitude allow accurate estimation of signal component amplitudes?

3.6 A periodic signal has only three spectral components, located at 40 Hz, 50 Hz and 75 Hz. Assuming that the signal is sampled under coherent sampling conditions, which are the minimum observation interval and the minimum sampling frequency required to avoid aliasing and process the sampled data with an FFT algorithm?

3.7 A signal with fundamental frequency $f_1 = 50$ Hz and a harmonic content upper-bounded to the 100th harmonics is sampled with a sampling frequency $f_s = 13.5$ kHz. The signal is observed over an observation interval of 512 samples. A DFT algorithm is applied to the sequence of sampled data. Which is the position, in the obtained array of frequency domain components, where the closest element to the signal 55th harmonic component is located?

3.8 The sampling frequency of a measuring instrument is $f_s = 6.4$ kHz. A periodic signal is sampled, with fundamental frequency $f_1 = 50$ Hz. This signal features all harmonic components up to the 133rd. Is the sampled sequence affected by aliasing errors? In the case it is affected by aliasing errors, which are the harmonic components, in the original continuous-time signal, that originate aliasing, and which harmonic components, in the sampled signal, are affected by aliasing?

CHAPTER 4

AD AND DA CONVERSION

NICLAS BJÖRSELL
Department of Electronics, Mathematics and Natural Sciences, University of Gävle, Gävle, Sweden

4.1 INTRODUCTION

In many measurement applications, the acquired signals are analog while the measurement equipment is computer–based. For example, a digital multimeter (DMM) acquires an analog voltage and presents a discrete value on its display. Furthermore, it is not unusual that measured signals are post-processed before presentation; they can be filtered or represented in the frequency domain. Moreover, measured data are stored in a digital format. Thus, there is a need to convert the analog signals into digital signals. This is typically performed by an analog–to–digital converter (ADC).

Measuring signals is a fundamental part of a measuring system and another important function is to measure the response to a given stimulus. Consequently, a measurement system must also be able to generate signals. It can be a stimulus used to characterize a transfer function of a component or device. It is common to have a waveform generator that generates a signal in a digital format and then converts it into an analog signal, which is performed using a digital–to–analog converter (DAC). When converting signals from analog to digital or vice versa, some information will be lost. This is easily recognized by simply noting that the amplitude of a digital signal has a limited number of fixed values, while an analog signal may have an arbitrary value with infinite resolution. In this chapter, the conversion process will be described. In addition, limitations will be described, and methods to reduce the impact of these limitations in various applications will be discussed.

4.2 SAMPLING

The process of converting analog signals into digital signals can be divided into two parts. First, the signal will be measured at discrete time instants and then quantized

Modern Measurements: Fundamentals and Applications, First Edition.
Edited by Alessandro Ferrero, Dario Petri, Paolo Carbone, and Marcantonio Catelani.

into fixed amplitude levels. One can envision two sequential blocks that perform these functions: a sampler followed by a quantizer. The key parameters to consider when selecting an ADC or DAC are the resolution of the quantizer in number of bits and the sampling rate of the sampler. These two functions will be discussed in the following subsections, but we will begin with a basic example.

Example 4.1 Sampling a Single Tone This initial example will describe what happens when a time-continuous single tone signal with an amplitude equal to one is converted to a discrete time signal by means of sampling and quantization. The single tone is represented by a cosine function. For the frequency f_0, the signal $x(t)$ is given by

$$x(t) = \cos(2\pi f_0 t). \tag{4.1}$$

The analog signal $x(t)$ is sampled with the sampling frequency $f_s = 1/T_s$, where T_s is the sampling interval. The sampled signal $y[n]$ will then be

$$y[n] = x\left(nT_s\right) = \cos\left(2\pi f_0 nT_s\right), \tag{4.2}$$

where n is an integer. $y[n]$ is the discrete time representation of $x(t)$ but is still continuous in amplitude. The next step is to quantize the signal in amplitude. This means that the amplitude is divided into a number of fixed levels called transition levels. Each transition level is represented by a binary code, k, and the digital signal will be represented by the code closest to the value of $y[n]$). This is illustrated in Figure 4.1. In the lower figure, the deviation between the analog and digital signal is shown. This is called the quantization error.

The full process is illustrated in Figure 4.2 for two different number of bits $B = 4$ and $B = 8$. The lower plots shows the spectrum of the quantized signal.

In the sampled signal, $y[n]$ we do not know anything about the signal between the samples, and in the quantized signal we lose information about the amplitude due to round-off errors. In the remaining part of this section, we will study the effects of quantization and sampling separately.

4.2.1 Quantization

The relationship between the time discrete, continuous amplitude input signal $y[n]$ and the digital output code k from an ideal ADC approximates the staircase transfer curve shown in Figure 4.1. For an ideal ADC, the code transition levels T_k within the ADC range (V_{min}, V_{max}) are given by

$$T_k = Q(k-1) + T_1, \tag{4.3}$$

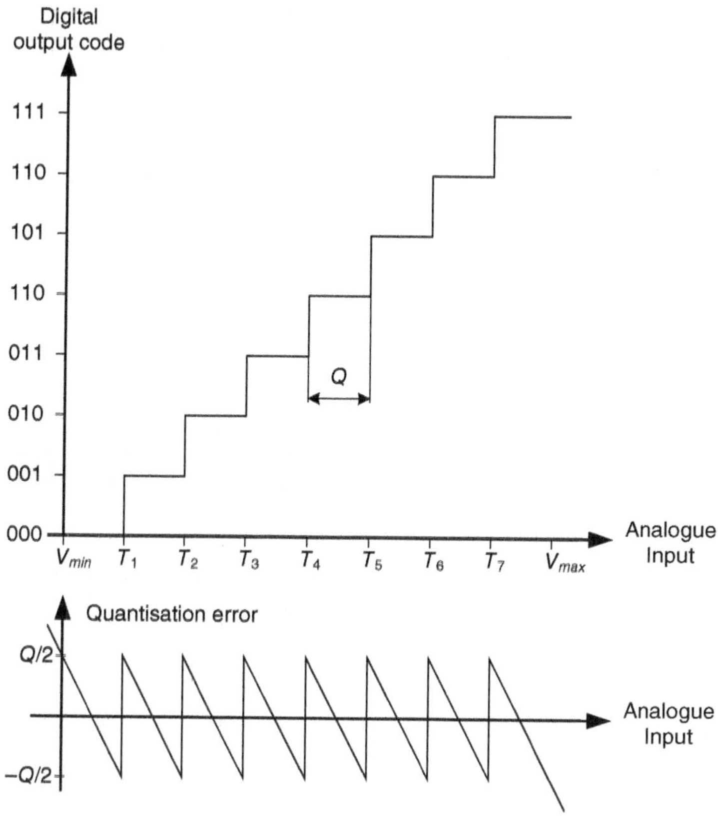

FIGURE 4.1 The relation between the analog input signal and the digital output code for a three-bit ADC (upper). The quantization error as a function of input amplitude is in the lower diagram.

where Q is the ideal width of a code bin; in other words, the full-scale range (FSR) of the ADC is divided by the total number of codes

$$Q = \left(V_{max} - V_{min}\right) / 2^{B}, \tag{4.4}$$

where B denotes the number of bits. Furthermore, T_1 is the ideal voltage corresponding to the first transition level, and $T_1 = V_{min} + Q$.[1] The code k spans $k = 0, ..., 2^B - 1$. The error generated by the deviation in the amplitude between the analog and digital signal is called the quantization error, $e(t)$. The quantization error can be considered to be uncorrelated to the input signal and is handled as a noise component of the

[1]Alternatively $T_1 = V_{min} + Q/2$, depending on the convention used: the mid-riser convention is used in the text and mid-tread in the footnote.

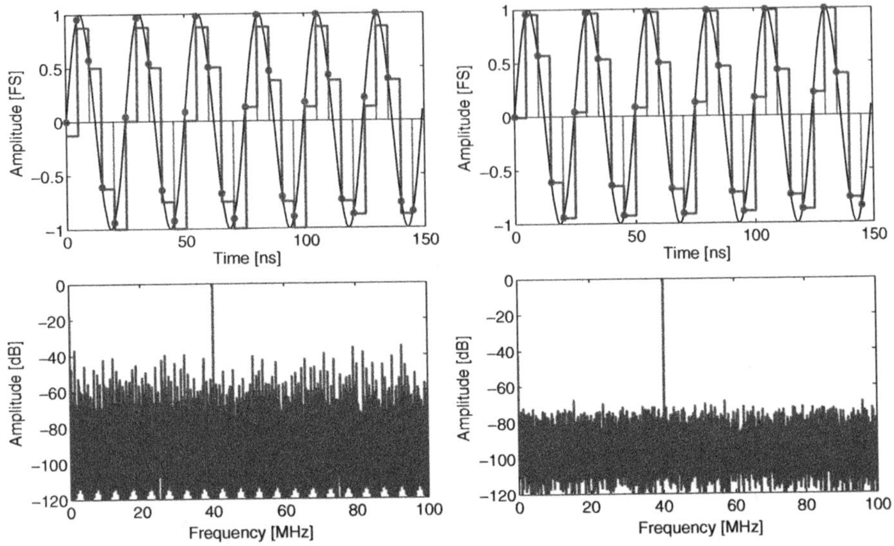

FIGURE 4.2 The time-continuous signal $x(t)$ (thin line) is sampled $y[n]$ (round markers) and quantized k (thick line) with resolution $B = 4$ (to the left) and $B = 8$ (right). The corresponding spectrum is in the lower plot.

output signal. The quantization noise is uniformly distributed, and the root mean square (RMS) value can be calculated as follows:

$$RMS(e) = \sqrt{\frac{1}{Q}\int_{-Q/2}^{Q/2} e^2 de} = \frac{Q}{\sqrt{12}}. \tag{4.5}$$

A block scheme for an ADC including sampler, quantizer and quantization error is shown in Figure 4.3. The noise will be reduced if the code bin width is decreased, that is, if the number of bits, B, is increased. The quantization error can easily be studied in the frequency domain. In this case, the quantization noise will be the noise floor. This is illustrated in Figure 4.2 for four- and eight-bit resolution, respectively. The interested reader can read more about quantization noise in e.g. Reference [1].

It is not only the quantization noise that affect how well we can represent the analog signal in the digital domain. The output from a non-ideal ADC will also be distorted due to non-linearities in the ADC. The distortion will occur at harmonic

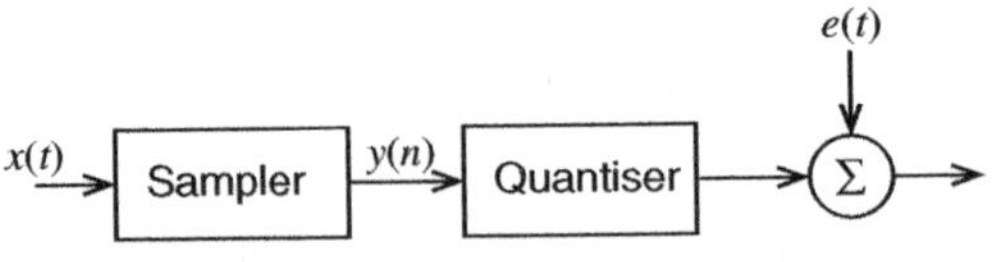

FIGURE 4.3 A block scheme of an ADC.

tones of the fundamental tone (e.g., f_0 in Example 4.1). Distortion will be further discussed in Section 4.4, but it is mentioned here to introduce the signal to noise and distortion ratio (SINAD). The SINAD is defined as follows [2]:

> For a pure sinewave input of a specified amplitude and frequency, the ratio of the RMS amplitude of the ADC output signal to the RMS amplitude of the output noise, where noise is defined to include not only random errors but also non-linear distortion and the effects of sampling time errors.

For an ideal ADC utilising the full scale, the SINAD can be calculated in dB as follows:

$$SINAD = 1.76 + 6.02B \qquad [\text{dB}]. \tag{4.6}$$

The proof can be found in the problems section at the end of the chapter. From the equation it is easy to see that for every bit resolution increases SINAD improves by about 6 dB.

4.2.2 Sampling Theorem

The shortcoming in sampling is that information about the analog signal between samples is lost. Consider the signal in Example 4.1; in Figure 4.4 the signal is sampled with three different sampling frequencies. When the signal frequency, f_0, is

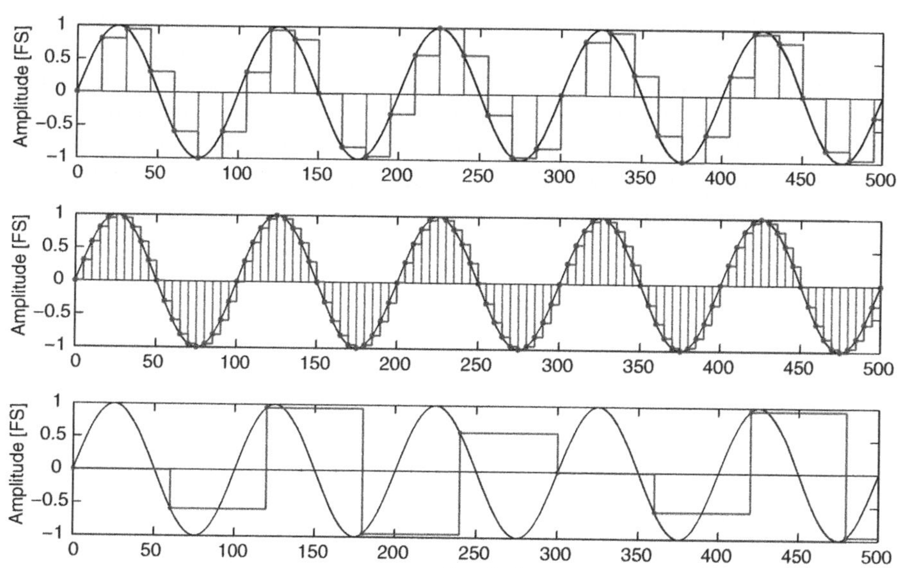

FIGURE 4.4 A sinewave is sampled at three different sampling rates; medium high (upper), high (middle), and low (lower).

low relative to the sampling frequency, f_s, the sampled signal becomes almost like a time-continuous sinewave. When the frequency is increased, the sampled signal becomes increasingly less like the continuous time signal, even if the signal is still a time-discrete sinewave. When the signal $x(t)$ is sampled at $f_s = 2f_0$, we have exactly two samples per period of $x(t)$; in a worst-case scenario, the sampling will be at the zero crossing and the sampled signal $y[n]$ will be a zero-valued DC level. If the frequency is further increased, the sampled signal will be observed as a signal of lower frequency. Sampling theory is covered in the previous chapter and we just recap that the Nyquist sampling theorem states that to correctly reconstruct a signal from its samples, the sampling frequency f_s needs to be greater than or equal to two times the highest frequency component in the waveform f_{max}. Nyquist sampling theorem has been modified by Shannon, who limited the lower bound to the bandwidth of the signal rather than the maximum frequency. This enables bandpass sampling, which will be introduced in Section 4.5.5. Let $x(t)$ be a band-limited signal with the frequency function $X(f)$ such that

$$X(f) = 0 \qquad |f| \geq f_N, \tag{4.7}$$

then $x(t)$, is uniquely determined by its sample content. See Section 3.1 for the full proof. A consequence of an incorrectly selected sample frequency is illustrated in Figure 4.5 (time domain) and 4.6 (frequency domain), where a continuous-time two-tone signal becomes a zero-valued DC signal when sampled.

4.2.3 Signal Reconstruction

There are at least two cases in which we would like to convert a digital signal into an analog signal. The first case is related to the ADC where a sampled signal is reconstructed by post-processing the output from an ADC. The second case is a generated digital signal that is converted via a DAC into an analog signal. Both cases have some fundamental theory in common. Assume that we have a discrete time signal $y[n] = x(nT_s)$ and we would like to generate a continuous time signal $\hat{x}(t)$ by using pulse-amplitude modulation

$$\hat{x}(t) = \sum_{n=-\infty}^{\infty} y(n)p\left(t - nT_s\right), \tag{4.8}$$

where $p(t)$ is the pulse function. The Fourier transform of (4.8) is

$$\hat{X}(f) = P(f)\,Y\left(fT_s\right). \tag{4.9}$$

The Fourier transform of a sampled signal is given by Poisson's sum formula (see Section 3.1 for details)

$$\hat{X}(f) = \frac{P(f)}{T_s} \sum_{k=-\infty}^{\infty} X\left(f - \frac{k}{T_s}\right), \tag{4.10}$$

where k is an integer.

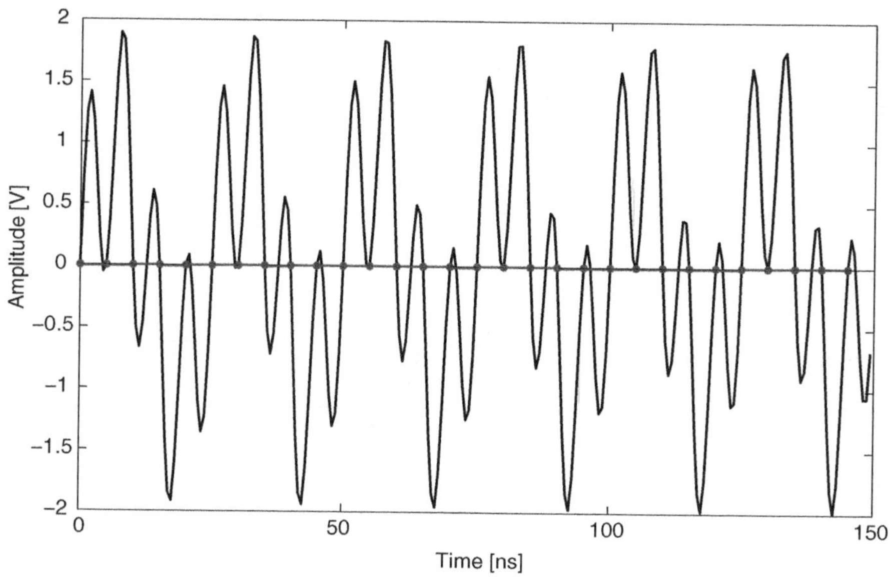

FIGURE 4.5 A two-tone signal where the highest frequency will fold down on the lowest frequency when sampled. The analog signal (solid line) has a peak value close to 2 while the sampled signal (round markers) looks like a zero-valued DC level.

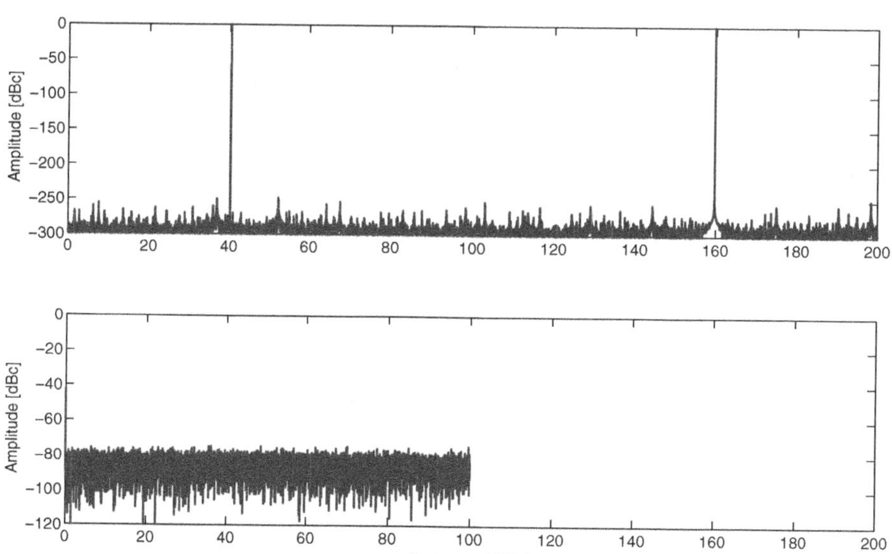

FIGURE 4.6 The spectrum of a two-tone analog signal (upper) and the corresponding spectrum for a sampled signal (lower) where the highest frequency is folded down on top of the lowest. Since the phase is 180 degrees phase shifted the signals will cancel out each other.

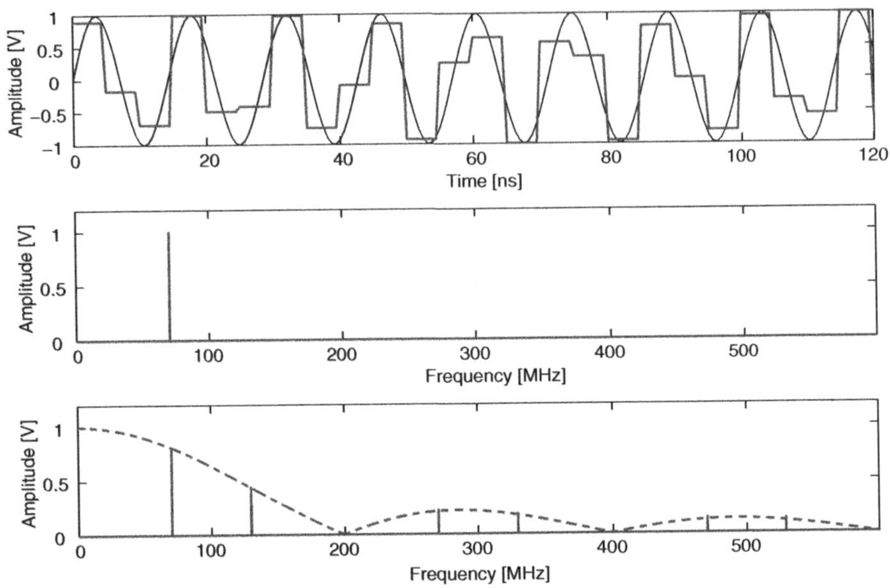

FIGURE 4.7 The upper diagram shows an expected single tone analog signal (thin line) which is generated by a digital signal (thick line). The spectrum for the analog signal is given in the middle but the output from a ZOH DAC will suffer from non-ideal signal reconstruction. Frequency components occur above f_N and the amplitude is corrupted by the sinc function (dashed line).

This means that the signal that we would like to generate, $X(f)$, will appear not only in the desired frequency range but also periodically at $\frac{1}{T_s}$. The zero-order hold (ZOH) is a mathematical model of the practical signal reconstruction performed by a conventional DAC. That is, it describes the effect of converting a discrete time signal to a continuous time signal by holding each sample value for one sample interval. In other words, $p(t) = 1$ and the corresponding frequency function will be

$$P(f) = sinc(f). \tag{4.11}$$

This is illustrated in Figure 4.7. To filter out additional frequency components, a low-pass filter is used after the DAC. Moreover, the amplitude of the analog signal decreases with frequency, which can be compensated when generating the digital signal.

If we, on the other hand, would like to perfectly reconstruct the analog signal based on the digital output from an ADC assuming that the Nyquist criteria is fulfilled, then $P(f)$ in (4.10) shall be chosen as

$$P(f) = \begin{cases} 0 & |f| \geq f_N \\ 1 & |f| < f_N \end{cases} \tag{4.12}$$

Taking the inverse Fourier transform $p(t) = F^{-1}[P(f)]$ for $f_s = 2f_{max}$ will produce

$$p(t) = \frac{\sin(\pi tT_s)}{\pi t/T_s}. \tag{4.13}$$

The original analog signal can then be perfectly reconstructed in theory, but an infinite amount of data is required according to (4.8).

4.2.4 Anti-Alias Filter

The alias effect described in Section 3.2.2 results in undesirable effects. It folds down signals outside the desired frequency range as well as noise. If we wish to avoid aliasing, the input signal must be forced to be band-limited to frequencies below the desired f_N. This can be accomplished by low-pass filtering the continuous time signal. This is performed using an anti-aliasing filter. The filter must be an analog filter; it is not possible to use a digital filter after the ADC because the undesirable frequency components above f_N cannot be separated from the desired components after the ADC.

4.3 ANALOG-TO-DIGITAL CONVERTERS

To emulate an analog signal as well as possible, we need high resolution and a high sampling rate. Unfortunately, it is difficult to obtain both at the same time. There are several different ADC architectures available on the market that are optimized for high resolution, a high sampling rate, or an optimum compromise between the two. One should also keep in mind that energy consumption and price are important parameters as well. In this section, an overview of the most common architectures will be provided. A more complete description can be found in ADC architectures textbooks (e.g., Reference 3). The following subsections provide a brief description that includes areas of application, advantages and disadvantages.

4.3.1 Flash ADCs

Flash ADCs are the fastest type of converters, operating at many mega samples per second (MS/s) to tens of giga samples per second (GS/s). They perform their multibit conversion directly. A resistor ladder is used to achieve a voltage division with $2^B - 1$ reference levels spaced Q apart. The analog signal is compared with each reference level by $2^B - 1$ parallel comparators. The outputs from the comparators are connected to a decoder-logic unit, which produces a parallel B bit output from the converter. The drawback is that Flash ADCs require stringent analog design to manage the large number of comparators and reference voltages required. Although flash converters are the fastest types available, their resolution is constrained by the available die size and by the excessive input capacitance and power consumption associated with the large number of comparators used. Their repetitive structure demands precise matching between the parallel comparator sections because any mismatch can produce static errors.

4.3.2 Pipelined ADCs

The pipelined ADC has become the most popular ADC architecture used to attain sampling rates ranging from a few MS/s to > 500 MS/s, with typical resolutions of 8–16 bits. It is derived from the subranging architecture. A subranging ADC can be considered to be composed of a sequence of sub-ADCs (SADCs) with low resolution (b bits), usually flash converters. The analog input signal is fed into a sample-and-hold amplifier (SHA), after which it is digitized by the first-stage SADC. The coarse b-bit most significant bit conversion is converted back to an analog signal using a b-bit sub-DAC (SDAC). Then, the SDAC output is subtracted from the SHA output, the difference is amplified, and this residual signal is digitized by a second-stage, b-bit SADC to generate the next bits of the total output. This procedure is repeated until the total number of bits, B, is achieved. Thus, the digital resolution is refined successively in each stage. However, one disadvantage of this process is that each stage requires some time to process the signal. This architecture is useful for resolutions of up to approximately eight bits. However, maintaining better than eight-bit alignment between the two stages (over temperature variations, in particular) can be difficult. There is no particular requirement for an equal number of bits per stage in the subranging architecture.

To increase the speed of the basic subranging ADC, the pipelined architecture has become very popular. The architecture is based on the subranging ADC, but an SHA is placed between the stages. All SHAs are clocked by the ADC clock signal. Thus, a new sample can be taken from the analog input while the previous sample is successively refined. This allows for a greater number of samples. The digital output will need several (the number of stages) clock pulses before an output signal is released. Thus, there is a latency in pipelined ADCs. However, after the initial latency, a new output is produced for each subsequent clock cycle. To ensure that the digital data from the individual stages corresponding to a particular sample arrives at the output simultaneously, shift registers must be added to each of the outputs of the pipeline stages. One of the disadvantages of pipeline architecture is that the comparators must be very precise to prevent error compounding down the pipeline. A pipeline architecture also includes error-correction logic, but we leave it to the interested reader to further study this area.

4.3.3 Integrating ADCs

Integrating ADCs typically feature low speed, low cost, and high resolution. An integrating converter integrates the input signal and correlates the integration time with a digital counter. The output of the counter is proportional to the amplitude of the sample, which can be demonstrated by allowing the quasi-stationary input signal charge/discharge a capacitor while a digital counter counts clock pulses. When the voltage over the capacitor reaches a predefined level, measured by comparator, a signal is sent to the counter to stop and output the counter value.

An improved design is the dual-slope converter; in this case, the input sample is integrated for a fixed time dictated by the digital value. When the counter overflows, a switch is thrown, a positive reference voltage is connected to the integrator, and the counter is simultaneously reset. When the integrator output reaches zero (fully discharged), a comparator switches its state, thereby latching the counter output.

The main disadvantage of such an architecture is that it can require $2B+1$ clock pulses to perform a full-scale conversion in a B-bit converter. The advantage of the dual-slope architecture is that the precision and nonlinearity issues that reduce the accuracy of the single slope version are cancelled out because the same circuitry is used for both the reference voltage and the sample voltage integration.

4.3.4 Successive Approximation Register ADCs

Successive approximation register (SAR) is a conversion technique that compares an analog input signal with a reference voltage. The reference voltage is the output of an B-bit DAC. Using the DAC output as a reference, this process approaches the final result as a sum of B weighting steps, in which each step is a single-bit conversion. In the first step, the most significant bit (MSB) in the DAC is one and the other bits are zero. If the input voltage is higher than the reference, the MSB is retained; otherwise, it is set to zero. Then, the next lowest bit is set to one and a new comparison is made. This process continues toward the least significant bit (LSB).

4.4 CRITICAL ADC PARAMETERS

It has already been noted that sampling rate and resolution are important ADC parameters and that it is not possible to achieve both high resolution and high sampling rate simultaneously. Furthermore, this has been proved in a continuously ongoing study on a majority of reported ADC designs, in which several ADC parameters are compared [4]. One of the factors compared is bandwidth (BW) versus SINAD, which is shown in Figure 4.8. Bandwidth is closely related to sampling rate (see Section 4.2.2), and SINAD is related to resolution; compare with Equation (4.6). The figure also shows that different values of jitter limits the overall performance. Jitter is the clock signal's deviation from true periodicity.

In addition to sampling rate and resolution, energy consumption is important in many ADC applications. However, these parameters are not the only critical ADC parameters. Depending on how an ADC will be applied, there are several additional parameters to consider. ADC data sheets provide information on the performance parameters, but it is also necessary to know which parameters are the most important for a specific application. In the IEEE standard 1241 [2], there is a table of parameters associated with different applications. A subset of typical measurement applications

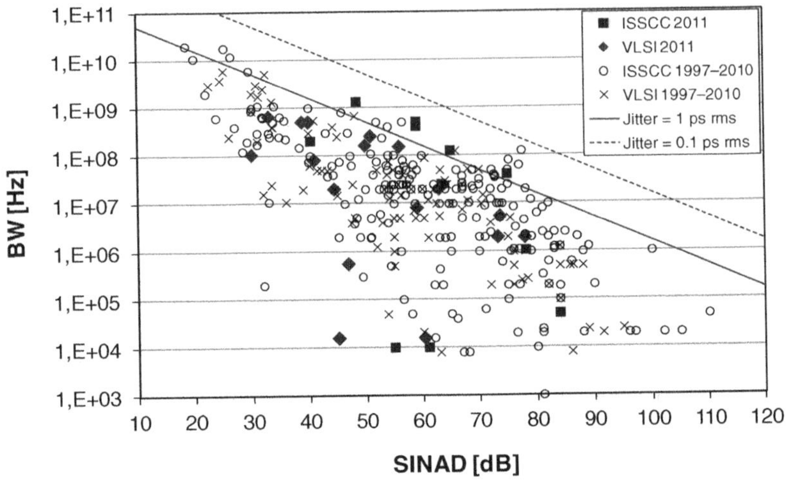

FIGURE 4.8 The trade-off between bandwidth and SINAD [4].

are given in Table 4.1. In the following sections, a brief description of some of the parameters will be provided. More thorough descriptions are provided in the standard [2].

4.4.1 Gain and Offset

Static gain and offset are the values by which the input values are multiplied and then to which the input values are added, respectively, to minimize the mean squared

TABLE 4.1 A subset of critical ADC parameters from IEEE std 1241 [2]

Typical applications	Critical ADC parameters	Performance issues
Data acquisition	DNL, INL, gain, offset, noise out-of-range recovery settling time, full-scale step response, channel-to-channel crosstalk	Channel-to-channel interaction, accuracy, traceability
Digital oscilloscope/ waveform recorder	SINAD, ENOB, noise, bandwidth, out-of-range recovery, word error rate	SINAD for wide bandwidth, amplitude resolution, low thermal noise for repeatability, bit error rate
Spectrum analysis	SINAD, ENOB, SFDR, noise	SINAD and SFDR for high linear, dynamic range measurements

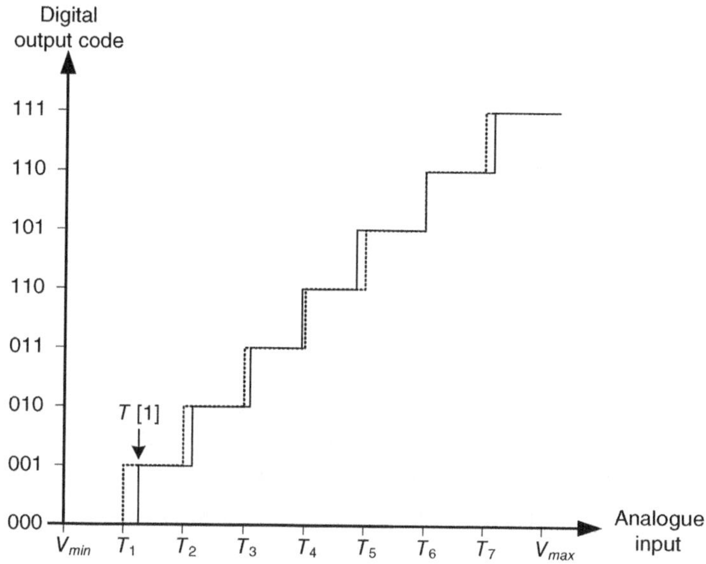

FIGURE 4.9 The relationship between the analog input signal v and the digital output code k from an ideal $B = 3$ bits ADC (dashed line) and practical ADC (solid line).

deviation from the output values. These values are called the independently based gain and offset.[2] The transfer characteristic can then be expressed as follows:

$$GT[k] + V_{os} + \varepsilon[k] = Q(k-1) + T_1, \tag{4.14}$$

where

$T[k]$	is the input value corresponding to the transition between codes k and $k-1$,
T_1	is the ideal value corresponding to $T[1]$,
V_{os}	is the output offset in units of the input quantity, nominally equal to zero,
G	is the gain,
$\varepsilon[k]$	is the residual error corresponding to the kth code transition.

The expression on the right side of Equation 4.14 gives the ideal code transition level (4.3), in input units, as a function of the binary output code k. Using conventional linear least-squares estimation techniques, the independently based static offset and gain minimizes the mean squared value of $\varepsilon[k]$ over all k.

4.4.2 Integral and Differential Non-linearity

Due to imperfections in all practical ADCs, the transfer curve is often somewhat distorted, which is illustrated by the solid line in Figure 4.9. The actual code transition

[2]There is also another set of values, terminal-based gain and offset, in which the gain and offset are based on terminal points.

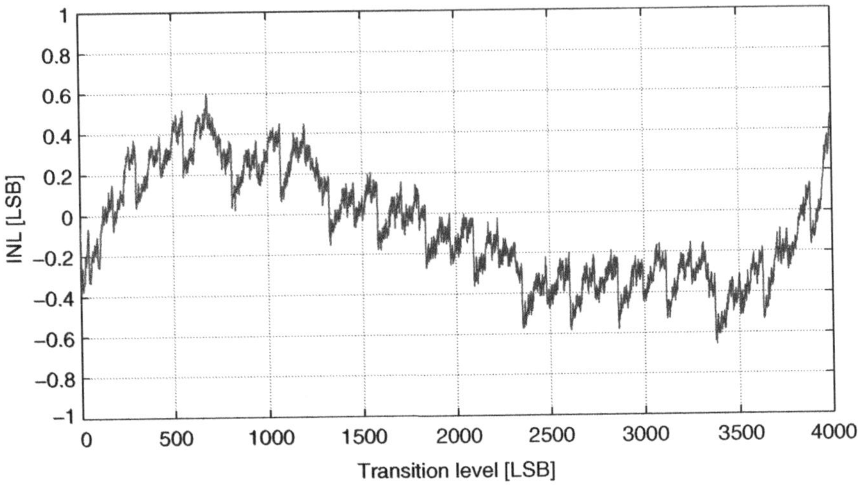

FIGURE 4.10 INL from a 12-bit ADC.

level $T[k]$ is the voltage that results in a transition from the ADC output code $k-1$ to k. Integral non-linearity (INL) is described as the difference between the ideal T_k in (4.3) and the actual $T[k]$ code transition levels of the ADC after a correction has been made for gain and offset errors.

$$INL[k] = \frac{T_k - GT[k] - V_{os}}{Q}. \tag{4.15}$$

Figure 4.10 shows an INL plot from a 12-bit ADC.

Differential non-linearity (DNL) is the difference, after correcting for the obtained static gain G, between a specified code bin width and the ideal code bin width Q, divided by the ideal code bin width. The DNL is given as follows:

$$DNL[k] = \frac{W[k] - Q}{Q}, \tag{4.16}$$

where $W[k]$ is the corrected width of code bin k. That is

$$W[k] = G(T[k+1] - T[k]). \tag{4.17}$$

The relation between $INL[k]$ and $DNL[k]$ is

$$DNL[k] = INL[k+1] - INL[k]. \tag{4.18}$$

4.4.3 Total Harmonic Distortion and Spurious-Free Dynamic Range

Commonly used measures of dynamic performance are often based on single-tone measurements; for example the spurious-free dynamic range (SFDR), total harmonic distortion (THD), and SINAD. Both THD and SFDR are defined in the frequency domain and the definitions are [2]

> For a pure sinewave input of specified amplitude and frequency, the root-sum-square (RSS) of all the harmonic distortion components, including their aliases in the spectral output of the ADC. Unless otherwise specified, THD is estimated by the RSS of the second through the tenth harmonics, inclusively. THD is often expressed as a decibel ratio with respect to the RMS amplitude of the output component at the input frequency.

> For a pure sinewave input of specified amplitude and frequency, the ratio of the amplitude of the ADC's output averaged spectral component at the input frequency to the amplitude of the largest harmonic or spurious spectral component observed over the full Nyquist band.

4.4.4 Effective Number of Bits

The SINAD was defined in Section 4.2.1 and is closely related to the effective number of bits (ENOB). It is used to compare the performance of an actual ADC to that of an ideal ADC. For an input sinewave of specified frequency and amplitude, after correcting for the gain and offset, the ENOB is the number of bits of an ideal ADC for which the RMS quantization error is equal to the RMS noise and distortion of the actual ADC.

4.5 SAMPLING TECHNIQUES

Thus far, this chapter has mainly focused on limitations in various aspects when converting between analog and digital signals. Information between samples is lost, round-off errors occur due to quantization, and imperfections in the ADC reduce performance even further. However, today's ADCs cover the needs of most applications. Moreover, there are several techniques that can further improve performance. These can be useful in applications with high demands or in reducing cost. In this section, we will introduce some commonly used methods: oversampling, sigma-delta conversion, dithering, time-interleaved sampling, undersampling, harmonic sampling, time-equivalent sampling and post-correction methods.

4.5.1 Oversampling

As discussed in Section 4.2.1, the quantization noise depends on the resolution and can be reduced by increasing the number of bits in the ADC. It is also possible to reduce the noise by increasing the sampling frequency and post-processing the data. This method is called oversampling because the signal is sampled at a higher frequency than that prescribed by the minimum requirements at twice the bandwidth of the signal.

The quantization noise power is spread over the frequency range $[-f_N..f_N]$. Consequently, if a higher sampling rate is used, the quantization noise power will spread over a wider frequency range; thus, the power spectral density will decrease because the area below the noise floor is constant. Every time the sampling frequency is doubled, the power spectral density of the quantization noise will be reduced by a factor of 2, which is equal to −3 dB. The increased sampling rate will both decrease the noise floor and increase the frequency range. However, the frequency above f_N is not requested and can be filtered out. Thus, the quantization noise in this range is removed, but the signal(s) inside the desired frequency range remains. Consequently, the SINAD will be reduced by 3 dB when doubling f_s. Compare with the equation for the SINAD (4.6); doubling the sampling frequency gives the same result as increasing the number of bits with $\frac{1}{2}$. In a more general case, the sampling frequency is multiplied by an oversampling ratio, M. The performance improvement corresponds to $\frac{log_2(M)}{2}$ bits. For example, if $M = 4$, we need one bit less. Because the signal is sampled at a higher sample rate, we will have more samples than we need. The unwanted samples are thrown away by a decimation filter.

4.5.2 Sigma-Delta, ΣΔ

The ΣΔ-converter is an ADC architecture and could have been introduced in Section 4.3, but because its conversion method is based on oversampling, it is appropriate to present this architecture after discussing oversampling. As an introduction to ΣΔ-converters, we assume that oversampling can be further improved. The trick is to filter the quantization noise so that the noise is distributed over the frequency range such that it is low in the frequency range of interest and high outside. The oversampling technique, as described thus far, distributes the noise uniformly over the whole frequency range $[-f_N \ldots f_N]$.

A ΣΔ design is shown in Figure 4.11. ΣΔ-converters consist of an integrator and a comparator with a feedback loop that contains a 1-bit DAC followed by a digital decimation filter. Through a series of iterations, the integrator, comparator, DAC, and summing junction produce a serial bit stream that represents the oversampled input voltage. Once digitized, the over-sampled signal goes through a digital filter to remove frequency components at or above the Nyquist frequency. A decimator then removes the oversampled data.

In order to study the effects on quantization noise, the ADC in Figure 4.11 is replaced with the ADC block diagram in Figure 4.3. The integrator and the ADC

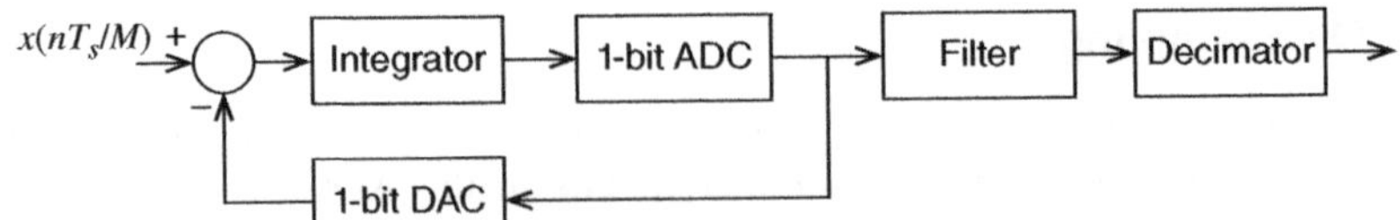

FIGURE 4.11 Architecture of a ΣΔ-converter.

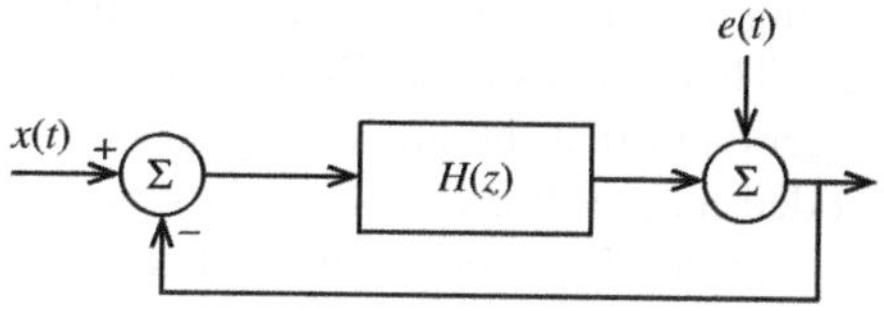

FIGURE 4.12 Block diagram of a ΣΔ-converter.

blocks are clustered into one block, $H(z)$ (the block diagram shown in Figure 4.12), that will act as a low-pass filter. One can find the transfer function from the input signal to the output, $H_{Signal}(z)$, as well as the quantization noise to the output, $H_{Noise}(z)$. These transfer functions are given by

$$H_{Signal}(z) = \frac{H(z)}{1 + H(z)}. \tag{4.19}$$

$$H_{Noise}(z) = \frac{1}{1 + H(z)}. \tag{4.20}$$

One can see that the transfer function for the signal will act as a low-pass filter, while the noise will be high-pass-filtered. Thus, the noise will be distributed to high frequencies that later will be filtered out by the digital filter before the decimator. See also Reference 5.

4.5.3 Dither

Distortion resulting from quantization can also be reduced by using a technique called dithering. The somewhat counter-intuitive idea of dithering is to add noise to the signal prior to quantization. There are many methods for doing this. The dither can be broadband noise; however, depending on how much noise must be injected, the SNR of the converter may be unduly sacrificed. Two methods are available to circumvent this problem. First, the dither can be generated with a pseudo-random digital number generator. These digital data are passed through a DAC and summed with the input of the ADC being tested. In the digital outputs of the ADC, the digital signal sent to the DAC is subtracted from the converter response. In this way, the noise summed into the analog input is digitally subtracted from the digital output, causing the SNR performance to return to normal. This technique is ideal for large dither signals. The other method is to generate the noise in such a manner that it occurs out of the frequency band of interest.

The main purpose of dithering is to randomise the DNL pattern of a non-ideal uniform quantizer. It can also be used to increase resolution in slowly varying signals. A slowly varying signal can be considered constant for a number of samples. In the absence of noise, all of these samples will have the same quantization error. Averaging the output samples will not reduce this error because the samples are all the same. However, if a zero-mean dither signal is added to the input prior to quantization, then the output will no longer be constant for all samples. Taking the mean of the output samples now has a meaning and might yield a result with a higher resolution than that of the quantizer itself. This can easily be illustrated by an example. Suppose the transition levels are integers and that the analog signal is 4.4 without added dither signal. It would be encoded as the discrete value 4. When we add a dither signal the amplitude sometimes exceeds 5 and occasionally becomes less than 4. Averaged over a longer period the value approaches 4.4. One may note that we again have a trade-off between bandwidth and resolution because averaging can be considered as low-pass filtering.

4.5.4 Time-Interleaved

Interleaving multiple ADCs is usually performed to increase the effective sampling rate, especially if there are no off-the-shelf ADCs available that fulfil the desired sample rate, linearity, and AC requirements of such applications. The main idea is to use two or more sub-ADCs (SADCs) in parallel, where each ADC has a lower sampling rate than the required one. For example, assume that there are two SADCs with the sampling rate $f_{s,SADC}$. The complete time-interleaved ADC (TI-ADC) will then have twice the sampling rate $f_{s,TI-ADC} = 2f_{s,SADC}$ of the individual SADCs. The SADCs will take every second sample, where one of the SADCs will be delayed by one sample. However, time-interleaving data converters is not an easy task because even with perfectly linear components gain/offset mismatches and timing errors can generate undesired spurs in the output spectrum. If one ADC shows an offset and the other a gain error, the digitized signal represents not only the original input signal but also an undesired error in the output. An offset discrepancy and gain mismatches show up as anomalies in the output. For interleaving designs, it is therefore necessary to choose ADCs with integrated gain and offset correction or include external circuitry that corrects these mismatches. Moreover, an INL of ± 1 LSB is quite common for individual ADCs; however, in an interleaving system, such errors can easily double. The appearance of non-linearity introduces distortion into the system, which degrades dynamic parameters such as SINAD and ENOB. Most of the errors discussed above can be overcome using calibration procedures in the time domain, careful circuit design and layout, a suitable selection of data converters, and digital post-processing. The interested reader can find more information in References 6 and 7.

4.5.5 Undersampling

When undersampling a signal, one makes use of the aliasing effect in a positive way. That is, frequencies above f_N are folded down to frequencies within the Nyquist band. However, when using undersampling, different techniques are used to make sure that the signals do not fall on top of each other. One way to do this is to replace the low-pass anti-alias filter with a bandpass filter for the frequency range of interest. This method is sometimes called bandpass sampling. The filter bandpass should be chosen such that all frequencies in the passband are folded down to unique frequencies in the 1st Nyquist band. The filter would not only suppress out-of-band signals, it would also suppress noise that would otherwise be folded into the 1st Nyquist band.

4.5.6 Harmonic Sampling

When measuring harmonic signals, undersampling can be used as well. For example, when measuring total harmonic distortion, it is recommended to measure 10 harmonics. For a high-frequency component, this implies a bandwidth of 10 times the fundamental tone, which can be difficult to achieve with sufficiently high resolution. However, in practice, this means that 11 frequencies will be measured. Because when the fundamental tone is known, all of the over-tones are also known. By the appropriate selection of the sampling frequency, all tones can be folded down to unique positions in the 1st Nyquist band [8].

4.5.7 Equivalent-Time Sampling

If the input signal is repetitive, equivalent-time sampling can be used to overcome limitations in sampling rate. Equivalent-time sampling is a process by which consecutive samples of a repetitive waveform are acquired and assembled from multiple repetitions of the waveform to produce a record of samples representing a single repetition of the waveform. The method is illustrated in Figure 4.13, which also serves as example to describe the method. By choosing the appropriate repetition rate of the input signal, four periods of the input waveform are recorded in a single record; then, upon rearranging the samples with a simple algorithm, a single period of the input signal is obtained, which is effectively sampled at four times the real-time sampling rate [9].

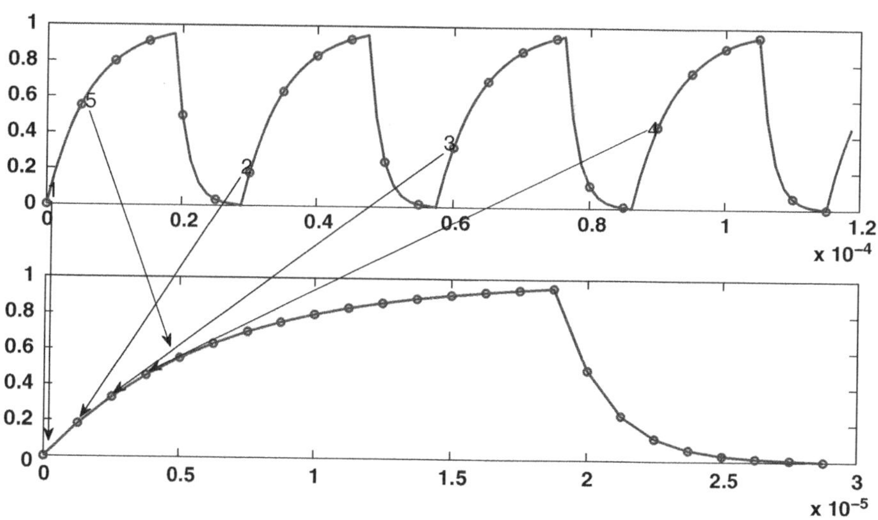

FIGURE 4.13 The principle of equivalent-time sampling. The upper plot is the sampled signal and the lower is the extracted signal.

4.5.8 Model-Based Post-correction

To compensate for non-ideal behavior in ADCs, such as that associated with gain, offset, INL, and DNL, one can use post-correction methods [10, 11]. Whenever the errors are systematic, it is possible to mitigate them by mapping the output to a corrected value. Post-correction can be divided into two different methods. One method is to use an inverse model, and the other is to add a correction term. When using a look-up table (LUT) for post-correction, the two methods are often denoted replacement and correction, respectively. The inverse model corresponds to replacement. The output code from the ADC is a table index. The code addresses a memory, where the memory value of that address is an estimate of the analog input. The index can also be compounded by one or more previous samples. If the current and previous samples together constitute an address in the correction table, it is called state–space correction. If the current sample and the difference between the current and previous sample (the slope) constitute an address, it is called phase–plane correction. However, such a modeling approach has a serious limitation; the error model requires burdensome experimental work for its identification. In model-based post-correction, the post-correction term is computed using a mathematical model. A parametric model requires less memory and does not need to be trained for every combination of present and previous samples. A well-assigned model is able to describe scenarios for which it is not trained [12].

4.6 DAC

The theory behind signal reconstruction was discussed with respect to the sampling theorem. In practice, there are different architectures for converting digital signals into analog signals. In the sections below, some of the most common architectures will be presented. A more complete description can be found in DAC architectures textbooks (e.g., Reference 3).

A DAC is fed with a code k; the output full-scale range (FSR) is divided into uniform intervals, with normal separation Q. The number of code levels is 2^B. The code 2^B is generally a virtual code that does not exist but serves as a good definition of the DAC. Thus, a 0-10 V, 10-bit DAC will theoretically never actually generate 10.0 V; the ideal output will be approximately 9.99 V. The key parameters of a DAC are to a great extent the same as those of an ADC, such as INL, DNL, SFDR, and SINAD. Additionally, for a DAC, glitches may occur when switching from one code to another.

4.6.1 Binary-Weighted

The binary-weighted DAC utilises a number of resistors (which can also be other elements such as current sources or capacitors) that are binary-weighted. It contains an array of B binary-weighted resistors, with B digital inputs feeding a summing amplifier to produce an analog output. This DAC architecture is not inherently monotonic and is actually quite hard to manufacture successfully at high resolutions. In addition, the output impedance of the voltage-mode binary DAC changes with the input code.

4.6.2 Kelvin Divider

A Kelvin divider architecture consists of 2^B equal resistors. The resistors are of equal value, so the input must be encoded. The output is determined by decoding 1 of 2^B switches to tap into a particular location on the resistor string. Compare with the ADC flash architecture. The Kelvin divider architecture has the advantage of having an inherently monotonic voltage output and low glitch (as only two switches operate during each code transition). It is also linear if all of the resistors are of equal value. The major disadvantage of this architecture is the large number of resistors and switches required.

4.6.3 Segmented

The voltage across the decoded resistor in a Kelvin divider circuit can be further subdivided to build a voltage-segmented DAC. This subdivision can be achieved through a second Kelvin divider circuit (in which case, the architecture is called a Kelvin–Varley divider). The output of the overall DAC remains monotonic as long as the individual segments are monotonic. Monotonicity is achieved when the offset of the buffer amplifiers is sufficiently small. Segmentation has the added benefit of reducing the number of resistors required to achieve a given resolution, allowing for smaller die sizes. Thus, it is common for high-resolution DACs to be segmented. Overall linearity is still determined by resistor matching.

4.6.4 R-2R

One of the most common DAC building-block structures is the R-2R resistor ladder network. It uses resistors of only two different values, the ratio of which is 2:1. A B-bit DAC requires $2B$ resistors, and they are quite easily trimmed. There are also relatively few resistors to trim.

4.6.5 PWM DAC

The device used to generate an analog voltage using a digital Pulse-Width-Modulated signal is known as a PWM DAC. PWM DACs are widely used in very low-cost applications, where accuracy is not a primary concern. Standalone DACs, however, are used for applications requiring higher accuracy. For example, a digital microcontroller outputs a PWM signal to a low-pass filter. The PWM pulse train's digital value becomes an analog voltage when it passes through the RC filter. Over a given period of time, the analog output is proportional to the PWM pulse's high durations.

4.7 CONCLUSIONS

A large extent of the measurements we perform will be processed by computer–based equipment. Thus, it is important that we can transfer the typically analog signals to digital without corrupting the data. However, converting analog signals to digital results in that the measured signal changes. In this chapter, we have shown the limitations in the converting process. The most common is the trade–off between sampling rate and resolution, but also other parameters should be considered. In general you can choose an ADC architecture that suits your application and usually it is possible to find an ADC with desired performance. But in those cases it is not possible, we have also demonstrated techniques for improving the performance of an existing ADC, such as dithering, time-interleaving, under- and oversampled ADCs. In the end of the chapter an overview of the close related DACs is given.

FURTHER READING

For those who want to immerse themselves further in the analog-to-digital converter and vice versa the following literature is suggested.

ADC architectures are well described in:

van de Plassche, R. *CMOS Integrated Analog-to-Digital and Digital-to-Analog Converters*, Kluwer Academic Publishers, Boston (2003).

de la Rosa, J.M., del Rio, R. *CMOS Sigma-Delta Converters: Practical Design Guide*, Wiley-IEEE Press (2013).

Louwsma, S., van Tuijl, E., Nauta, B. Time-interleaved Analog-to-Digital Converters, Springer eBooks (2011).

The latter also contains some post-correction methods for ADC imperfections such as gain, offset and timing errors.

Quantization and dithering can be studied in:

Widrow, B., Kollar, I. Quantization Noise, Cambridge University Press, Cambridge (2008).

For those who do want to read more about the standard for ADC testing, I recommend a magazine paper which gives a good overview or the complete standard.

IEEE, IEEE Standard for Terminology and Test Methods for Analog-to-Digital Converters. IEEE Std 1241–2010, 1–139 (2010).

Linnenbrink, T.E., Blair, J., Rapuano, S., Daponte, P., Balestrieri, E., De Vito, L., Max, S., Tilden, S. J., ADC testing, *IEEE Instrumentation and Measurement Magazine*, 9, 39–49 (2006).

REFERENCES

1. Widrow, B., Kollar, I. *Quantization Noise*, Cambridge University Press, Cambridge (2008).
2. IEEE, *IEEE Standard for Terminology and Test Methods for Analog-to-Digital Converters*. IEEE Std 1241–2010, 1–139 (2010).

3. van de Plassche, R. *CMOS Integrated Analog-to-Digital and Digital-to-Analog Converters*, Kluwer Academic Publishers, Boston, (2003).
4. Murmann, B., *ADC Performance Survey 1997–2012*, [Online]. Available at http://www.stanford.edu/ murmann/adcsurvey.html
5. Aziz, P.M., Sorensen, H.V., vn der Spiegel, J., An overview of sigma-delta converters, *IEEE Signal Processing Magazine*, **13** , 61–84 (1996).
6. Elbornsson, J., Gustafsson, F., Eklund, J.E. , Analysis of mismatch effects in a randomly interleaved A/D converter system, *IEEE Transactions on Circuits and Systems I: Regular Papers*, **52** (3), 465–476 (2005).
7. Elbornsson, J., Gustafsson, F., Eklund, J.E., Blind equalization of time errors in a time-interleaved ADC system, *IEEE Transactions on Signal Processing*, **53** (4), 1413–1424 (2005).
8. Nader, C., van Moer, W., Björsell, N., Händel, P., Wideband radio frequency measurements: from instrumentation to sampling theory, *IEEE Microwave Magazine*, **14** (2), 85–98 (2013).
9. Remley, K.A., Williams, D.F. Sampling Oscilloscope techniques, *National institute Standards Technology*, Boulder, CO, Tech Rep. 47W–7209 (1989).
10. Arpaia, P., Daponte, P., Rapuano, S. A state of the art on ADC modelling, *Elsevier Computer Standards and Interfaces*, **26**, 31–42 (2003).
11. Balestrieri, E., Daponte, P., Rapuano, S., A state of the art on ADC error compensation methods, *IEEE Instrumentation And Measurement Technology Conference*, 711–716 (2004).
12. Björsell, N., Händel, P., Achievable ADC performance by post-correction utilizing dynamic modeling of the integral nonlinearity, *EURASIP Journal on Advances in Signal Processing*, p. 497187–1 (2008).

EXERCISES

4.1 An analog signal $x_a = \cos(2\pi 25t) + 2\cos(2\pi 110t)$ is sampled with a sampling frequency $f_{sampling} = 200$ samples per second.

1. What are the frequencies in the resulting discrete–time signal $x_d(n)$?
2. Assume that a third unwanted frequency component with a magnitude of 0.5 at 175 Hz distort the signal. How will that affect the spectrum of the sampled signal?

4.2 An analog signal is sampled with sampling frequency 320 Hz. The sampled signal has spectral components at 4, 10, 117, 158, and 159 Hz. Then the sampling frequency is changed to 312 Hz, which gives a sampled signal with spectral components at 10, 12, 117, 150, and 153 Hz. What spectral components were in the analog signal?

4.3 A signal $y(n) = \sin(2\pi f_0 \frac{n}{f_s})$ will be generated by an eight-bit DAC. The sampling frequency, f_s, is 160 kHz and the signal frequency, f_0, is 60 kHz. The eight sample long input sequence to the DAC is 127, 218, 0, 218, 128, 37, 255, 37. The output of the DAC is a zero-order-hold analog signal. Sketch the spectrum of the analog output signal.

4.4 Effective number of bits (ENOB) is closely related to SINAD.

1. Give a derivation of the equation for SINAD (4.6).
2. What will happen if the signals peak value is less than full scale?
3. A 12-bit ADC has a SINAD = 71 dB. What will be the ENOB of the ADC?
4. Assume that oversampling can be used. What will the ENOB be for the same ADC if the sampling frequency is doubled?

CHAPTER 5

BASIC INSTRUMENTS: MULTIMETERS

DANIEL SLOMOVITZ
Department of Electrical Engineering, University of the Republic,
Montevideo, Uruguay

5.1 INTRODUCTION

Since the first decades of the last century, *multimeter* has been a common word in the technical electrical vocabulary. This device is a very useful measuring instrument. Typically, the instrument can be used to measure current, voltage, and resistance without the need of carrying a large set of different measuring devices. It covers large ranges in each quantity, from very low values, such as a few μA and millivolts to higher ones up to 20 A and 1000 V. In the resistance ranges, it covers from some ohms to tens of megohms. First types of multimeters were based on analog devices such as the D'Arsonval moving coil galvanometer. It can measure low direct currents (dc) in the order of 50 μA at full scale with an internal resistance of some hundreds of ohms. That means that full scale requires only some millivolts. To get other ranges and quantities, different resistors are associated in series-parallel connections. Also alternate voltages (ac) can be measured by this instrument, using rectifier diodes. For technicians, it was a marvellous instrument that included a whole laboratory in a small portable device. Nowadays, digital electronics has substituted analog devices and new quantities can be measured with improved accuracy. Not only traditional alternate voltage is available, but also alternate currents. Regarding other quantities, capacitance, frequency, temperature, and transistor checking are present in modern digital multimeters, and what is more amazing, from less than US$ 10. This aggressive competence has led to the disappearance of analog multimeters from the market. The best analog multimeter ever existed did not exceed 1% of accuracy. On the other hand, electronic and digital techniques have allowed the development of very high accuracy digital multimeters (DMM) with accuracies in the order of 1 part in 10^6. The improvement is shocking. The core of DMMs is the analog to digital converter (ADC), which is explained in Chapter 4. Different types of ADC are used

Modern Measurements: Fundamentals and Applications, First Edition.
Edited by Alessandro Ferrero, Dario Petri, Paolo Carbone, and Marcantonio Catelani.

depending on the accuracy. Most of them are based on integrating devices reaching resolutions from 10 to 27 bits. Linearity error of the ADC is one of the most important uncertainty sources in the calculation of the DMM accuracy. Chapter 1 analyzes how to evaluate the uncertainty according to standard methods. From this technique, manufacturers derive the accuracy specifications of their DMMs. As any measuring instrument, DMMs must be periodically calibrated. Standards of different quantities are necessary for that process, in particular voltage and resistance standards. They are discussed in Chapter 2.

5.2 HISTORY

Long before the Christian Era, in China, the compass was known. Some historians date this discovery in the 27th century BC, at the time of Emperor Huang-Ti. The term magnetic comes from the ancient Greek city of Magnesia where natural stones with magnetic properties were discovered. In Western countries, the compass was reinvented in 1269, four millennia later! In the south of Italy, Peregrinus (Pierre de Maricourt) wrote a letter to his friend Sigerus in France (Epistola of Magnete) in which he described how to find the poles in a magnetized sphere and how to make a compass.

Some centuries later, Oersted and Ampère showed the relationship between electricity and magnetism, specifically between electric current and magnetism. From the discovery of the movement of a compass needle due to an adjacent current started the development of galvanometers. This was the word used by Ampère for this measuring instrument to honour Galvani.

A substantial improvement in the sensitivity of galvanometers was introduced by Schweigger. He discovered that the effect can be increased if a multiturn coil is used instead of a single straight conductor (see Figure 5.1). Magnetic needle was placed inside the coil, so that currents passing above and below produce forces in the same direction, adding their values. He invented the so-called Schweigger multiplier. The development of the galvanometer had a great advance in the 19th century due to the telegraph. This invention required sensitive receivers, even more for the submarine telegraphy.

In 1858, Thomson (Lord Kelvin) invented the marine galvanometer with a beam of light instead of the needle and a magnetic shielding which turned the operation free of external conditions. Figure 5.2 shows a diagram of this galvanometer. Another new idea was to exchange the location of the magnet and coil. Sturgeon in 1836 was the first to propose a galvanometer where the mobile element was the coil and the fixed element the magnet. This allowed using heavier magnets, dramatically increasing the magnetic field and therefore the sensitivity of the instrument.

Further improvements of the magnetic circuit were proposed by Jacques-Arsène d'Arsonval (director of a biophysics laboratory) and Marcel Deprez (engineer). In 1882 they proposed to add a magnetic core into the coil. A cylindrical piece of iron was placed inside the coil, reducing in this way the air space of the magnetic circuit to the minimum required for the coil movement. The sensitivity was increased as

FIGURE 5.1 Schweigger multiplier.

well as the linearity, because the field direction was always collinear with the coil plane. Modern analog multimeters are based on moving-coil galvanometers using that principle. Figure 5.3 shows an example of a modern analog multimeter.

The first multimeter patent registration was done by Donald Macadie (London), applied on May 20, 1922 [1]. He called his invention "Electrical Measurement

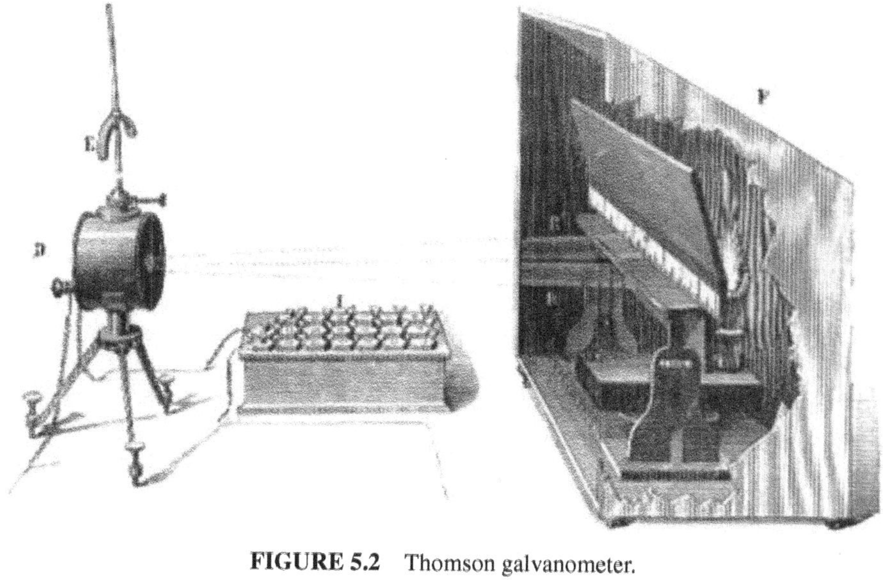

FIGURE 5.2 Thomson galvanometer.

FIGURE 5.3 Modern analog multimeter.

Instrument". It included, in a single portable instrument, all devices for measuring current, voltage, and resistance. Figure 5.4 shows the front panel and Figure 5.5 the electrical diagram. It is very similar to modern analog multimeters, except for the ac quantities that it did not have. It was based on a moving coil galvanometer, which had different resistors to implement the voltage and current ranges. The resistance ranges needed the battery *p* to supply the current through the circuit. A compensation for the battery depletion was done by the potentiometer *c*. For voltage ranges, series resistors (e, f) were added to the galvanometer, and for current ranges, shunt resistors *b*.

Few years later, commercial units appeared in the market, as the AVO meter (see Figure 5.6).

Digital multimeters in the market appeared in the 1970s. The FLUKE 8000A (Figure 5.7) together with the $2\frac{1}{2}$ digit HEATHKIT IM-1202 (Figure 5.8) and the SINCLAIR DM1 (Figure 5.9) were very popular instruments at that time. HEATHKIT sold kits with a set that included all parts and instructions to assemble their instruments. Even complex instruments such as oscilloscopes were included in its offer. The last two units used Nixies displays, as PHILIPS PM2420 digital multimeter, while FLUKE used red LEDs. Some 8000A units are still working. The availability of DMMs was a very significant step in improving the measurement accuracy for technicians and small laboratories due to their low cost.

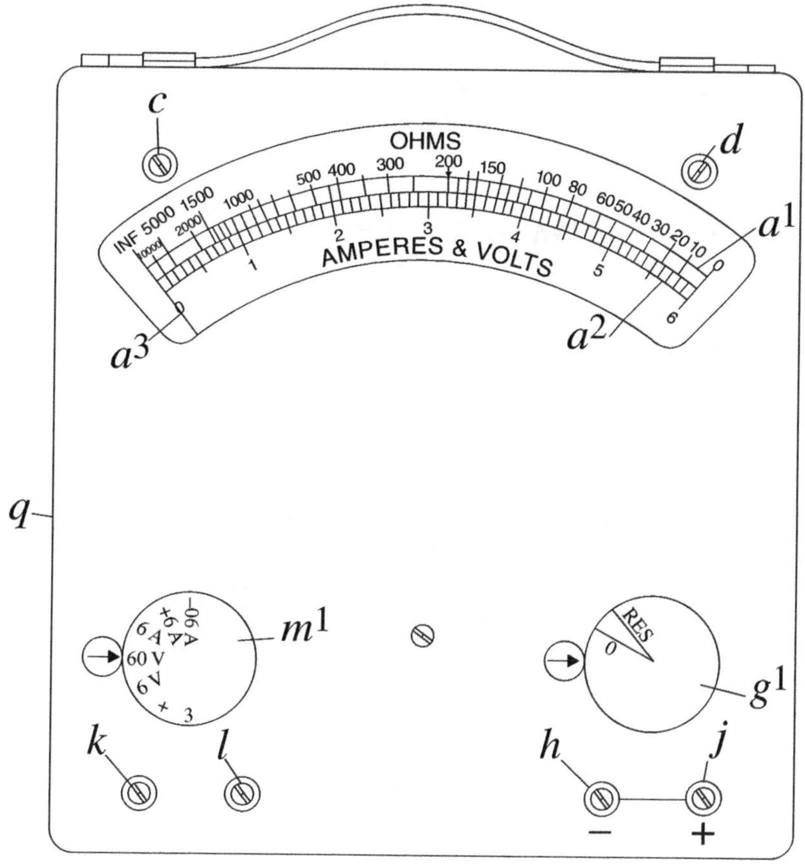

FIGURE 5.4 Front panel of the Macadie multimeter.

5.3 MAIN CHARACTERISTICS

To define the performance of a DMM, there are some key parameters to take into account. Although the following list is not complete, it includes most relevant points to consider when we need to select a unit.

5.3.1 Ranges

The first characteristic to verify is if the DMM can measure the quantity and the range we need. Most low cost DMMs only have dc and ac voltages, dc currents and resistance quantities. More expensive ones add ac current and special quantities. Also the number of ranges for the same quantity increases following the price. Traditionally, the range selection is performed by a rotating switch. Each position corresponds to a particular quantity and range (see Figure 5.10). Other models use keys and touch screen.

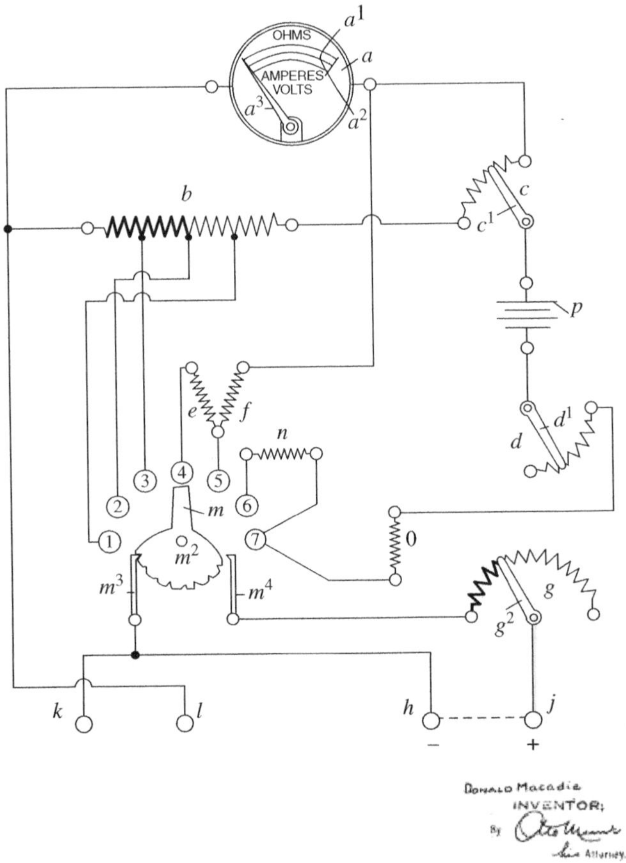

FIGURE 5.5 Electrical diagram of the Macadie multimeter.

If the measurand exceeds the full scale value, a warning message appears in the display indicating *overflow*. New DMMs have automatic range selection. They change the range according to the input value. The user only has to select the quantity, that is, voltage, current, resistance, or other. Some models can also verify if the test leads are at the appropriate binding posts. If a voltage range is selected but test leads are connected to current input, an alarm signal appears. It avoids dangerous situations because of the very low impedance that the DMM has in its current input, which may lead to a short circuit. At high current ranges, another caution must be taken. Some DMMs that have 10 A or 20 A range limit the maximum time of the measurement to a few seconds. If the measurement lasts more than this period, the instrument could have a permanent damage. A warning sign on this limitation is shown close to the 20 A terminal. The triangle with an exclamation mark inside, as Figure 5.11 shows, means: risk of danger, important information, see the manual.

FIGURE 5.6 One of the first multimeters in the market.

5.3.2 Number of Digits and Resolution

Resolution is defined as the *smallest change in a quantity being measured that causes a perceptible change in the corresponding indication* [2]. It only depends on the numbers of digits that the display has. More precise DMMs need to have more digits in its display. To show 1 V with 0.1% uncertainty, at least 4 digits are necessary.

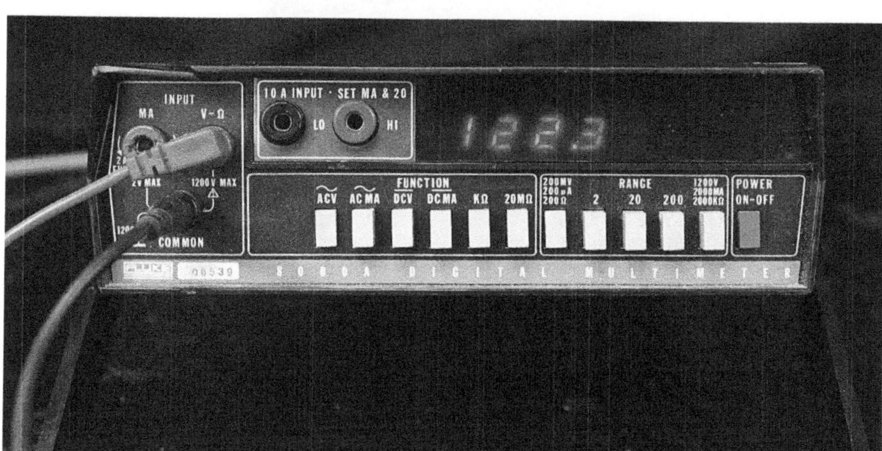

FIGURE 5.7 First FLUKE digital multimeter.

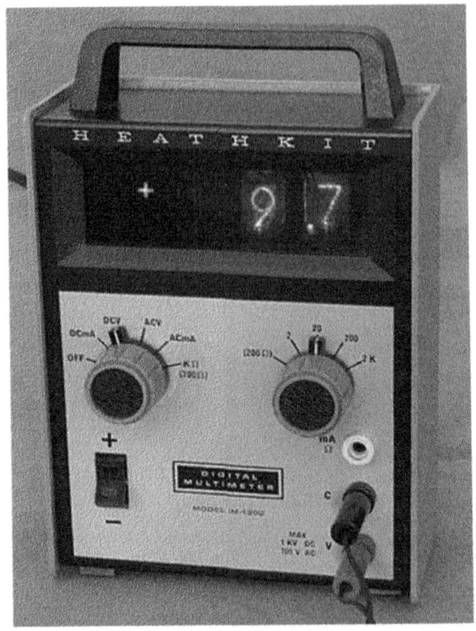

FIGURE 5.8 Heathkit 2 $\frac{1}{2}$digit multimeter.

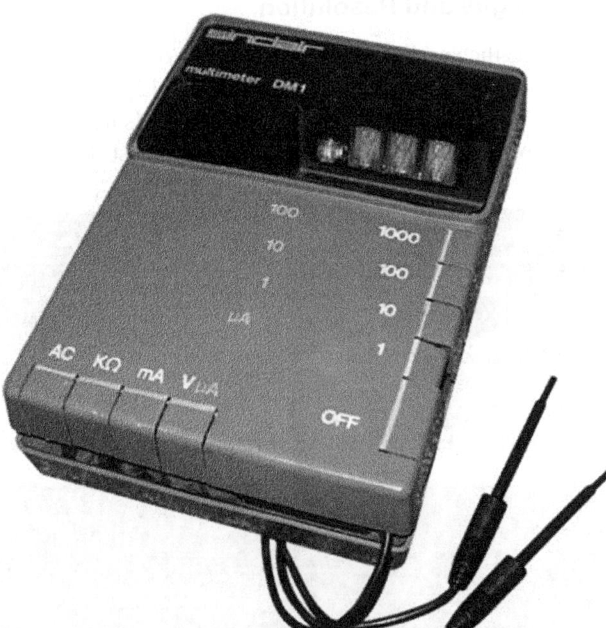

FIGURE 5.9 First SINCLAIR digital multimeter.

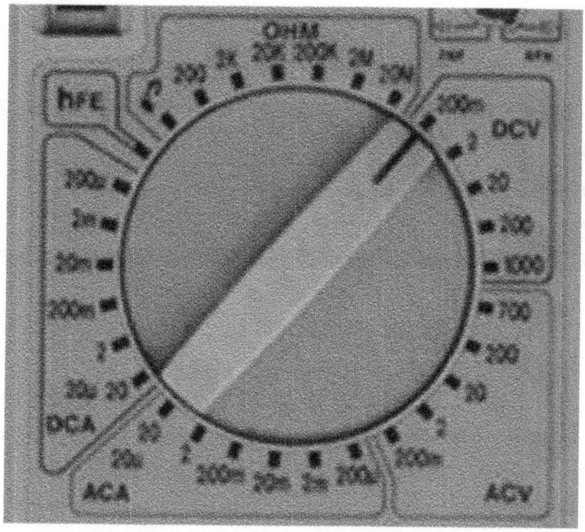

FIGURE 5.10 Rotating range selector.

The display must be able to show 1.000. Only 3 digits are not enough because the reading would be 1.00 and we cannot know the value at the millivolt level. With this resolution, any value between 0.995 and 1.005, with a variation of ±0.5%, has the same display indication of 1.00. The simplest way people generally characterize the precision of a DMM is from its number of digits, but really it refers only to its resolution. It would have 3, 4, 5, or more digits; however, it also can have fractional number of digits, that is, three and a half digits. What does it mean? That the most significant digit (MSD, the one at the left end) has only two values: 0 (or blank) or 1. That is, the highest value the DMM can show is 1999 for a $3\frac{1}{2}$ digit model. This number is also called *count*. A $6\frac{1}{2}$ digit DMM can reach 1999999 on its display. The position of the decimal point is irrelevant. The number of digits only refers to the

FIGURE 5.11 Multimeter terminals.

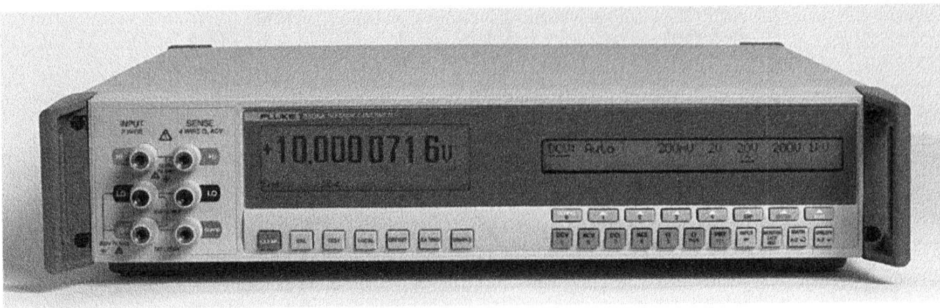

FIGURE 5.12 FLUKE model 8508A with 8 $\frac{1}{2}$ digits (By courtesy of FLUKE).

amount of digits the display has. New units have maximum values where the MSD can reach values of 3, 4 or others. In this case, it is usual to name it as an N $\frac{3}{4}$ DMM where N is the number of normal digits that vary between 0 and 9, that is, if full scale is 399999, the DMM is called as 5 $\frac{3}{4}$. With the same number of digits, a N $\frac{3}{4}$ instrument has better resolution than an N $\frac{1}{2}$ one. As an example, a voltage of 2.5 V is displayed as 2.50 in a 3 $\frac{1}{2}$ instrument, but 2.500 in a 3 $\frac{3}{4}$ one, increasing the resolution 10 times. Nowadays, the number of digits goes from 3 $\frac{1}{2}$ for the cheapest DMM up to 8 $\frac{1}{2}$ digits for expensive ones. Figure 5.12 shows one DMM with the largest number of digits.

5.3.3 Accuracy

The definition of accuracy is: *quality which characterizes the ability of a measuring instrument to provide an indicated value close to a true value of the measurand* [2]. It must not be understood as precision. Precision is: *closeness of agreement between indications or measured quantity values obtained by replicate measurements on the same or similar objects under specified conditions*. In this way, one measuring instrument can be very precise but inaccurate. This is the case when it always shows the same value when the measurement is repeated, but this value is far from the true one. What manufacturers generally assure is the *accuracy*, which is the maximum error they guarantee, but for some applications it would be useful to know the *precision*. For example, if we have to match resistors in a differential amplifier, it is not relevant the exact value of each resistor. The important point is the difference between them, so that the precision of the DMM. Accuracy specification is generally done as a percentage of the measuring value plus a number of digits. This last value sounds strange. It means that the last significant digit (LSD, at the right end of the display) can vary from the true value. For example, accuracy specifications of 0.5% of reading plus 4 digits in a 3 $\frac{3}{4}$ DMM, measuring 2 V, means that the real value would be between $2 \pm (2 \times 0.5/100 + 0.004 = 0.014)$. What is the reason for this complex way to inform the accuracy? The problem is to limit the instrument error at low scale values. If only a percentage of the display value is declared as accuracy, it reduces to zero when the measuring value is zero. No manufacturer would accept that obligation. To cover errors around low scale values, other specification is added. In the previous example, if the display value is 0.002, the accuracy specification would be ± 0.004, which is 200% of the reading. Obviously, it is not convenient to use this range. A lower one is better, that is, a millivolt range. Low cost DMMs have basic accuracy around 1%, depending on the quantity and range, while expensive ones can reach few parts in 10^6. The resolution must be in accordance with the accuracy. It is not coherent that a DMM had an accuracy of 0.01% and a display resolution of only $3\frac{1}{2}$ digit. The uncertainty produced by the low resolution is so large that avoid taking any advantage from the high accuracy. In this case, it would have a good measuring system, but could not show the result accordingly. On the other hand, there are DMMs with resolutions much higher than their accuracies. The reason of that is based on the precision specification. As stated previously, it may be useful, in some applications, to know the value of a measurement with low resolution uncertainty, although it had a significant deviation from the real value.

5.3.4 Loading Effects

Any measurement affects in some way the measurand. It is not possible to measure an object without making some change on it. Also, a simple visual inspection shoots a huge amount of photons on the observed objet. Has this perturbation a significant effect on the result of the measurement? As most questions, it has two answers, yes and no. It depends on the particular measurement we are doing. If we are measuring the length of a room by using a laser-meter, the effect of the photon shocks on the wall will be negligible. However, if we are trying to get the position of an atomic particle and neglect this effect, the result will be totally wrong. In electricity there are same problems. To measure a current that is passing through a conductor, some interference will be done on the circuit. One method is to cut the circuit and insert an ammeter. This kind of instrument has a low internal impedance, but not zero. Some resistance will be added, enough to reduce the circulating current by a significant amount in some applications. For example, in the 20 A range, a DMM can have an internal resistance of 40 mΩ (leads included). This means that with a current of 10 A there will be a voltage drop of 0.4 V. If the circuit has an emf of 3 V, the inclusion of the ammeter will reduce the actual current by more than 10%, so that a large error will appear in the result. A similar situation occurs when a voltage range is selected. The internal impedance of the DMM is not infinite, as an ideal voltmeter must have. The error produced by the load of the DMM will depend on the impedance of the circuit under test. Typical internal impedances in the voltage ranges are around 10 MΩ, although some models reach 10 GΩ in some ranges (generally, 10 V or lower). A circuit with a Thevenin impedance of 1 MΩ will reduce the actual voltage around 10% when the DMM is connected. These examples show that the load of the measuring instrument must be taken into account in all measurements, and an estimation of its influence on the circuit must be calculated. As a general rule, the larger the voltage impedance and the lower the current impedance, the better. Although most DMMs manufacturers declare the impedance of the voltage ranges, unfortunately, they do not declare on the impedance of the current ranges. Even, the impedances of the leads must not be neglected. Also this value is not declared in the instrument manuals. In some instruments, their impedances are higher than the internal impedance of the instrument. Regarding other quantities like resistance, frequency, or capacitance, the internal impedance of the DMM is not the only relevant parameter. In ac voltage measurements, not only the input resistance of the DMM but also the input capacitance must be known. At high frequencies, the impedance of the capacitance can be of the same order than the resistance. This effect is more important in high accuracy DMMs, which can have a large bandwidth. Usual capacitances between inputs are in the order of 100 pF. Larger capacitances, up to 1000 pF, exist between the low input and ground or guard in mains-powered equipment. In measuring high resistances, another parameter of the test probes and leads must be considered: their isolation. Some DMMs can reach resistance ranges as high as 200 MΩ. Cables with 2 GΩ of isolation will produce an error of 10%. If high accuracy resistance measurement must be done at high value, perhaps there are no testing probes with enough isolation to match the accepted maximum error. In that case, it is better to use shielded cables connecting the shields to the GUARD terminal of the DMM.

5.3.5 Guard

Most DMMs only have, for voltage ranges, the HIGH and LOW terminals, but more expensive models include the GUARD terminal (see Figure 5.13). This terminal is internally connected to an electrostatic shield that surrounds the instrument. In this

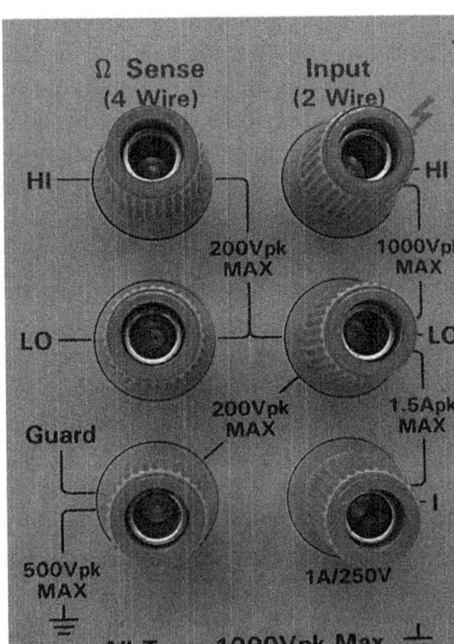

FIGURE 5.13 Guard terminal.

way, any external interference is blocked by the guard, avoiding disturbances on the measurement. The question is: where must the guard terminal be connected? One simple point is ground, but in this case the impedance between ground and guard terminals must be taken into account. This impedance is equivalent to a resistor in parallel with a capacitor. The resistance value generally is very high (more than 1 GΩ), but the capacitance impedance can be low at high frequencies. Typical values are around 1000 pF. In cases the capacitance impedance affects the measurement, the guard terminal must be connected to different places than ground, that is, a Wagner ground. It is a potential point equal to the potential of the LOW terminal, but isolated from it. This potential point can be generated by a follower amplifier, with high input impedance, connected between the LOW and the GUARD terminals.

5.3.6 Four Terminals

Test lead resistance may prevent to measure very low resistances. In this case, advanced models have the four-wire method. This means that the DMM has four terminals to measure a single resistor (see Figure 5.14). Two of these terminals impose a current though the resistor under test, while the other two terminals (sense) measure the voltage drop at the resistor ends, as Figure 5.15 shows. In this way, the DMM shows the ratio between voltage and current so that there is no lead resistance effect. This method applies to low value resistances, but also to medium values (hundreds of ohms) if high accuracy is needed.

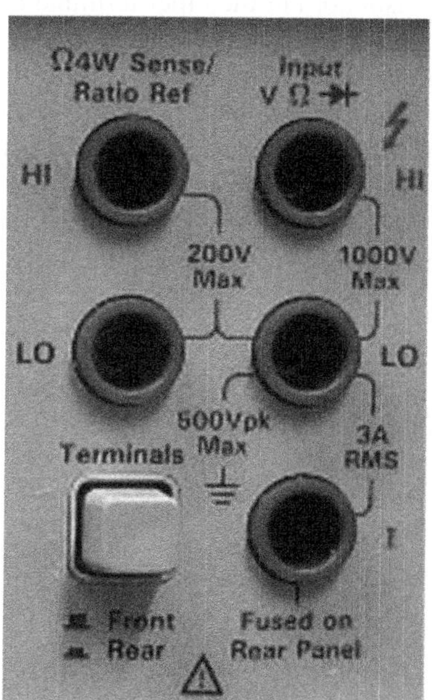

FIGURE 5.14 Four-terminal DMM.

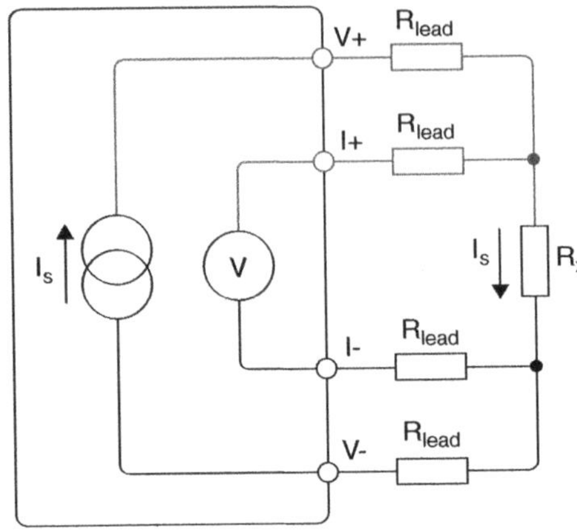

FIGURE 5.15 Internal circuit of a four-terminal DMM.

5.3.7 Accessories

Many external devices can be added to a DMM to extend its ranges such as high voltage probes, current clamp, and temperature probes. Most DMMs limit the voltage ranges up to 1000 V. If higher voltages are necessary to measure, a high voltage probe is needed (see Figure 5.16). It can increase the maximum value up to 40 kV. However, it can only be applied to limited energy circuits (Category I), like televisions or photo copy machines. It is not intended for high energy networks. It has one high voltage input terminal with a large isolator but the other terminal has no special isolation. It is connected to the common input of the DMM. In this way, only one terminal of the input has high voltage isolation. The circuits where the probe is used must be referred to ground potential. Internally, these probes have a resistor divider, one high value high voltage resistor and one low voltage output resistor. The input impedance of the DMM constitutes part of the low voltage arm, so that its value must match the probe specifications. Usually, a 10 MΩ DMM input impedance is required. An instrument with 1 MΩ impedance will have very large errors. In ac measurements,

FIGURE 5.16 40 kV voltage probe.

the target application is for power frequency. If high frequencies are needed, see the manufacturer manual, not only to check if the bandwidth is appropriate, but also if the voltage range is enough. As frequency increases, the maximum voltage decreases. A probe that can stand 20 kV at 50 Hz would reduce that specification to 10 kV at 1 MHz.

Another useful probe is the current probe. This kind of probe can measure currents circulating though a cable without the need of cutting the circuit to insert the measuring instrument. It senses the magnetic field created by the current around the conductor and generates a voltage or current proportional to the first one. It works based on the Ampère law. It relates currents with magnetic fields. For ac applications, the probe has a winding that acts as the secondary winding of a transformer. The primary winding is the cable under test. It has one turn. Then, for a probe 1000:1, the secondary winding has 1000 turns. Most of these probes have a magnetic iron core to improve the accuracy. Although the output signal of that transformer is a current, generally it includes a low value shunt resistor to get a current to voltage transducer. In this way, the selector of the DMM must be in the ac voltage ranges. This type of current probe does not work under dc condition. For this kind of application, another working principle must be used, like the Hall effect. It allows the direct measurement of the magnetic field, regardless of whether it is an ac or dc field. These probes have internal electronic circuits with a battery for its power supply. The Hall sensor detects the magnetic field caused by the measuring current, and generates a voltage proportional to that field. When iron cores are used, as usual, it is better to nullify the total magnetic field by using an auxiliary winding that imposes a controlled current. In this way, the effect of the non-linearity of the core is greatly reduced. The accuracy of these probes is around 3%, and the bandwidth 20 kHz. To get the total uncertainty of a measurement, the accuracy of the DMM must be added. Figure 5.17 shows the model 80i ac–dc current probe.

To measure temperatures, many DMMs have a special input for temperature probes. Most common probes are based on thermocouples. This transducer converts temperature to voltage. More precisely, it senses the temperature difference between

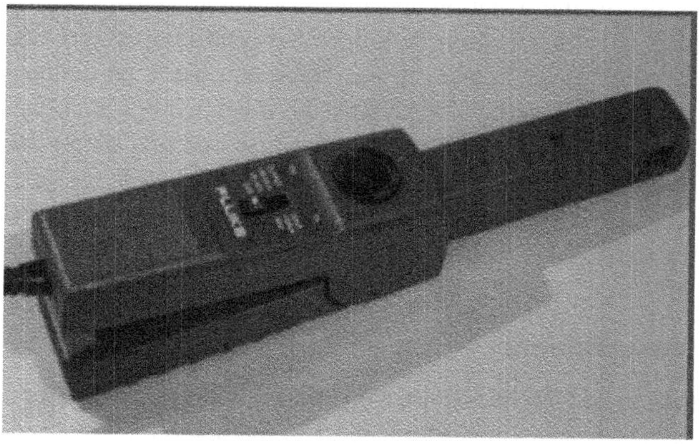

FIGURE 5.17 Current probe FLUKE 80i (By courtesy of FLUKE).

FIGURE 5.18 Thermocouple temperature probe (By courtesy of FLUKE).

both extremes of the probe (differential sensor). In this way, to get the temperature of the remote end, it is necessary to know the temperature of the end near to the DMM, generally equal to the ambient temperature. DMMs which have temperature facilities also have an internal electronic circuit that compensates the reading for the ambient temperature. That is, it converts a differential sensor into an absolute sensor. In some instruments, a special input jack exists to avoid errors from thermoelectric emfs generated at the contact between the plug and the jack. A commercial temperature probe is shown in Figure 5.18.

The temperature range for thermocouples goes from very low temperatures up to thousands of degrees Celsius. However, the probes delivered with DMMs have, generally, temperature limits from -40°C up to 300°C. This limitation is due to the DMM, but also to the isolation and linearity of the sensor. Thermocouples are classified according to the relation they have between temperature and voltages in different classes (i.e., J, K, T, and others). Each one has a particular temperature-voltage coefficient, so that the DMM must be designed for it. Low cost DMMs are usually matched for types J or K. Also they assume that the probe has a linear relationship between voltage and temperature, which is not true for large ranges. The temperature accuracy is around 1°C for most models. If higher accuracy is needed, DMM is not the right instrument.

5.3.8 AC Measurements

The bandwidth of most DMMs is limited up to few kilohertz. They are not instruments designed for high frequency measurements. The main application of ac ranges is for power frequency (50 Hz, 60 Hz) and eventually its harmonic content. For ac voltage and current ranges an ac to dc converter is necessary because the analog-to-digital converter of the DMM has only dc response. Many different circuits can convert an ac signal to a dc one. They belong to two classes according to their response, true RMS (root mean square) and aav (average absolute value). The first ones compute ac voltage according to its definition.

$$V_{ac} = \frac{1}{T}\int_0^T v^2(t)dt. \tag{5.1}$$

In this way, it works well regardless of the signal waveform. On the other hand, other DMMs use cheaper circuits to convert ac signals, most of them based on aav converters, implemented by a rectifier. The equation of the model of this converter is

$$V_{ac} = \frac{\pi}{2\sqrt{2}}\frac{1}{T}\int_0^T |v(t)|dt. \tag{5.2}$$

It performs the average of the absolute value of the signal. This is done by the integral in (5.2).

The coefficient $\pi/2\sqrt{2}$ is necessary to get the appropriate value when the ac signal has sinusoidal waveform. In this sense, with a sinusoidal ac input of 1 V RMS, the output will be 1 V dc. The absolute value is done by a full-wave rectifier, but instead of using diodes alone, DMMs add electronic amplifiers to eliminate the conduction voltage of the diode. In this way, these circuits emulate ideal diodes with null cut-in voltages. In analog multimeters no electronic devices exist, so that rectification is done by real diodes. In this type of meter, the lowest ac voltage range is limited to 10 V. Lower ranges cannot be achieved because of the diodes cut-in voltages, between 0.2 V and 0.7 V depending on the type. On the contrary, DMMs have ac voltage ranges in millivolt values. Figure 5.19 shows an example of an ideal full-wave rectifier. At the end, an R-C low pass filter performs the average of the signal. At its output V_o, a dc voltage equal to the aav value of the input V_{in} is obtained.

However, if the signal has another waveform, such as pulse, or has harmonic distortion, the result will be erroneous. The entity of errors produced by the aav-based DMMs depends on the type of distortion the signal has. An usual current waveform of an energy saving lamp is shown in Figure 5.20 which has a large harmonic distortion. Its peak value is 5.5 times its RMS value, instead of 1.41 if the waveform was sinusoidal.

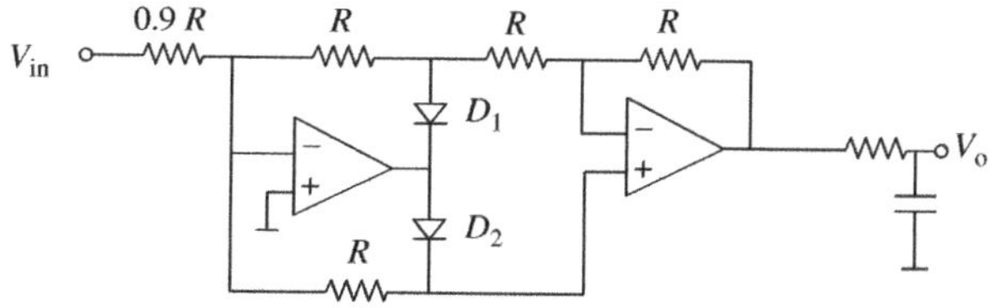

FIGURE 5.19 Rectifier circuit.

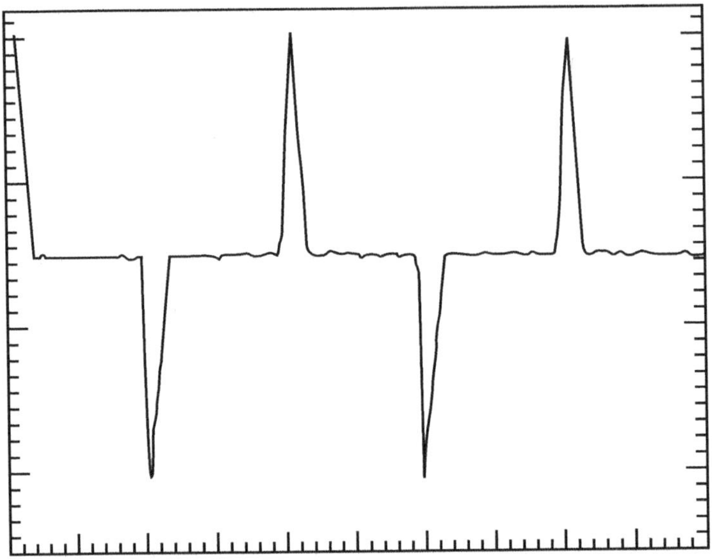

FIGURE 5.20 Distorted current waveform produced by an energy saving lamp.

A DMM with aav response will produce an error of −60% when measuring this current. This example shows the importance of knowing the internal type of the ac-to-dc converter, and to select the appropriated instrument according to the distortion of the signal.Table 5.1 shows the errors for typical waveforms. All values are based on 1 V peak voltage signal. The *Error* column represents the error of a DMM having an ideal aav converter.

Figure 5.21 shows a similar comparison for pulse waveforms. The aav error is calculated against the duty cycle of the signal. An error as large as −50% appears with duty cycles of 0.1. Under complex waveforms it is not easy to estimate the error, but a safe solution is to use a true RMS DMM in all cases a significant distortion is suspected.

Observing these results, an obvious question arises. Why do so many DMMs use aav converters instead of true RMS converters? The answer is very simple: cost. To implement an aav converter, a couple of diodes, an OpAmp and few passive components are enough. However, for true RMS converters, more complex and expensive circuits are necessary, as shown in Figure 5.22. It is the block diagram of

TABLE 5.1 Typical errors of an aav response DMM under distorted waveforms

Waveform	Vrms (V)	Vaav (V)	Error (%)
Sinusoidal	$1/(\sqrt{2})$	$1/(\sqrt{2})$	0
Rectangular	1	1.11	11
Triangular	$1/(\sqrt{3})$	0.555	−4

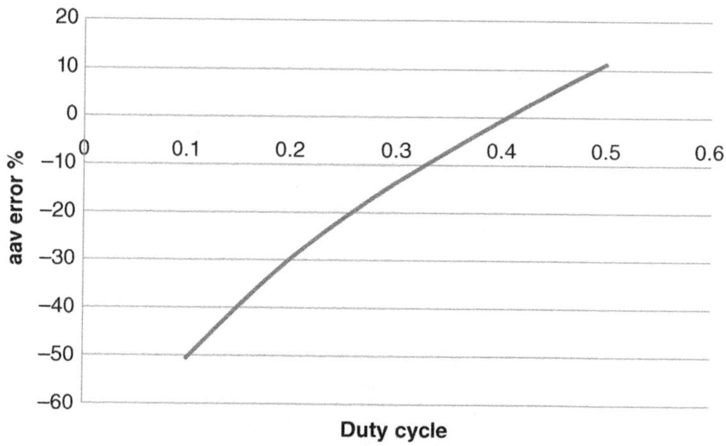

FIGURE 5.21 Error of an aav response DMM with pulse signal against its duty cycle.

the AD 637 integrated circuit. A multiplier-divider block generates an output equal to v_{in}^2/V_o, being V_o the dc voltage output of the converter. This signal passes through a low pass filter to get its average value V_o. In this way,

$$V_o = \langle {v_{in}}^2 \rangle / V_o \tag{5.3}$$

that is

$$V_o = \sqrt{\langle {v_{in}}^2 \rangle} \tag{5.4}$$

which is the true RMS value of the input signal.

Many DMMs that compute the true RMS value inform on this capability in their front panel. Figure 5.23 shows an example.

5.3.9 Safety

An important specification for the user is safety. DMMs are used as voltmeters to measure voltages in power networks. In this application, high power exists so there is

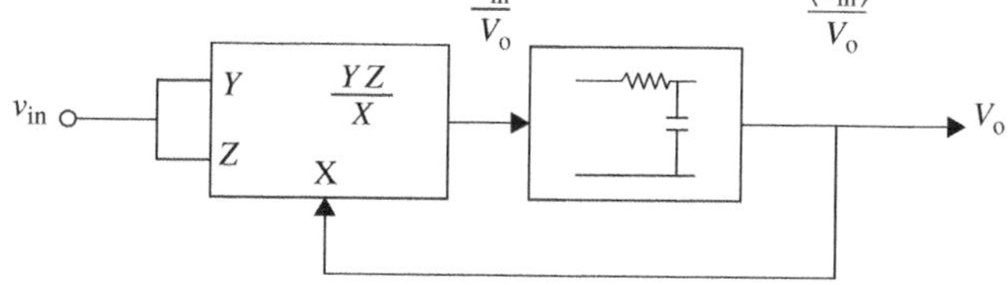

FIGURE 5.22 Circuit for computing RMS values.

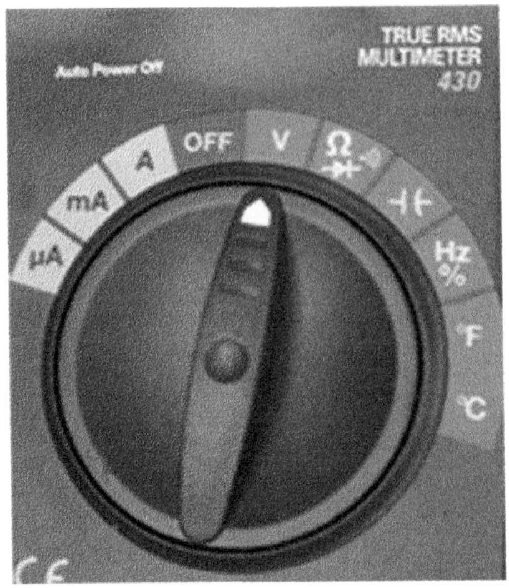

FIGURE 5.23 Front panel of a DMM with true RMS capability.

a risk of shock and burn. Usually, personal risk is associated with high voltages, but the real problem is current and power. The human body can be seriously damaged if a few milliamps circulate through it. Figure 5.24 shows a normalized diagram of time–current curves [3]. It is divided into four zones, from no effect to fibrillation. Current as low as 100 mA may produce death if it circulates more than a couple of seconds. Current depends on the voltage and resistance. The resistance of the human

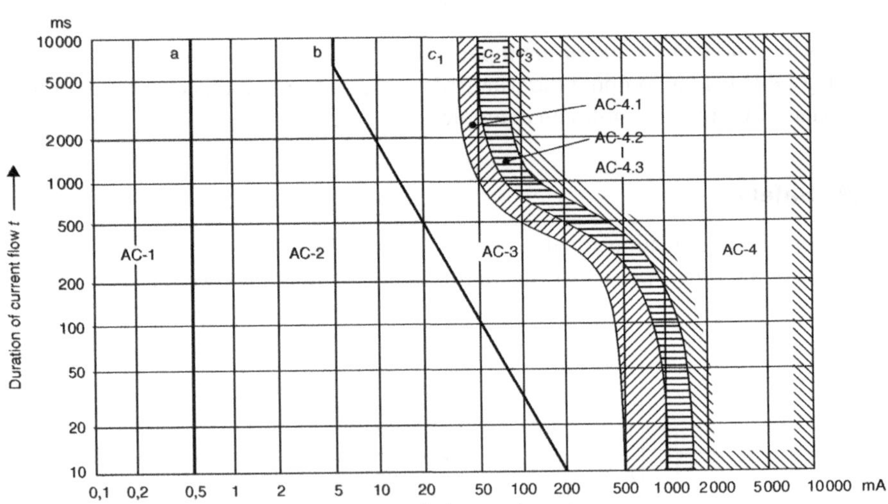

FIGURE 5.24 Time-current safety curve. IEC 60479-1 ed.4.0 Copyright 2005 IEC Geneva, Switzerland. www.iec.ch.

TABLE 5.2 Instrument categories according to safety requirements

Overvoltage installation category	Working voltage (dc or ac RMS to ground)	Peak inpulse transient (20 repetitions)	Test source
CAT I	600 V	2500 V	30 Ω
CAT I	1000 V	4000 V	30 Ω
CAT II	600 V	4000 V	12 Ω
CAT II	1000 V	6000 V	12 Ω
CAT III	600 V	6000 V	2 Ω
CAT III	1000 V	8000 V	2 Ω
CAT IV	600 V	8000 V	2 Ω

body strongly depends on the skin surface. If it is dry, resistances are around 2 kΩ, but under wet condition, that value can be much lower. This shows that voltages as low as 50 V can produce serious effect under wet conditions. On the contrary, high voltage sources up to tens of kilovolt, but with very low energy, are not dangerous. This is the case of high voltages sources of some electronic devices, where the maximum current, under short circuit condition, is not larger than 2 mA.

In power network measurements, another risk factor appears. It is the energy in case of a short circuit. These networks have very small impedances, so that a 220 V power outlet can supply thousands of amperes under short circuit conditions. Burn and shock risks are associated with short circuits during the measurement, but how can a short circuit be produced? One way is by using inappropriate test leads. These accessories may touch each other initiating a dangerous short circuit. Another way is because of the occurrence of a transient overvoltage in the network. These transients exist, and are generated by far lightning strikes or switching in the network. Larger voltage transients can appear up to 8 kV. There are standards that propose tests to prove the DMM can withstand these overvoltages without failure.Table 5.2 shows different DMM categories included in IEC 61010 [4].

Category I instruments are designed to measure electronic equipment, where no transients exist and all sources have low energy. Category II refers to instruments that will be used to measure single-phase circuits with medium internal impedances, like outlets at more distance than 10 m of a category III source. Category III is related to three-phase distribution, and category IV to measurements on the energy meter placed in the power network. For a detailed description, see the referred standard. Each category has different working voltages depending on the impulse peak voltage that the instrument can withstand.

Most DMMs show the safety category in their front panels, as Figure 5.25 shows. In this way, there are units of category II that can stand impulse peak voltages of 6 kV, for using with working voltages up to 1000 V, while others can be used only up to 600 V, corresponding to impulse peak voltage of 4 kV.

Ambient conditions, like temperature, humidity, and pressure can affect the safety conditions. Care must be taken with temperatures larger than 40 °C, humidity larger than 80% or altitude higher than 1000 m.

FIGURE 5.25 Symbol of category class.

5.3.10 Calibration

During their lives, DMMs as well as any other measuring instrument must be periodically calibrated. Calibration consists in comparing the DMM readings to the value of a standard. DMM can measure many quantities, so many standards are necessary. For example, for DCV ranges, a voltage standard is necessary, but for resistance ranges, standard resistors are needed. The difference between the reading of the DMM and the value of the standard is the error of the instrument and it is stated in the calibration certificate. Also, the uncertainty of the error must be provided, which gives the dispersion that the error has. These tasks are done by calibration laboratories, normally under the accreditation of the ISO 17025 standard. To reduce the amount of standard artifacts needed to perform a calibration of all DMM quantities, special pieces of equipment, called calibrators, are used. They can be set as voltage sources (dc and ac) and current sources (dc and ac) and can even emulate resistors, capacitors, and temperature probes. Calibration must be done in regular periods, generally 1 or 2 years. It is objective evidence that the DMM is under its manufacturer specifications. When the DMM is covered under an ISO 9000 quality system, calibration must be done periodically as an obligation. This standard states that all instruments used in a process under this ISO certification must have a valid calibration. A usual question is if a new instrument needs a calibration certificate. In some cases, especially for cheaper models, the calibration is more expensive than the cost of the instrument, and it is cheaper to buy a new unit than calibrate the old one. The answer depends on the information the manufacturer delivers with the DMM. In high precision DMMs, it is common to see an individual calibration certificate, done in the accredited manufacturer laboratory. But, in inexpensive DMMs, the only information delivered with the unit is a general manual that states the specification limits. This is not a calibration certificate. It refers to the error limits the manufacturer assigns to that model, but no individual test result is shown for each specific unit. Perhaps, some new units are outside their specifications. In this case, a calibration is needed, although the instrument was new. With the use, internal parts can change, and periodical calibrations are necessary in regular periods.

5.3.11 Selection

To select the best DMM for certain application, all parameters previously mentioned must be taken into account. First of all, which is the most common use that we need? It is not convenient to select the instrument taking into account uncommon applications. DMMs are designed as a general measuring device and it would be costly to choose an expensive unit that has accuracies or ranges that we never will use. Ranges and accuracy are two of the first parameters to be considered. *Portability* is another important parameter. Is it necessary to use the meter in the field? In this case, it is helpful a DMM that can admit shocks and has a carrying case adapted to handle manipulation. Is the DMM intended to measure distorted signals? In this case, a true RMS DMM must be selected. Another important point is safety. If measurements in high power circuits are necessary, a high category class must be chosen.

5.4 CONCLUSIONS

Multimeters have more than 100 years of history and they have changed a lot over time, from the analog-based instruments to the modern digital ones. Nowadays, they have many functions, not only voltage, current, and resistance ranges. Modern digital multimeters (DMM) can measure capacitance, frequency, temperature, diodes, and transistors, reaching accuracies of 1 part in 10^6 in the basic units. On the other hand, they are more difficult to use than their simpler analog counterpart. Generally, DMMs have a set of menus that allows to adjust many parameters, such as offset, resolution, sampling rate, accuracy, and communication settings, among others. This leads to a complex operation that requires more technical knowledge from the users. For all type of multimeters, it is necessary to take into account the loading effect that the instrument has on the device under test. Voltmeters reduce the impedance of the tested circuit and ammeters increase it. Special voltage measurements require GUARDING to avoid the influence of noise or other parasitic effects. GUARD terminal is normal in medium cost DMMs. For low value or high precision resistance measurements, four-terminal configuration eliminates the influence of the lead and connection resistances. To extend voltage and current ranges there are different accessories such as high voltage probes and current transformers. Their characteristics must be taken into account not only to ensure that the measured values are within their ranges, but also to evaluate the total uncertainty of the measurement. In some cases, these accessories can increase a lot the uncertainty result. Safety is another important characteristic of any type of instrument, but in the case of multimeters this is more relevant because these instruments are intended to be used in a wide range of installations. Particularly, personal risk must be taken into account when measuring on power networks. The large short-circuit power that this kind of installation has can create very dangerous situations under mishandling or bad selection of the instrument. For this use, DMMs require to fulfill special tests that assure their safe operation even under overvoltages and peaks superposed to the normal power supply. Finally, multimeters as well as any other measuring instrument must be periodically recalibrated. This task must be done, preferably, by accredited laboratories that have traceability to National Standards.

FURTHER READING

Several texts can be found that provide interesting insights into the basic instruments in general and the multimeters in particular.

An overview of the basic instruments used in electrical measurements in the first part of the twentieth century can be found in:

Stout, M.B. *Basic Electrical Measurements*, Prentice-Hall, Englewood Cliffs, NJ (1960).

Kinnard, I.F. *Applied Electrical Measurements,* John Wiley & Sons, New York; Chapman & Hall, Ltd., London (1956).

A more modern approach to electrical and electronic instruments can be found in:

Oliver, B.M., Cage, J.M. *Electronic Measurements and Instrumentation*, McGraw-Hill, Inc., London, UK (1975).

Kularatna, N. *Digital and Analogue Instrumentation: Testing and Measurement*, The Institution of Electrical Engineers, London (2003).

Sydenham, P.H. *Overview of Electrical and Electronic Technique. Handbook of Measuring System Design,* John Wiley & Sons, New York (2005).

REFERENCES

1. *US Patent 1593024* (July 20, 1926).
2. *International Vocabulary of Metrology—Basic and General Concepts and Associated Terms (VIM),* 3rd edition (2008). available at http://www.bipm.org/utils/common/documents/jcgm/JCGM_200_2012.pdf
3. *IEC standard 60479-2* (2007).
4. *IEC standadrd 61010-2* (2012).
5. Slomovitz,D. *Electrical Measurements* [In Spanish]. IEEE-R9, ISBN 07803-8395-8 (2004).
6. Fluke Corporation. *Calibration: Philosophy in Practice*, ISBN-13: 978-0963865007, ISBN-10: 0963865005, (1994).

EXERCISES

5.1 If a DMM is used to measure a dc voltage in a circuit,

(a) what is the error due to the load effect of the input impedance of the voltmeter if this value is 10 MΩ and the Thevenin impedance of the circuit is 350 kΩ?

(b) What is the maximum Thevenin impedance to get a load error lower than 0.01%?

(c) To avoid the load error, the 10 GΩ input impedance is selected in the voltmeter. However, in this case the specifications state that a current of 30 pA can circulate through the input terminals of the voltmeter. Under these conditions, what is the approximate Thevenin impedance to get an error lower than 0.1% when measuring 10 mV dc?

5.2 To measure the current that circulate through a load of about 8 Ω connected to a voltage source of 12 V approximately, a DMM is inserted as ammeter in the 2 A range. The manufacturer specifications state that the maximum voltage drop under full scale is 700 mV. What will be the error in the measurement due to the insertion of the ammeter?

5.3 What will be the percentage error, when measuring a resistor with a DMM, due to the heating of the unit under test in case this is a thermistor of 7.5 kΩ with a thermal coefficient of -3.5 %/K and a dissipation constant of 1 mW/K, if the current source of the DMM applies 100 μA?

5.4 To extend the voltage range of a DMM up to 2000 V, an external resistor is added. The DMM is used in its maximum range, which is 700 V with an internal impedance of 10 MΩ.

(a) What is the value of that resistor to get a reading of 200.0 when 2000 V is applied?

(b) To reduce this value to 10 MΩ, a parallel resistor is connected to the DMM. What is the value of this parallel resistor?

5.5 The accuracy specification of a $4\,\frac{1}{2}$ digit DMM, when measuring 132.08 V ac, is 0.5% of reading + 12 digits. What is the voltage interval where the real voltage value is?

5.6 The RMS value of non-sinusoidal voltage waveforms must be measured by a DMM that computes the average-absolute-value (full-wave rectifier system).

(a) What is the error in case of a nearly square waveform?

(b) What is the error in case of a half-wave rectifier waveform?

5.7 What is the advantage when measuring resistance using the four-terminal-method instead of using the two-terminal-method?

5.8 In which cases is it necessary to use low thermal EMF test leads?

5.9 How often is it necessary to calibrate DMMs?

CHAPTER 6

BASIC INSTRUMENTS: OSCILLOSCOPES

JORGE FERNANDEZ DAHER
Independent Consultant, Cuba, Shangrila, Canelones, Uruguay

6.1 INTRODUCTION

Although one can think this is a quite new instrument, the cathode-ray oscilloscope was invented by the end of the nineteenth Century. During the first quarter of the twentieth Century the main improvement was the development of a better and more precise time base. Later, World War II pushed for more improvements in this instrument because of the needs for better radars and communications. The final big step came with the introduction of digital oscilloscopes by the 1970s.

The oscilloscope is a very important and powerful instrument, which is useful for engineers and scientists when designing and testing components and circuits. It is also very useful for maintenance technicians who need to determine and measure the waveform of different electrical and non-electrical signals. The oscilloscope can measure input voltage signals, but with the help of different transducers or sensors practically any physical phenomena can be measured. A transducer or sensor is a device which converts specific physical stimuli into an electrical signal so that it can be measured by an oscilloscope. Engineers and technicians need to measure different phenomena like pressure, level, temperature, sound, or brain or heart waves. This is just a list of possible physical quantities that you can find in your work but there are many others. It is only necessary to find the appropriate transducer or sensor so that the oscilloscope can measure the signal.

The modern oscilloscope is based on fast digitizer circuits, which use different analog to digital conversion techniques. In Chapter 4 there is a good and wide description of the different methods for conversion. The fastest oscilloscopes use flash analog-to-digital converters (ADCs), whose working principle is described in Chapter 4. Although digital oscilloscopes presented a huge improvement compared to

Modern Measurements: Fundamentals and Applications, First Edition.
Edited by Alessandro Ferrero, Dario Petri, Paolo Carbone, and Marcantonio Catelani.

cathode-ray oscilloscopes, they also had some problems. In this chapter we describe the characteristics and limitations, like undersampling, of this device. In this book there are some chapters dealing with topics related to how digital oscilloscopes work. For example, data acquisition is described in Chapter 4 while sampling methods and errors found are described in Chapter 3. Digital signal processors (DSPs) are widely used in oscilloscopes so the topics covered in Chapter 3 will help understand capabilities and limitations found in digital oscilloscopes.

References presented in this chapter include, apart from many publications from manufacturers, a pair of IEEE publications as recommended reading. *IEEE Standard Specification of General-Purpose Laboratory Cathode-Ray Oscilloscopes* was published in 1970 by the Instrumentation and Measurement Society [4]. The second publication is *IEEE Standard for Digitizing Waveform Recorders,* revised in 2001 [5], which deals with descriptions, performance characteristics and error sources.

6.2 TYPES OF WAVEFORMS

A waveform is the shape of an electrical signal. This shape provides the information you need for your work. You can see if the measured voltage is constant or if it varies with time and how fast it changes in time. You can also see if the amplitude of the signal is varying with time and how fast it does so. In the real world there are many different waveforms so in this book we will mention the ones you can find more frequently.

6.2.1 Sinewave

This waveform is typically related to alternating voltages or currents. Its amplitude can be constant as in the case of the voltage you can find in any power outlet or it can change with time. This waveform is usual in power sources.

6.2.2 Square or Rectangular Wave

This waveform changes periodically from one value to another one. In the case of the square waveform the signal stays the same time in both values and in the case of rectangular waveform the two times are different. This waveform is usual in digital systems.

6.2.3 Triangular or Sawtooth Wave

This waveform varies linearly with time going up and down between two values. The two ramps can be equal or different. The sawtooth wave is a special case where one of the ramps is practically vertical. This waveform is usual in television circuits.

6.2.4 Pulses

A pulse is a sudden change in the amplitude of the signal and then back to the previous value. A step is also a sudden change in the amplitude but it remains in the last value. Both can be non-periodical signals. They are also known as single shots or transients. These waveforms can also occur periodically. A train of pulses is a typical signal in digital circuits or in communications like in parallel data buses.

6.3 WAVEFORM MEASUREMENTS

During their work engineers and technicians need to measure different properties of the signals present in circuits. We are going to see the basic properties of a signal and how to measure them.

6.3.1 Amplitude

One of the properties to measure is the voltage difference between two points in the circuit. Generally one of them is ground but not always and in this case it is necessary to do a differential voltage measurement. We will discuss this deeper in the measurements techniques section of this chapter. The amplitude can be measured from the maximum peak to the minimum peak of the voltage and in this case it is called peak-to-peak voltage. It is also useful to measure the root-mean-square (RMS) voltage and in this case the oscilloscope measures the voltage, squares it, finds the average voltage and takes the square root.

6.3.2 Phase Shift

When you have two similar waveforms, the oscilloscope can measure the time difference between them. This is measured in degrees or radians. This measurement is typical when analysing filters or amplifiers.

6.3.3 Period and Frequency

In the case of a periodic signal it is important to measure its period or its frequency. A period is the amount of time the signal takes to complete one cycle and it is expressed in seconds. The frequency of the signal is the number of times the signal repeats itself in one unit of time. If the unit of time is seconds then the frequency is expressed in hertz. Both magnitudes are reciprocals of each other. So 1/period equals the frequency and 1/frequency equals the period.

6.4 TYPES OF OSCILLOSCOPES

In general nowadays it is difficult to find analog oscilloscopes so we will concentrate on the modern digital ones. Anyway, there are some common characteristics in all

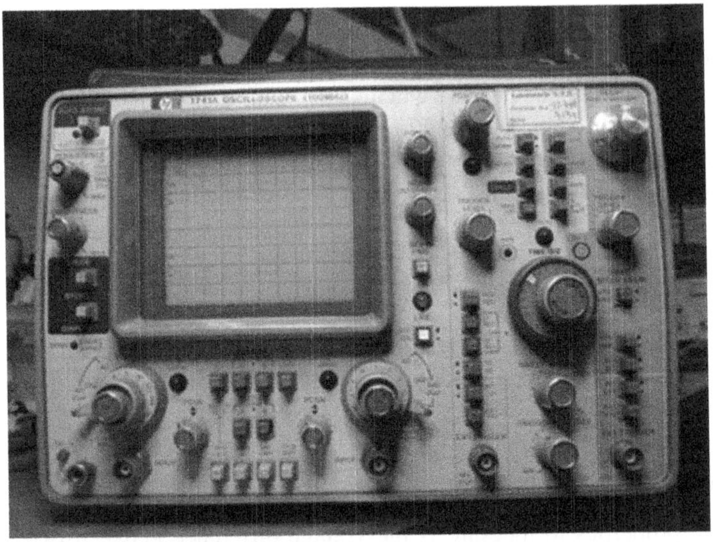

FIGURE 6.1 Hewlett Packard 1741 analog oscilloscope.

types of oscilloscopes. The front panel of all oscilloscopes is divided in many sections like the channel or vertical section, the display or screen, the horizontal or time base controls, and the trigger controls. In the case of digital oscilloscopes you may also have soft keys whose function can vary depending on the software menu selected on the screen. Figure 6.1 shows Hewlett Packard 1741 analog oscilloscope which was capable of storing waveforms in the screen phosphor. Figure 6.2 shows Hewlett Packard 54502 digital oscilloscope, one of the first powerful ones 20 years ago.

Figure 6.3 shows Tektronix 11402 digital oscilloscope, which was more similar to a computer than to an oscilloscope in its user interface. Every function in the oscilloscope was activated through soft keys.

FIGURE 6.2 Hewlett Packard 54502 digital oscilloscope.

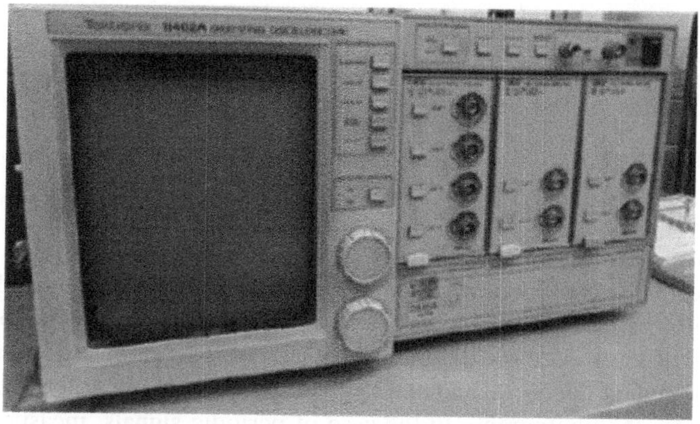

FIGURE 6.3 Tektronix 11402 digital oscilloscope.

Figure 6.4 shows Tektronix TDS 210 digital oscilloscope, which was one of the first powerful and economical models 15 years ago.

Analog oscilloscopes use cathode ray tubes to display the waveform of the signals connected to the channel inputs. The signal is connected to the vertical section of the front panel and it goes through amplifiers and signal conditioning circuits to adapt the voltage present at the input to the requirements of the vertical circuits. The internal time base circuits generate sawtooth signals which make the electron beam move from the left side to the right side of the screen making the successive sweeps needed to see the waveform on the screen. The phosphor on the inner side of the screen illuminates when hit by the electron beam. The trigger system makes the display look stable. Once the trace is completed, the trigger circuits wait for a new event, like a rising voltage crossing a given value, to start a new trace. The phosphor of the screen keeps illuminated for some time so the intensity of the trace gives you an idea of how frequent the signal passes by the same points. This feature makes this oscilloscope useful when trying to detect transients, which will be

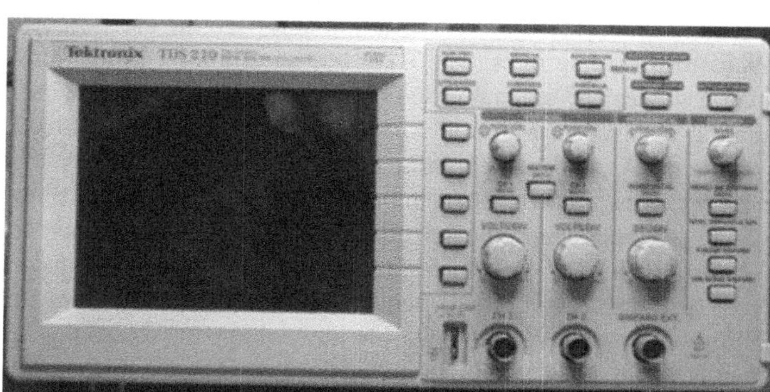

FIGURE 6.4 Tektronix TDS 210 digital oscilloscope.

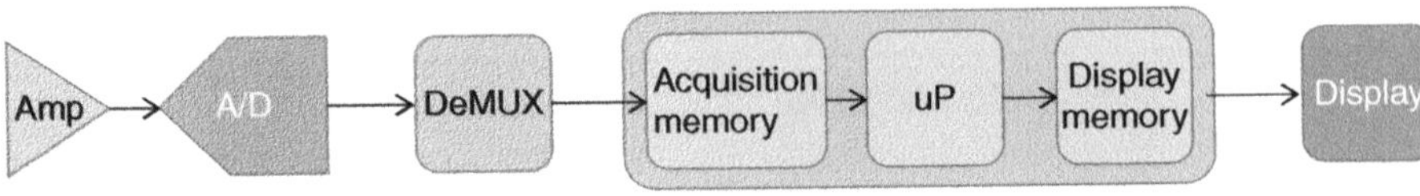

FIGURE 6.5 Digital sampling oscilloscope (DSO) architecture.

shown for a short time and not as bright as the periodic signal shown on the screen. Anyway, the analog oscilloscope cannot freeze the image indefinitely and it does not allow for automatic measurements in the waveform. In the past it was solved taking photographs of the screen for recording waveforms. Measurements could be done manually looking at the horizontal and vertical divisions in the grid on the screen or later measuring the photograph. In the case of periodic signals, measuring was not so difficult because the image on the screen, although it was always updated, did not change giving you enough time. Only in the case of transients the solution was to take a photograph. The bandwidth of analog oscilloscopes was limited in part by the cathode ray tube. There was a limitation in the deflection speed of the electron beam and so this imposed a limit in bandwidth. Some analog oscilloscopes showed two different signals connected to two input channels. One way to show the two signals is with one single electron beam and multiplexing the input signals. True dual beam oscilloscopes were not so common. Many years ago, when technology allowed for fast acquisition, analog-to-digital conversion and relatively large memory integrated circuits, and digital oscilloscopes appeared in the market. The signal is connected to the input channels, goes through signal-conditioning circuits, passes through analog-to-digital converters, and is displayed on the screen. The information can also be stored in memory indefinitely. Figure 6.5 shows a basic digital sampling oscilloscope (DSO) architecture. Now that the information is digital you can perform a series of processing tasks and measurements very easily. It can even be transmitted through digital communication ports, saved to flash drive or to CDs or DVDs. When digital oscilloscopes appeared, the bandwidth of the instrument was no longer dependent on the display system as it was with analog scopes.

Digital phosphor oscilloscopes (DPOs) use a parallel-processing architecture like the one shown in Figure 6.6. With this different approach the oscilloscope achieves very high waveform update rates. The DPO has an amplifier in the vertical channel as in analog scopes and an analog-to-digital converter as in digital oscilloscope. The information is saved in a digital phosphor database and then directly sent to the display system. As this transmission is parallel, this architecture can achieve faster processing times so there is an important increase in speed. DPOs are very similar to analog scopes since the refresh speed is very high compared to digital oscilloscopes. Handheld oscilloscopes generally are digital oscilloscopes with the advantage of being portable and battery-operated. Another advantage of this type of oscilloscopes is that you can measure any voltage not referred to earth since the instrument is isolated from the mains power system. There is no ground reference in handheld oscilloscopes. Generally handheld oscilloscopes are not as powerful as larger ones. PC-based oscilloscopes are based on data acquisition circuits, which can

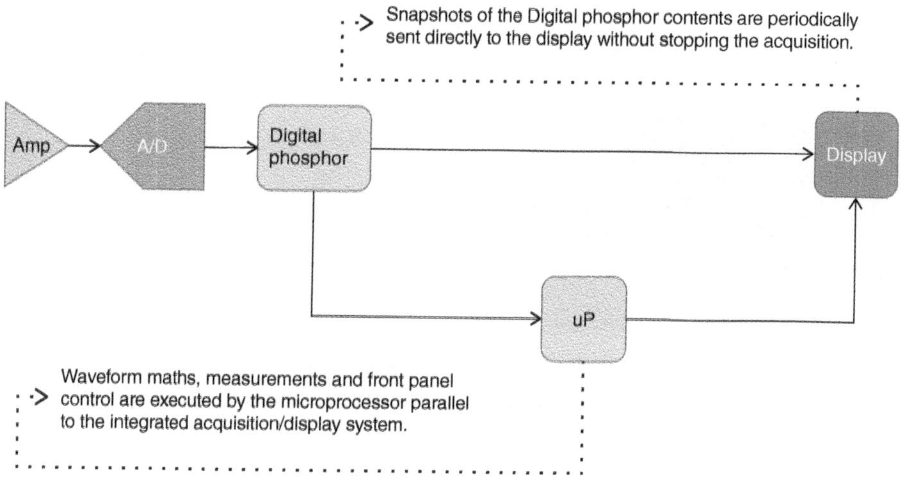

FIGURE 6.6 Digital phosphor oscilloscope (DPO) architecture.

be directly connected to the PC bus internally, or to the PC by a serial or parallel port externally. This hardware is controlled by software installed on the PC. Mixed signal oscilloscopes have inputs for analog signals and digital signals and are able to display both types of waveforms simultaneously. Generally the number of digital channels is larger than the number of analog channels. They can trigger on any type of signals or any combination of them. In Figure 6.7 you can see the input connectors on the front panel of a mixed signal oscilloscope.

6.5 OSCILLOSCOPE CONTROLS

Practically all oscilloscopes have a front panel divided into different sectors with knobs and buttons to operate the instrument. Digital oscilloscopes have also what are called soft keys. These are buttons with no fixed function assigned. Their function depends on the menu being selected. In some cases you may even have a mouse on the screen and there are also oscilloscopes with touch screens. In any case there are at least some basic sections on the front panel, the vertical controls, the horizontal

FIGURE 6.7 Mixed signal oscilloscope (Copyright © Tektronix. All Rights Reserved. Reprinted with permission).

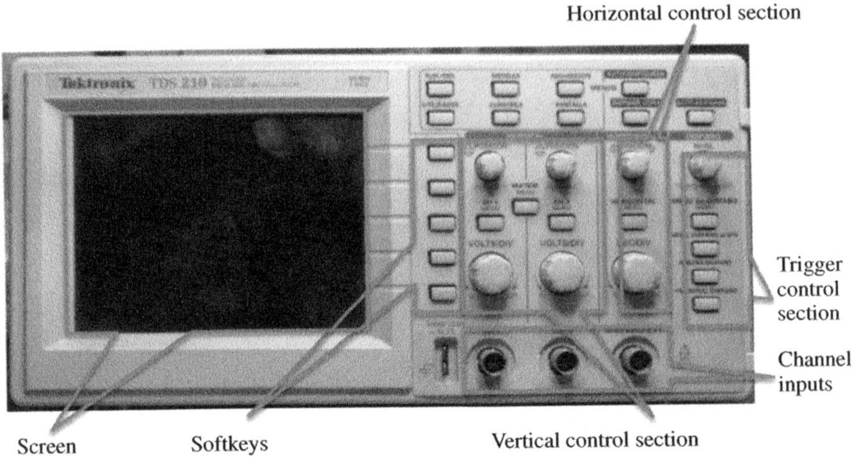

FIGURE 6.8 Typical digital oscilloscope front panel.

controls, and the trigger controls. Even analog oscilloscopes have these basic groups of controls. Figure 6.8 shows the front panel and the different sections. Figure 6.9 shows a basic description of the screen and the axis.

When you are using an oscilloscope you always need to adjust the following parameters so that you can see any waveform on the screen: the gain of the amplifier in the vertical section which is marked with the volts per division on the screen, the time base which is marked with seconds per division, and the trigger level and mode so that you can see a stable image. In digital oscilloscopes there is a very useful control generally named auto scale which makes the oscilloscope find by itself the best selection for each of the three controls mentioned above. Most of the time this is the best option. Even if you may need to adjust any of the controls this is the easiest way to begin.

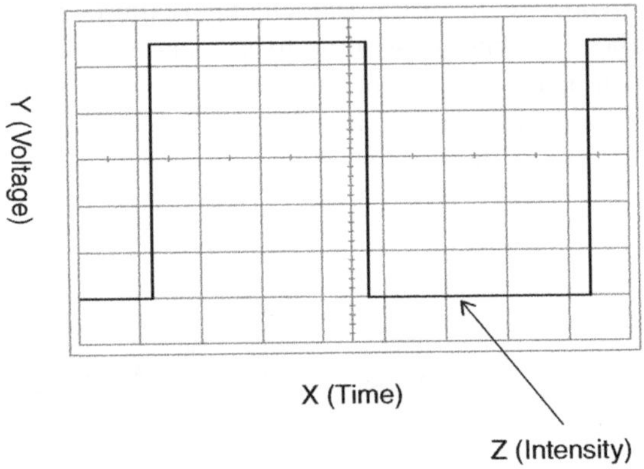

FIGURE 6.9 Oscilloscope screen.

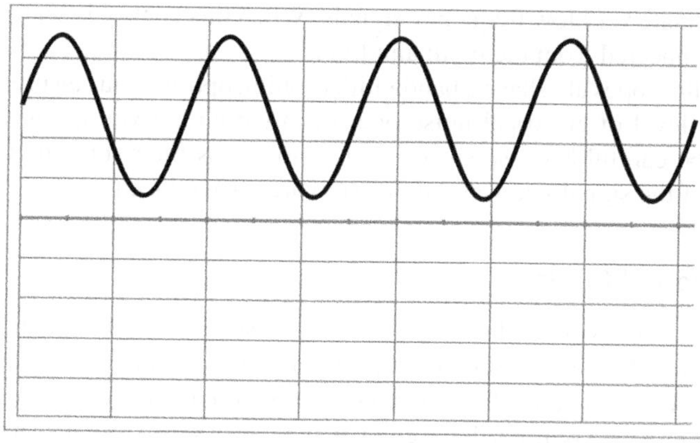

FIGURE 6.10 DC coupled signal.

6.5.1 Vertical Controls

In this section you have two different controls, the gain and the offset. The gain or volts per division control adjusts the gain or the attenuation of the input amplifiers so that you can accommodate the waveform on the screen. It is always recommended to use as much gain as possible so that you can see the peak-to-peak signal almost occupying the whole screen. The offset control or position control moves up and down the waveform on the screen without changing its amplitude. This control is used when you have to display two waveforms and you prefer not to overlap them. There is another control named coupling which changes the way the signal is connected to the input amplifier. When the coupling is DC the signal goes directly to the input amplifier (see Figure 6.10). When it is AC then the DC component of the signal is blocked so you will see the waveform centered on zero volts or ground (see Figure 6.11). In this coupling mode, the input channel acts as a high pass filter

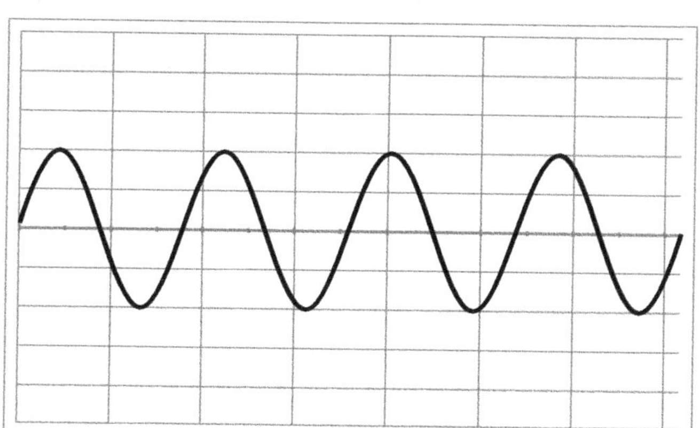

FIGURE 6.11 AC coupled signal.

so it may attenuate very low frequency signals. When this control is set to ground the signal is disconnected from the input amplifier.

Some oscilloscopes also have a bandwidth limiting option. You can use this control to reduce the level of electrical noise present in the circuit you are measuring but you have to be careful because since this is a low pass filter some high frequency components of the signal you want to measure are also reduced.

6.5.2 Horizontal Controls

These controls allow you to change the horizontal scale of the image on the screen. The control named base time sets the seconds per division showed on the screen. In this way you can zoom parts of the signal or show more than one cycle in the case of periodic signals. There is also a delay control, which is used to provide an offset to the signal along the horizontal axis. In digital oscilloscopes the horizontal controls are related to sample rate and record length. The time between two consecutive samples is called sample interval. There are some types of acquisition modes.

6.5.2.1 Sample Rate and Mode The sample rate of a digital oscilloscope is the number of samples it can acquire in one second. It is recommended that the sample rate must be at least three times the bandwidth (see Figure 6.12). Generally the specifications refer to the maximum sample rate. There are many factors that can modify this value. In simple digital models there may be only one analog-to-digital converter shared by all the channels so the maximum sample rate can only be achieved when using one channel. If you want to show more than one channel the sample rate is reduced according to the total number of channels used. Figure 6.12 shows the importance of an adequate selection of sample rate. If the signal you are measuring has fast transients and you select a low sample rate you can miss those transients.

In the sample mode the oscilloscope captures one sample every sample interval. In digital oscilloscopes there are two sampling modes, real-time and equivalent-time sampling. When the frequency components of the signal you want to measure are well bellow the sampling frequency of the oscilloscope, the real-time sampling mode can capture enough waveform points so as to reproduce the original signal. When the signal you want to measure has high frequency components that are too fast for real-time sampling, you can use equivalent-time sampling. In this mode the oscilloscope builds the waveform with successive sweeps, sampling part of the signal

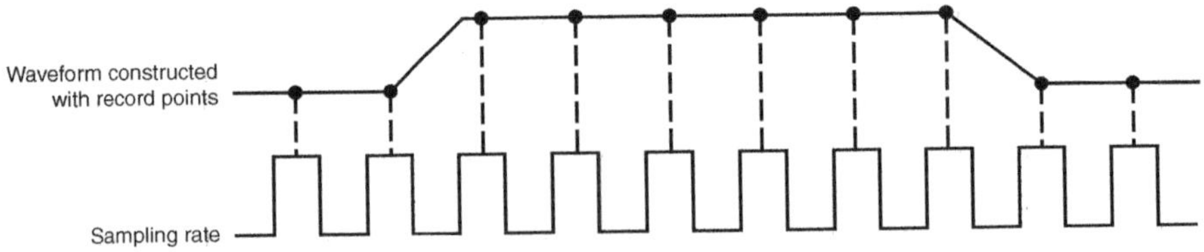

FIGURE 6.12 Sample rate and waveform reconstruction (Copyright © Tektronix. All Rights Reserved. Reprinted with permission).

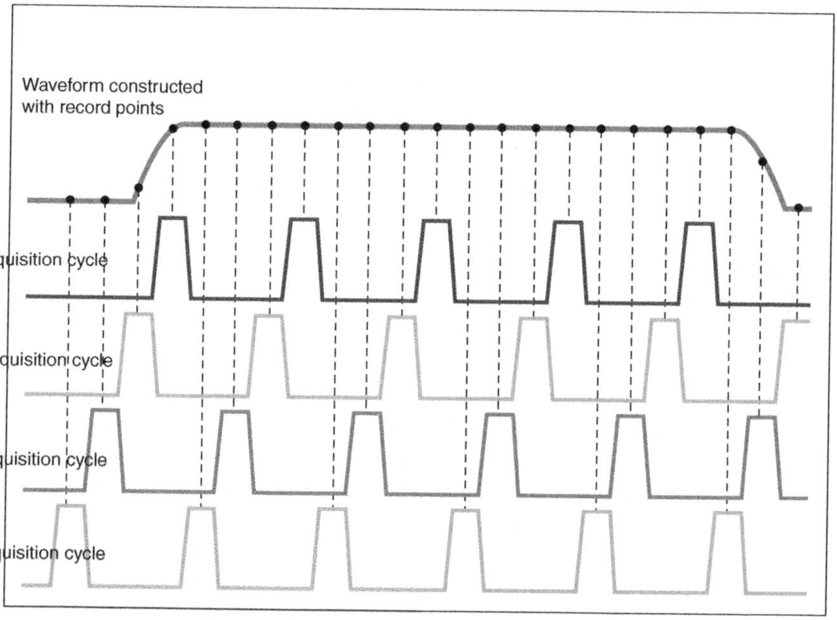

FIGURE 6.13 Equivalent-time sampling.

in each sweep. Then it builds the original signal interlacing the samples captured in each sweep (see Figure 6.13).

6.5.2.2 Peak Detect Mode In this mode the oscilloscope captures two samples in each sample interval, the maximum and the minimum value. This mode is useful for capturing narrow pulses that are spaced far from each other. This can be done because in this mode, the acquisition system runs at the fastest speed so as not to miss any quick variation of the signal, irrespective of the time base set for the horizontal axis.

6.5.2.3 Average Mode In this mode the oscilloscope works in the same way as in sample mode but it will show on screen the average value of all samples in one sample interval during successive acquisitions. This mode also reduces noise without limiting bandwidth like in analog oscilloscopes but you need to make several acquisitions.

6.5.3 Trigger System

The trigger system in an oscilloscope allows the user to synchronize the horizontal sweep in the correct moment so as to see a stable waveform in the case of periodic signals. When you need to capture transients this system also allows you to trigger the acquisition at the desired moment. The trigger level control lets you select the

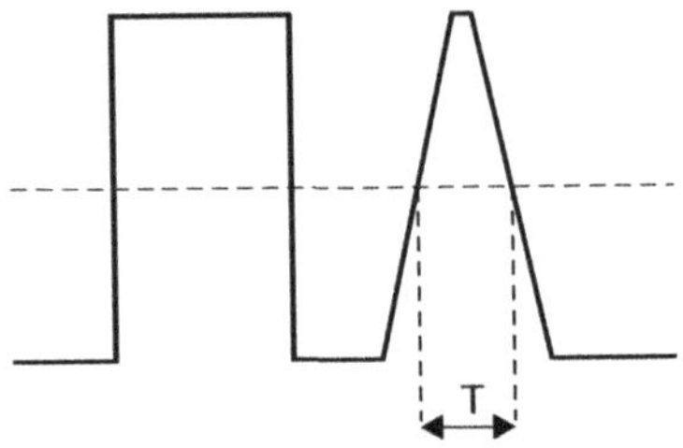

FIGURE 6.14 Glitch triggering (Copyright © Tektronix. All Rights Reserved. Reprinted with permission).

voltage level of the triggering signal at which you want the horizontal sweep to start. The signal used for triggering can be any of the input channels, the power source signal, or another external signal connected to a special input. One of the controls in this system lets you select the source for the trigger. You can also choose between different modes of triggering. In the case of analog oscilloscopes the most commonly used trigger mode is edge triggering. In this mode trigger starts when the input voltage reaches the trigger level you have set with the level control. You can also select to trigger on a rising edge or on a falling edge. In digital oscilloscopes there are many other types of trigger. They depend on certain specific characteristics of the input signal, conditioning the trigger to amplitude, width, polarity, logic state, or patterns. One example is the one called glitch triggering. In this mode the oscilloscope can trigger on a pulse whose width is greater than or less than a specific time interval set by you. In Figure 6.14 you can see how it works. When the trigger system sees a pulse narrower than T then it triggers the sweep. This is very useful when trying to capture random pulses that do not happen very often.

With edge triggering the oscilloscope will very probably miss the pulses.

There is another control that allows you to add some delay to the trigger systems setting the horizontal position of the trigger. With this delay you can see the signal during pre-trigger. This feature is useful when you need to know what caused some glitch or pulse or any behavior of the signal. The digital oscilloscope can show the signal during pre-trigger very easily because the acquisition system is always sampling the input signal and the trigger only tells when to start recording the waveform to memory. Analog oscilloscopes also had pre-trigger capabilities but with a very limited amount of time using a delay line in the input channel of the vertical system.

There is another characteristic of the trigger system called mode. Generally there are two modes in analog oscilloscopes, auto and normal. In digital oscilloscopes and also in analog ones but with memory, there is one more mode called single trigger or single shot. When selecting the auto mode you will always see a waveform on the screen no matter if the oscilloscope is triggering. In the normal mode the oscilloscopes only make a horizontal sweep if the trigger conditions are met, otherwise the horizontal system is stopped. The normal mode can be difficult to use while adjusting the oscilloscope controls because you do not see anything on the screen unless the oscilloscope is triggered. It is recommended that you start with auto mode while you make all adjustments and then switch to normal mode so that you see only the waveform you want. The trigger system also has the same coupling control as the vertical system. The trigger signal can be AC or DC coupled. There are other options for coupling the trigger signal like high frequency rejection, low frequency rejection or noise rejection. These options can help stabilize the image when there are certain instabilities in different frequency ranges or noise. Another important feature of the trigger system is the holdoff control. It is an adjustable time interval after a valid trigger during which the oscilloscope cannot trigger. This function is very useful when measuring complex waveforms that can produce valid triggers at many points during one cycle.

6.5.4 Display System

The display system in an oscilloscope includes a display and the knobs and buttons used for modifying the way it shows the waveforms. The controls can be different in analog and digital oscilloscopes but some of them are common to both. The screen has a grid with vertical and horizontal lines as major divisions and some tick marks as minor divisions (see Figure 6.15). The labels in vertical controls like volts/divisions and horizontal controls like seconds/division refer to the major divisions. The intensity control adjusts the brightness of the waveform. In analog oscilloscopes you have to increase the brightness when measuring very fast signals. In digital oscilloscopes this is not necessary. In analog oscilloscopes the focus control makes the trace sharper and the rotation control adjusts the alignment of the trace with the horizontal lines. In digital oscilloscopes there are other controls related to colour, grading, intensity of the grid, and turning on or off the information showed on the screen.

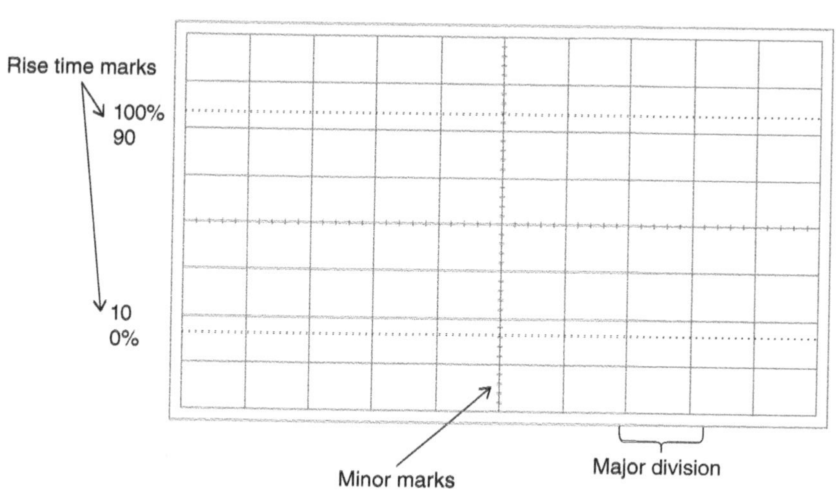

FIGURE 6.15 Grid and main marks.

6.6 MEASUREMENTS

In analog oscilloscopes the only way of making measurements was manually using the grid on the screen. In digital oscilloscopes there is a wide selection of measurements functions that are performed automatically. You only need to select which measurement and which waveform. The result is shown on the screen. The range of measurements depends on the complexity and power of the specific oscilloscope model so only the most common ones will be presented here.

Amplitude or voltage measurements are one of the basic functions provided by this instrument. If you want to measure voltages manually, you can use the grid on the screen and the scale set in the vertical control in volts/division. It is always recommended that you adjust the vertical gain so that the waveform occupies most of the graticule minimizing the influence of errors. Some oscilloscopes have a very useful tool called cursors. When you want to measure voltages you can activate horizontal cursors that can be moved up and down. The oscilloscope measures the voltage difference between the two cursors allowing you to measure any voltage apart from peak-to-peak or RMS.

6.6.1 Peak-to-Peak Voltage

This measurement evaluates the voltage difference between the highest value and the lowest value of the waveform.

6.6.2 RMS Voltage

As previously said, the root-mean-square (RMS) voltage is evaluated by measuring the voltage, squaring it, finding the average value and taking its square root. This is useful when evaluating power in a circuit

6.6.3 Rise Time

This measurement evaluates the time a waveform takes to go from 10% to 90% of its peak-to-peak voltage. Sometimes these percentages can also be changed arbitrarily. This is explained in Figure 6.16.

6.6.4 Fall Time

This measurement is the same as the previous one but going from 90% to 10% of the peak-to-peak voltage. This is also explained in Figure 6.16.

6.6.5 Pulse Width

This measurement can be performed on positive or negative pulses. When measuring positive pulses the oscilloscope evaluates the time the waveform takes to go from 50% of the peak-to-peak voltage to the maximum voltage and back to the 50% level.

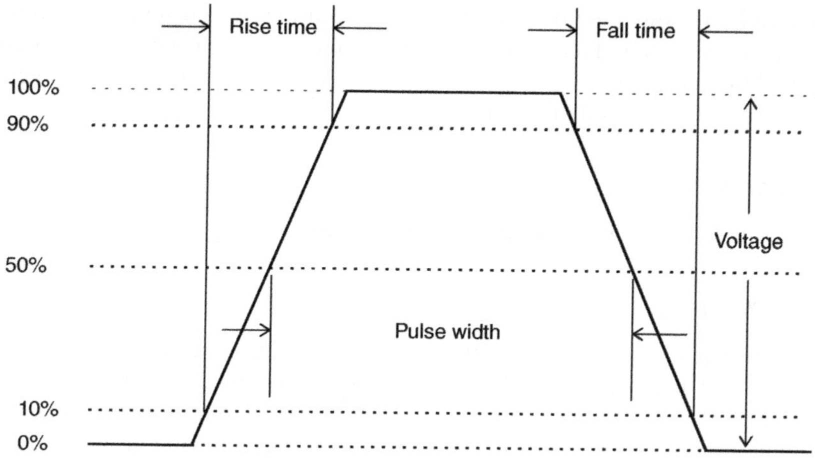

FIGURE 6.16 Rise and fall time and pulse width parameters.

When measuring negative pulses the procedure is the same but it analyzes the signal portion going to the minimum voltage (see Figure 6.16).

When you need to measure pulses you have to work on the trigger setting to have a stable signal. Using the holdoff control is very important in order to trigger only when valid conditions are met and measuring the correct portion of the signal. Figure 6.17 shows the effect of using the holdoff control to avoid triggering in parts of the waveform where the image would be unstable. Sometimes it may be necessary to zoom on the horizontal axis to increase accuracy or to be able to see any details in the pulse.

If you need to perform time and frequency measurements you can use the horizontal axis of the grid and the value set in the time base control in seconds/division. You can do it manually just looking at the screen. If cursors are available then you can use vertical cursors, moving them to the right or to the left and the oscilloscope

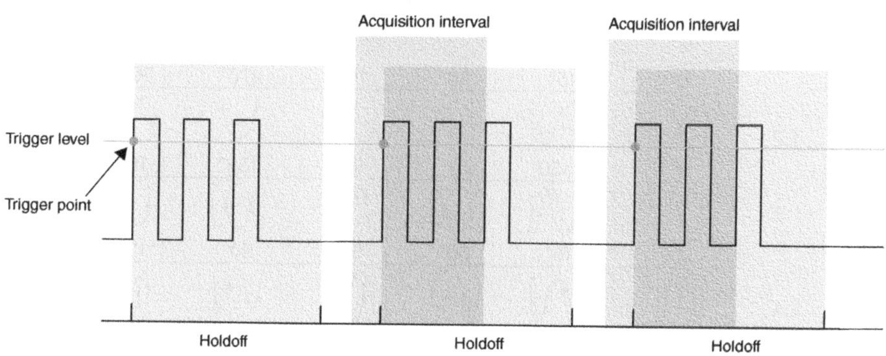

FIGURE 6.17 Holdoff control (Copyright © Tektronix. All Rights Reserved. Reprinted with permission).

automatically measures the time difference between the two cursors. As in voltage measurements you should try to expand the time zone to be measured so as to occupy as much of the screen as possible to minimize errors. Frequency can be calculated by the reciprocal of the period of the signal.

6.6.6 Period

This measurement evaluates the time it takes to complete a cycle in periodic signals.

6.6.7 Frequency

This measurement evaluates the frequency of the signal that is equal to the reciprocal of the period. There are many other measurement functions like duty cycle, delay, phase, mean, area, overshoot, maximum, and minimum. Many oscilloscopes allow performing more than one measurement simultaneously, showing all values on the screen updated in real time.

6.6.8 Phase Shift Measurements

Oscilloscopes have a very useful function for measuring phase shift between two signals called XY mode. You need to connect one of the signals to any vertical channel and the other signal to the horizontal system. On the screen you will see what is called a Lissajous pattern. Looking at the pattern you can calculate both phase shift and frequency ratio. Figure 6.18 shows some possible Lissajous patterns showing these two parameters.

6.6.9 Mathematical Functions

Depending on the model, digital oscilloscopes can perform some mathematical functions on the waveform data. Some of them are the Fourier transform, integration,

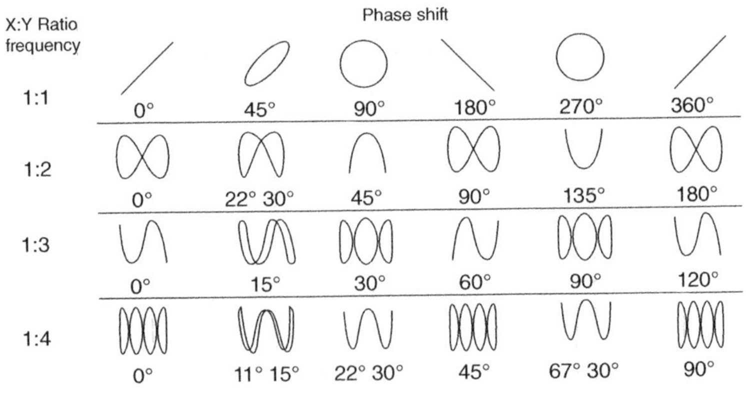

FIGURE 6.18 Lissajous patterns (Copyright © Tektronix. All Rights Reserved. Reprinted with permission).

basic functions like adding or subtracting waveforms or even using waveforms stored in the oscilloscope memory.

6.7 PERFORMANCE CHARACTERISTICS

There are many oscilloscope properties that vary with the power and complexity of each model. These properties have a strong influence on the oscilloscope performance and they may determine the suitability of any oscilloscope to perform a specific measurement. It is important that technicians understand the meaning of these properties because they determine the appropriate selection of the instrument.

6.7.1 Bandwidth

This characteristic determines the range of frequencies that the oscilloscope is able to display. This is specified as the frequency at which a sinusoidal signal is attenuated −3 dB compared to the true amplitude (see Figure 6.19). There is a basic rule of thumb for determining the needed bandwidth for measuring a signal accurately. The bandwidth should be at least five times the highest frequency component of the signal. In Figure 6.20 you can see a step voltage measured at different bandwidths. It is clearly seen that when the bandwidth is too low, the high frequency components of the signal are attenuated.

6.7.2 Rise Time

This characteristic is fundamental when using the oscilloscope in digital systems trying to measure pulses. The oscilloscope rise time must be short enough to allow capturing fast transitions of the signal. It is clear that rise time and bandwidth are

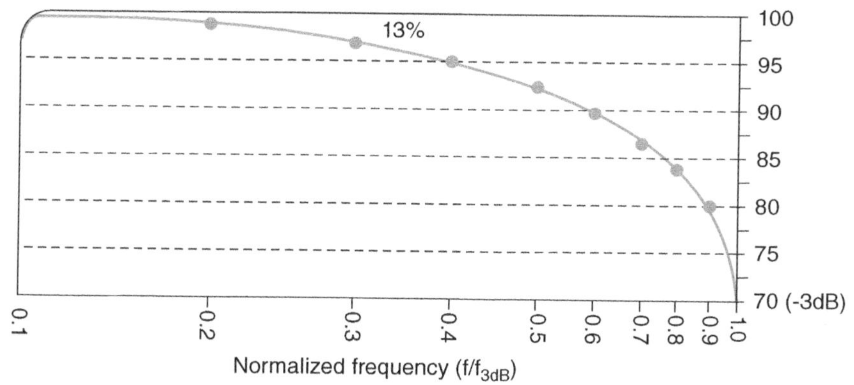

FIGURE 6.19 Oscilloscope bandwidth (Copyright © Tektronix. All Rights Reserved. Reprinted with permission).

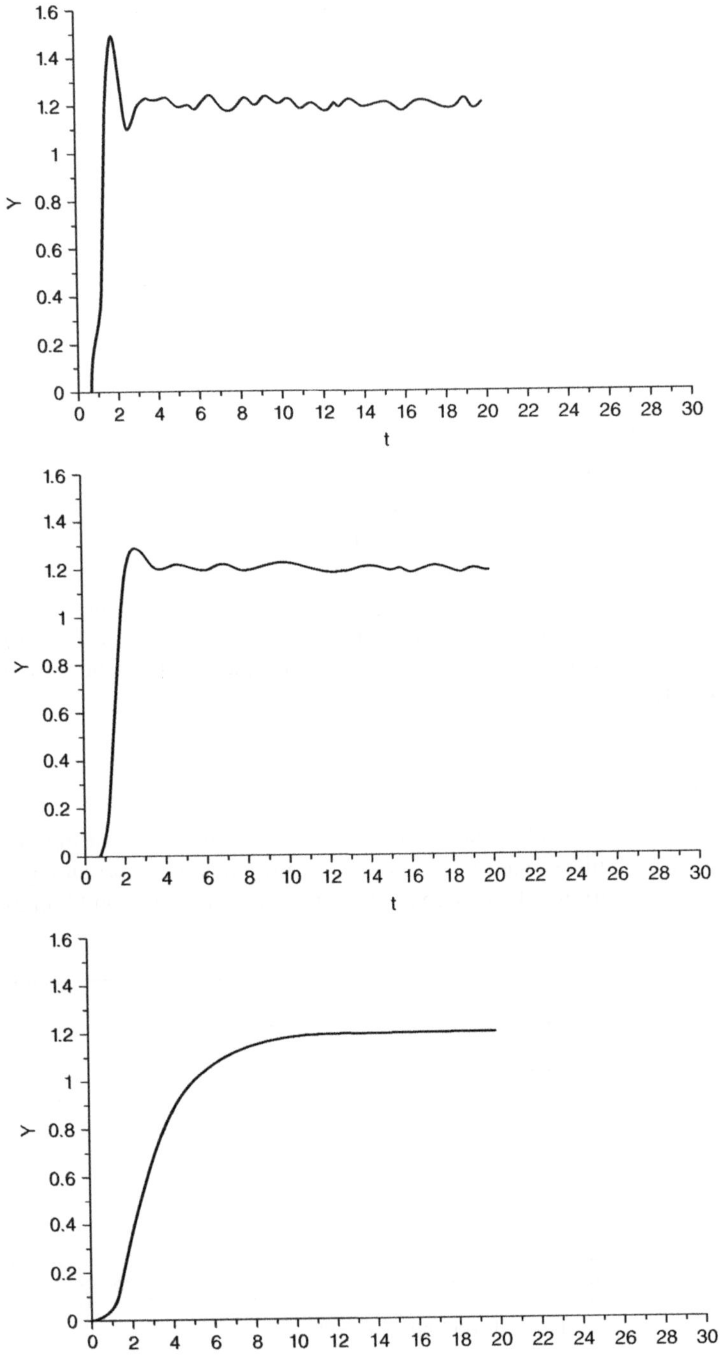

FIGURE 6.20 Effect of bandwidth.

related characteristics in an oscilloscope. There is also a basic rule of thumb for determining the necessary rise time for measuring a signal accurately. The oscilloscope rise time must be five times smaller than the fastest rise time of the signal.

6.7.3 Channels

The channels of the oscilloscope are the different inputs to the vertical amplifiers. Oscilloscopes have different number of channels and in some cases, like mixed signal oscilloscopes, also different types of channels. When selecting an oscilloscope you must take into account the number of signals to be shown simultaneously. You need as many channels as simultaneous signals. If you are going to measure both analog and digital signals then you must consider using a mixed signal oscilloscope instead of a basic digital oscilloscope.

6.7.4 Vertical Resolution

This characteristic is related to the amount of bits of the analog-to-digital converter and it will affect the accuracy of this conversion. The more bits it has the more accurate is the conversion from a signal voltage to a digital value. This characteristic is expressed in bits. The most simple and economic models start with eight bits but you can find resolutions of 14, 16 or even more bits.

6.7.5 Gain Accuracy

This characteristic indicates how accurately the vertical system amplifies a signal voltage and it is expressed in percentage error.

6.7.6 Horizontal Accuracy

This characteristic indicates how accurately the horizontal system shows the timing of signals. It is expressed in percentage error.

6.7.7 Record Length

This characteristic refers to the number of records the internal memory of the oscilloscope can store. The memory of the instrument is limited so at a certain sample rate you will be able to save a predetermined interval of time. At high sample rates the interval time decreases; at low sample rates the interval time increases (see Figure 6.21). The time interval can be determined dividing the record length by the sample rate. In some oscilloscopes you can change the record length to allow you to use the full memory capacity for the acquisition of a complex digital signal. In case of a simple periodic signal you may decrease the record length and allow capturing more waveforms.

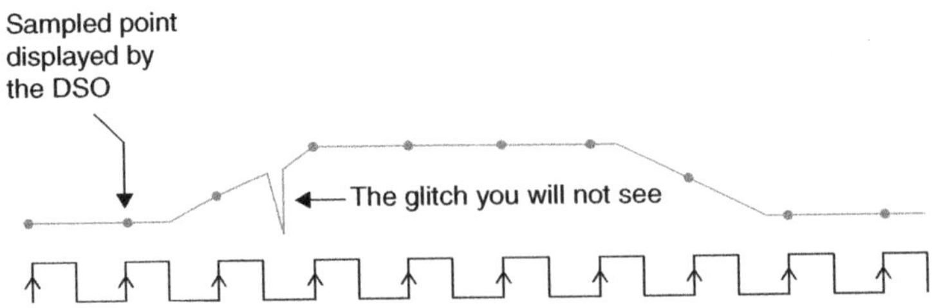

FIGURE 6.21 Importance of adequate sample rate (Copyright © Tektronix. All Rights Reserved. Reprinted with permission).

6.7.8 Update Rate

This characteristic is related to the rate at which a digital oscilloscope can acquire and refresh the display of a waveform. The oscilloscope has a dead time between two successive acquisitions and if the signal you are measuring has some variations or events during that time you will miss it. This is explained in Figure 6.22. It is clear that the faster the update rate the shorter the dead time and the higher chances of seeing infrequent events of the signal. As in other parameters you have to check the specifications of each model since the update rate may change depending on the mode the oscilloscope is working. This characteristic is expressed in waveforms/second. It is affected by the internal architecture of the oscilloscope. In digital storage oscilloscopes using serial configurations the update rate can go from 10 to 5000 waveforms/second while

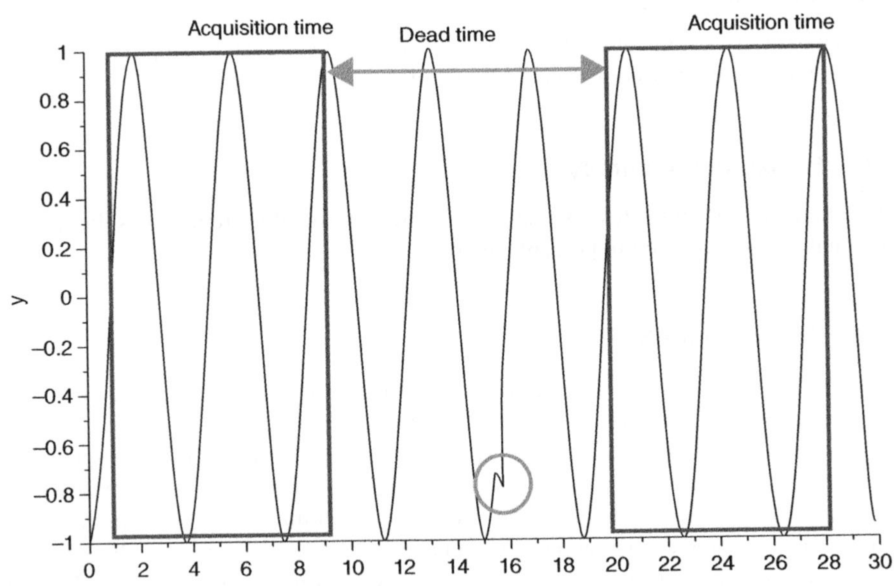

FIGURE 6.22 Waveform update rate.

FIGURE 6.23 A wide range of interface connections.

in digital phosphor oscilloscopes using parallel configurations the update rate can go up to millions of waveforms/second.

6.7.9 Connectivity

Nowadays it is very easy to connect an oscilloscope to any digital system. The need for saving valuable data, sharing the information with others, creating reports on measurements and even performing other measurements and analysis over the acquired data is satisfied with the connecting capabilities of modern oscilloscopes. There are many standard interfaces for connecting these instruments to control networks like RS-232, IEEE-488, USB and Ethernet. With appropriate software and development systems you can not only download data from the oscilloscope but also control it from a remote computer. In more sophisticated development systems you can also create virtual instruments where an oscilloscope can be part of it acquiring a signal and sending the data to the system for further analysis. Figure 6.23 shows the different interface connectors found in a Tektronix TDS 210 oscilloscope.

6.8 OSCILLOSCOPE PROBES

The probes are also an important part of the measurement system because they connect the circuit or equipment under test to the oscilloscope. The selection of the adequate probe is as important as the selection of the oscilloscope. For example, if you are using an oscilloscope with a 1 GHz bandwidth and your probe's bandwidth is only 500 MHz then you are limiting the performance of the whole system. A probe has to be as transparent as possible when feeding the signal to the input amplifier. It would be desirable for a probe to have linear phase behavior, no attenuation of the amplitude, equal or larger bandwidth than the oscilloscope, high noise immunity, and practically no load to the circuit being measured.

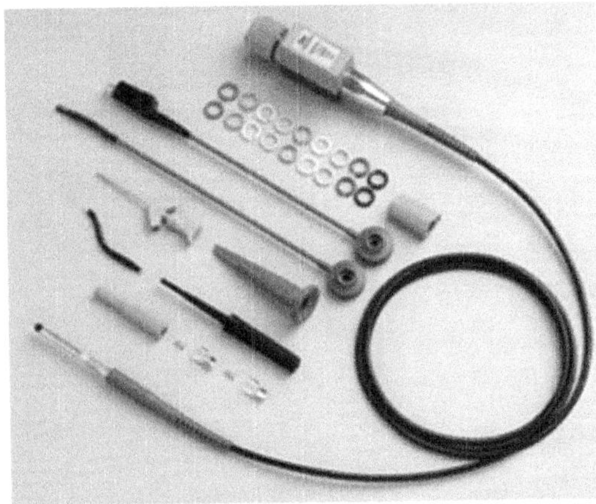

FIGURE 6.24 Passive probe (Copyright © Tektronix. All Rights Reserved. Reprinted with permission).

6.8.1 Passive Probes

This type of probe is the most common, inexpensive, and easy to use. They have only passive components and they do not need any power supply to operate correctly (see Figures 6.24 and 6.25). Generally they use connectors and wires; resistors and capacitors are added when they need to provide some attenuation of the signal. When the amplitude of the signal is low a 1X probe is enough but when you may have higher voltages in the circuit under test you need a 10X probe. One common version of passive probe has a switch in itself to change the attenuation from 1X to 10X. The typical bandwidth ranges for passive probes go from 100 MHz to 500 MHz.

These probes have a low frequency adjustment control used to compensate and match the probe capacitance to the oscilloscope input capacitance. In the front panel

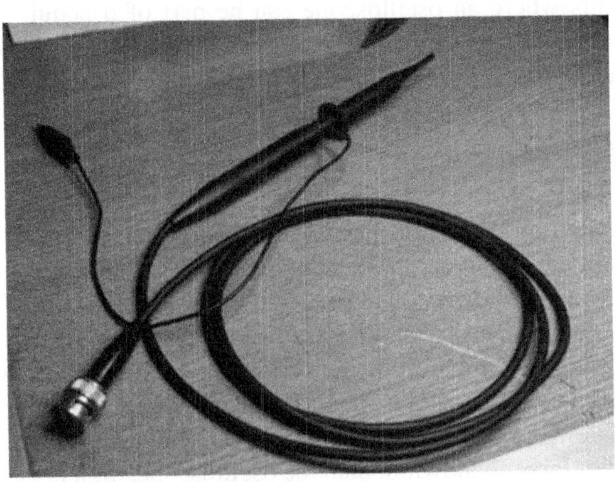

FIGURE 6.25 Example of passive probe.

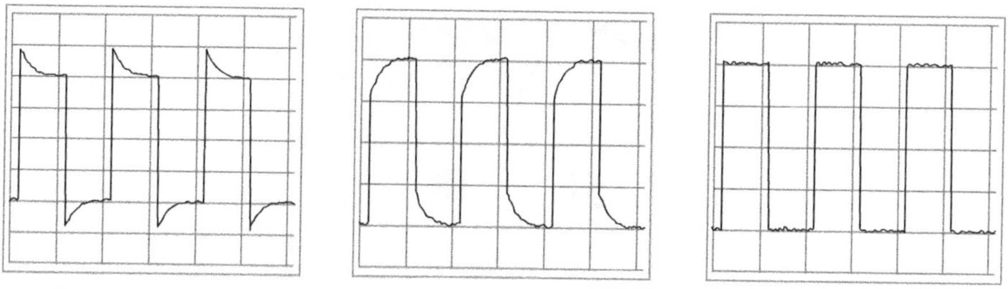

FIGURE 6.26 Probe compensation.

of the oscilloscope you can access a calibration signal output for this purpose. It generates a square wave with a frequency around 1 MHz. When you connect an uncompensated probe to this calibration signal the square wave looks with some deformation in the corners. You need to adjust the compensation control in the probe until you see a perfect square wave with square corners. In Figure 6.26 the different waveforms obtained are shown.

In the case of high voltage probes up to some tens of kilovolts (see Figure 6.27) the compensation system is more complicated than a single adjustment. In Figure 6.28 you can see controls for adjusting different resistors and capacitors.

6.8.2 Active Probes

These probes need a power supply to operate and the oscilloscope provides this through the input connector. The probe has amplifiers to condition and amplify the signal. They can work up to higher frequencies than passive probes and they are very suitable for demanding applications (see Figure 6.29). The active probes have some advantages over passive probes. They include low loading of the signal source, high

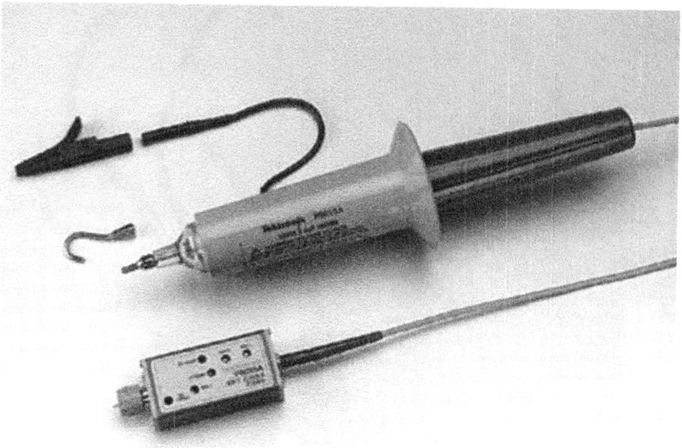

FIGURE 6.27 High voltage probe (Copyright © Tektronix. All Rights Reserved. Reprinted with permission).

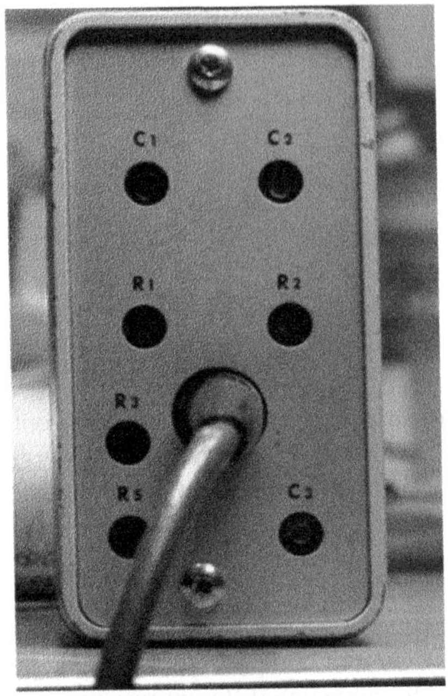

FIGURE 6.28 High voltage probe compensation.

resolution, automatic recognition by the oscilloscope, better overall combination of resistive and capacitive loading and higher bandwidths. There are some disadvantages like being more expensive, less rugged and lower working voltage ranges than passive probes. Active probes can perform single-ended or differential measurements. If you use passive probes and you need to perform a differential measurement you need two probes and inputs. An active differential probe has a differential amplifier that subtracts the two signals so you only need to use one channel in the oscilloscope.

FIGURE 6.29 Different active probes (Copyright © Tektronix. All Rights Reserved. Reprinted with permission).

6.9 USING THE OSCILLOSCOPE

6.9.1 Grounding

Using a good ground connection is fundamental when using oscilloscopes. The two reasons for grounding are safety of the operator and the achievement of the best possible accuracy in the measurements. The ground connection is done through the three-conductor power cord plugged in a grounded power outlet. In case of any fault or any accidental contact with high voltages present in the circuit under test the current flows through the protective conductor to ground since this is the path with lowest resistance. In this way the operator is safe and no current will flow through his body. If the circuit under test and the oscilloscope share the same ground there is no possible circulating current between the two ground connections that could result in noise and interference during measurements. This will guarantee the best possible accuracy during the tests. When working with integrated circuits it is also good practice to ground yourself. This avoids building up electrostatic charges in your hands that could result in damage to the circuit. You should wear a wrist grounding strap.

6.9.2 Calibration

It is recommended that you calibrate the oscilloscope on a regular basis. Once a year can be a good selection but as in all instruments the calibration period has to be adjusted according to instruments characteristics and conditions of use. Modern digital oscilloscopes have one internal function for self-calibrating the signal path. This action is recommended when the ambient temperature has changed more than 5°C from the last self-calibration. Generally the oscilloscope can monitor temperature and it can even suggest self-calibration sending a message on the screen. This procedure will take a few minutes and the accuracy of the signal channel will be again under the instruments technical specifications.

6.10 CONCLUSIONS

This chapter analyzed one of the most common instruments in use in laboratories or fieldwork. It started with some history showing the characteristics of analog and digital oscilloscopes and how they have evolved through decades. It presented a brief description of the different waveforms one may find in normal engineering work. The chapter also describes different waveform characteristics and how to measure them using oscilloscopes. A wide variety of analog and digital oscilloscopes, old and new ones, produced by different manufacturers was shown in pictures. By describing their building blocks it is easy to understand limitations of each type of oscilloscope. It gives a general idea of how to use oscilloscopes by explaining each of the most common controls found in their front panels. It is important to note that it is not only necessary to understand how to use controls but also understand how each type of oscilloscope works. This chapter showed all the performance characteristics of oscilloscopes and how they can affect the measurements performed with them. With the new digital oscilloscopes the use may have become easier than in the old analog ones but a good understanding of sampling rate and its effect on waveform reconstruction is fundamental to avoid wrong interpretation of waveforms. Oscilloscopes do not work alone; they must use one fundamental accessory, probes. This chapter showed the different types of probes available and their characteristics to be able to use the correct one for each different application. As in all instruments safety when using oscilloscopes is important so this chapter explained general considerations when grounding them. Again as in all instruments calibration is fundamental to make correct measurements so this chapter gave some recommendations on this.

FURTHER READING

As we mentioned at the beginning of this chapter, we recommend reading two standards from IEEE, which give a deep knowledge on oscilloscope performance and characteristics. They deal widely with oscilloscope performance and characteristics. Analog oscilloscopes have practically disappeared from laboratories and this is the reason why we did not write much about them in this chapter. Anyway, for those interested in learning more about them, we can suggest reading the book *Understanding and Using the Oscilloscope* by Clayton Hallmark, published in 1974 by Foulsham–Tab Limited. If your needs are understanding present-day digital oscilloscopes then we recommend you read the book *Digital Storage Oscilloscopes* by Ian Hickman, published in 1997 by Elsevier. There is also a good paper published in the *IEEE Instrumentation and Measurement Magazine* in December 2006 about the evolution of oscilloscopes named The history and technology of oscilloscopes by J. Miguel Dias Pereira. It is also very useful to read the many application notes and white papers written by instruments manufacturers. These are independent from oscilloscope models and brands so they are a very good source of information on how to use hem in different fields. Apart from the ones already mentioned in the references we can add a few more. *12 Things to Consider when Choosing an Oscilloscope* from Tektronix is a useful guide when selecting an oscilloscope going through the basic parameters and a clear explanation of how each of them can impact in the performance. *Probes and Probing* from Teledyne Lecroy deals with different topics about oscilloscope probes and how the use of them can affect our measurements. *Evaluating Oscilloscope Fundamentals* from Keysight discuss different types of oscilloscopes, performance characteristics and also helps with using probes. *Evaluating Oscilloscope Sample Rates vs. Sampling Fidelity* from Agilent discuss the well-known problem of choosing the correct sampling method and frequency depending on the type of signal to be measured. *Evaluating Oscilloscope Vertical Noise Characteristics* from Agilent helps you understand noise signals and how to measure them. It also explains how to make measurements when there is noise and how to minimize its effect.

REFERENCES

1. *XYZs of Oscilloscopes. Primer*, Tektronix.
2. *AN 1606, Oscilloscope Fundamentals*, Agilent Technologies.

3. *Oscilloscope Fundamentals. Primer*, Rohde & Schwarz.

4. IEEE Standard Specification of General-Purpose Laboratory Cathode-Ray Oscilloscopes. *IEEE Transactions on Instrumentation and Measurement*, **IM-19**(3) pp. 180–194. (August 1970).

5. IEEE Std 1057-1994 (R2001), IEEE Standard for Digitizing Waveform Recorders, Sponsor: Waveform Measurements and Analysis Committee of the IEEE Instrumentation and Measurement Society.

EXERCISES

6.1 In your opinion which type of oscilloscope is easier to use at first sight, analog or digital oscilloscopes?

6.2 Mention at least three advantages in digital oscilloscopes compared to analog ones.

6.3 Explain the differences between AC and DC coupling in the vertical channel.

6.4 We have to measure the ripple of a battery charger. The output DC voltage is 100 Vdc and the ripple is around 1 Vac. How would you set the vertical controls to be able to measure ripple with the least possible error.

6.5 In which case would you use line coupling as a trigger mode?

6.6 Connect a signal generator to the scope. Generate a square wave voltage. How would you set your trigger system to display the front of the square wave?

6.7 Set your signal generator to generate a 40 Hz sinewave. Set the oscilloscope so that the sampling frequency is around 2 kS/s. Then increase the frequency of the signal generator to 2040 Hz keeping the same sampling frequency in the oscilloscope. Explain why you see the same sinewave on the scope screen even though you know you changed the measured signal.

6.8 When trying to display an unknown waveform how would you set the trigger mode at first? Why?

6.9 How can you verify that a 10x probe is correctly compensated?

6.10 Mention the factors that can impact on the vertical bandwidth in an analog scope.

6.11 Suppose you have to measure a single shot signal with an analog scope. How can you see the pre-trigger signal? And with a digital scope?

6.12 You have to measure a periodic signal formed by a train of four 10 ms rectangular pulses separated by 10 ms intervals and a 30 ms interval with zero voltage. Explain how to set the trigger system to have a stable display.

CHAPTER 7

FUNDAMENTALS OF HARD AND SOFT MEASUREMENT

LUCA MARI,[1] PAOLO CARBONE[2] and DARIO PETRI[3]
[1]Scuola di Ingegneria Industriale, Universitá Cattaneo – LIUC, Castellanza, Italy
[2]Dipartimento di Ingegneria, Universitá degli Studi di Perugia, Perugia, Italy
[3]Dipartimento di Ingegneria Industriale, Università degli Studi di Trento, Povo, Italy

7.1 INTRODUCTION

Being an infrastructural, widespread activity, performed by human beings since millennia, measurement is laden with stereotypes. A single sentence synthesizes several of them: "Philosophers commonly define measurement as the correct assignment of numbers to physical variables" [1, p. 367]. It is hard to believe that (some) philosophers can really support such a position about measurement, and this for a number of reasons. A short, critical discussion about them allows us to introduce the main features of the present chapter.

- First, particularly in the case of measurement it is important to maintain the distinction between the empirical entity subject to measurement—call it a "quantity," or more generally a "property," of the object under measurement—and its mathematical representation, for example a random variable. Hence, measurement should be characterized as the assignment (of values) to quantities or properties not to variables. Performing a measurement requires an experimental stage: a purely mathematical process is not a measurement.[1]

[1]This ambiguity has its origins in the Euclidean tradition, where "x measures y" was meant as "x divides y" (and therefore x can be exploited as unit for y: "The term measure is used conversely to multiple [...] Hence [if] A and B have a common measure [they] are said to be commensurable." [5, p. 9]). In this context the English term "measurement," or "mensuration" refers to procedural demonstrations of geometric propositions, such as "The area of any circle is equal to a right-angled triangle in which one of the sides about the right angle is equal to the radius, and the other to the circumference, of the circle" in the

Modern Measurements: Fundamentals and Applications, First Edition.
Edited by Alessandro Ferrero, Dario Petri, Paolo Carbone, and Marcantonio Catelani.

- Second, generally measurement does not assign numbers (e.g., 2) but quantity values (e.g., 1.23 meters): the latter include a number—called the "numerical quantity value"—and only in the specific case of measurements performed in absolute scale, for example, by counting, quantity values are numbers because the unit is "absorbed" in the definition of the property intended to be measured, that is, the measurand. Moreover, it is now commonly acknowledged that ordinal properties are also measurable (see, e.g., Reference [2, 1.26]), for which numbers do not necessarily play a specific role.
- Third, characterizing measurement as a correct assignment makes such undefined correctness a definitional criterion, thus feeding the traditional claim of the special effectiveness of measurement while still leaving it unjustified. Rather, measurement is acknowledged as generally affected by errors, as witnessed by its lack of complete accuracy, and an important task is to discuss how this "partial correctness" should be reported in terms of measurement uncertainty.
- Finally, why should measurement be limited to physical entities? Admittedly, the mentioned philosophers are not alone in this assumption: for example, the Encyclopedia Britannica opens its description of measurement by stating that it is "the process of associating numbers with physical quantities and phenomena." Of course, the question is crucial for what follows.

Measurement is a designed–on–purpose process, not a natural entity preexisting to the human endeavor and to be discovered in its independent features: more than a stable entity that we can only decide how to call, measurement itself is a moving target, the changes being triggered by social needs, available technologies, scientific models, philosophical conceptions. Hence, a quest for the "true meaning" of "measurement" is questionable, if not just meaningless [3]. The opposite position, a radical conventionalism assuming that everyone can call "measurement" whatever she/he believes it fit, would just prevent any social understanding, particularly by researchers from different disciplines. A possible way out might be building on tradition, by referring to the way the concept has been used in the past. On the other hand, historical events around measurability and measurement have been complex and controversial. While the nineteenth century witnessed the first efforts to study the measurability of non-physical (in particular psychophysical and then psychological) properties (with some unavoidable naivety or ideological bias [4]), a well-known debate within the British Association for the Advancement of Science (BAAS) in the first half of the twentieth century led to a clash between the communities of physicists and psychologists: they were unable to find a common ground as for the very possibility to identify a criterion of measurability, in front of the sharp assumption of the physicists that measurement applies only to additive properties [7]. Since then, the research toward the measurement of non-physical properties can be traced along two main, somehow contradictory, lines: on the one hand, to find reliable procedures for the additive evaluation of properties, as witnessed by techniques such

Archimedes' short treatise "Measurement of a circle" [6, p. 91-98]. Of course, no experimental activities are expected here.

as simultaneous conjoint measurement [8] and Rasch measurement [9, 10], where "measurement" is used as a more or less synonymous of "additive evaluation"; on the other hand, to formalize property evaluations in a consistent framework, of which additivity is interpreted as just one of the possible features. This was first presented in terms of "scales of measurement" [11] and then generalized to "representational theories of measurement" [12]. Interestingly, with the exception of some papers (particularly since Reference 13) these theories remained unused, if not unknown, in the traditional measurement science of physical quantities, and the two basic reference documents—the *International Vocabulary of Metrology* (VIM) [2] and the *Guide to the Expression of Uncertainty in Measurement* (GUM) [14]—do not even mention them. This summary reconstruction shows the ambiguities of the (at least recent) tradition, which can be then hardly exploited to ground a widely shareable position on measurement. We claim that this conundrum might be solved by emphasizing the structural and pragmatic features that the evaluation process of a (physical or non-physical) property should exhibit to be convincingly considered a measurement [3,15]. The analysis of the relations and the differences of the concept of measurability as applied to physical and non-physical properties has been undertaken in References [16–18]. The underlying idea is that measurement is not a purely empirical process, as instead sometimes supposed in the past, because its experimental outcome (e.g., the stable state of the measuring instrument produced by its interaction with the object under measurement) must be suitably interpreted in order to be represented as a value for the measurand. Hence, the measurability of a property implies that a metrological model for it is available. A major distinction is then recognized in the fact such a model is embedded in a "well-formed theory for broad domains of knowledge" [16, p. 42] or it just refers to an "underlying model" [3, p. 243], that loosely gives a more or less socially agreed, pre-metrological meaning of the property:

- the first case, called *strongly defined measurement*, or *measurement in hard systems*, or also *hard measurement* for short, is the usual one for physical properties, physics itself (together with geometry) being the "well-formed theory";
- by difference, the second case is termed *weakly defined measurement*, or *measurement in soft systems*, or also *soft measurement* for short, and it is the one which typically applies to many non-physical properties.

Hence, is soft measurement possible as such? Are there some recognizable commonalities in those processes that are called "measurements" in the physical and in the social sciences?

This chapter aims at exploring this issue, with a systematic, structural reference to the case of hard measurement, and is organized as follows. At first we analyze the main features a process might be expected to exhibit to be considered, properly, a measurement, and on this basis we propose an introductory, comparative overview on the concepts of hard and soft measurement, their similarities, and differences. Then we present the basic features of a measurement process, as usually thought of in physics and engineering, with the aim of providing an organized body of knowledge from which useful guidelines could be drawn for the design and the execution of valid measurements for complex properties, both in physical and non-physical domains. In particular a conceptual framework describing the structure of the different stages and activities involved in a generic measurement process is presented. This framework is then applied to illustrate the remarkable problem of the measurability of research quality.

7.2 A CHARACTERIZATION OF MEASUREMENT

Can a non-physical property be measured then? And how to avoid a conventional (e.g., by counting the occurrences of expressions such as "measurement of non-physical property" in the scientific literature and deciding according to their number) or a purely nominal (e.g., let us decide to call "measurement of non-physical property" such and such activity) answer? The problem is critical: measurement is an information production process, such as description, judgment by experience, and guess, but different than these it is generally considered able to produce information worth the public trust because it is "reliable, accurate and comparable" [19, p. 84]. Furthermore, this is the rationale socially given for the resources that are devoted to perform measurements and that would not be justified, for example, guess [15]. What is the source of this claim? Nowadays the terms "measurement" and "measure" are often used in non-physical sciences. Consider, for example, the following, interesting examples of properties that "psychologists claim to be able to measure": "general intellectual ability ("intelligence"); various specific intellectual abilities (verbal ability, spatial ability, and so on); the intensities of different kinds of sensations (loudness, brightness, etc.); the subjective probability of occurrence of various possible events (such as winning some gamble); the strength of attitudes toward social policies (e.g., euthanasia or abortion); the subjective value of various commodities (such as laptops or wilderness areas); degrees of personality traits (introversion, neuroticism, etc.); strength of association between a stimulus and the overt response elicited (such as Hull's habit "strength"); levels of skill (e.g., social skill or typing skill); and levels of achievement in various areas (such as spelling or arithmetic)" [20, pp. 3–4]. The actual possibility of measuring properties such as those exemplified above depends, first of all, on the features acknowledged to measurement. Let us review the most important candidate features, by inquiring about their being necessary or sufficient conditions of measurability. The analysis will reveal a complex, still evolving scenario.

7.2.1 Measurement as Value Assignment

In general terms, measurement is a both conceptual and empirical process aimed at providing information to effectively describe the empirical world. Such information is about the property of an object (e.g., the length of a table), which is represented, in the simplest cases, as a single property value (e.g., 1.23 meters) and, more generally, as a complex entity such as an interval of property values (e.g., 1.230 ± 0.005 meters). Hence, the basic entities involved in a measurement are

- *objects*, such as tables, including "phenomena, bodies, and substances" [2, 1.1], but also persons, software products, business organizations, production processes, ...; hereafter we will use the expression *object under measurement* to designate them[2];

- *general properties*, such as length, loudness, brightness, intelligence, readability, quality, ...;
- *individual properties*, or simply *properties* when not ambiguous,[3] such as the length of a given table or the loudness of a given sound; hence each individual property is the instance of a general property, obtained by considering that general property of a given object (for example, the length of a given table is the instance of the general property length as related to that table); the individual property intended to be measured is called the *measurand* [2, 2.1, 2.3][4];
- *property values*, such as 1.23 meters, which are intended to provide descriptive information on the current state of the object under measurement with respect to the measurand.

Hence, measurement can be operatively thought of as a process producing one or more property values that are attributed to the measurand with the aim of representing it. Also, any general property can be formally considered as a function that, once applied to a given object, is instanced into an individual property, that is

$$individual\_property = general\_property(object)$$

Prior to the execution of measurement, the individual property is just identified as being related to the given object, but the value representing it is unknown (or known with unsatisfactory accuracy). The measurement is aimed at establishing a suitable association between the individual property and a property value, customarily in the form of a functional relationship as:

$$general\_property(object) = property\_value$$

or more generally:

$$general\_property(object) = set\,(or\,probability\,distribution,\ or \ldots)\,of\,property\,values$$

meaning that the measurand is considered to be represented by the given property value or by a more complex entity, such as a set of property values "together with any other available relevant information" [2]. Information must be finally expressed according to conventional rules such as, for example, "1.23 m" (English format) or

[2]The expression "measured object" should be avoided because it is misleading: a property of the object, not the object itself, is the entity intended to be measured.

[3]Unfortunately there is not a standard terminology for distinguishing between general and individual properties. For example, individual properties are sometimes denoted as "manifestations of (general) properties" in the literature.

[4]There can be a difference between the property intended to be measured and the property that is actually measured. For example, we can be interested in measuring the length of an object at a specified temperature, but during the execution of measurement the temperature can be different from the specified one, or the measurement execution can alter the length of the object itself. As a consequence, the measurement result might refer to a property different from the measurand.

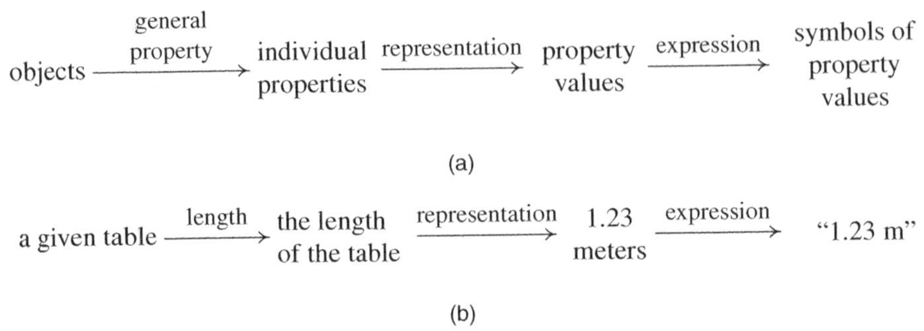

FIGURE 7.1 The basic entities involved in a measurement process and their relations: (a) general structure and (b) an example.

"1,23 m" (French, Italian, … format), that is, as a pair: numeral—symbol of a measurement unit. The diagram in Figure 7.1 highlights these entities and their relations.

As sketched in Figure 7.1, the first step of the process (from an object to an individual property of it) is preliminary to the execution of any measurement and requires the identification of a model in which the object is recognized to exhibit the general property of interest. On the other hand, the last step (from property values to a symbolic expressions for them) is just a linguistic matter. More interestingly, the core step of this process, that is, representation, can be performed by executing measurement. However, this characterization is still not specific enough to rule out situations such as "I believe that that object is long more or less 2 meters," which does not become a measurement result even if (peculiarly) rephrased as "is long exactly 2.0123456789 m." The way information is represented constitutes then a necessary but not sufficient condition for measurement: if measurement is a way to attribute values to properties, not all processes of value assignment—v-assignment for short henceforth—are measurements.[5]

Summarizing the above considerations, measurement is [15]

- a way to attribute values to individual properties, that is a v-assignment process;
- an informative v-assignment, since it is aimed at acquiring and conveying information on empirical objects;
- an informative property v-assignment; indeed any object exhibits multiple aspects (i.e., many properties) and what is measured is not the object as such but a property of it.

However, more specific features have to be looked for in measurement to characterize it as a specific kind of v-assignment.

[5]The concept 'value assignment' is sometimes denoted as "evaluation," as in the case of "evaluation of a variable" or "evaluation of a function." On the other hand, this is an ambiguous term particularly in non–physical sciences, where it is used also with an axiological connotation (e.g., in the sense of "good evaluation," sometimes with an ethical flavor) or implying a decision making process. These seem to be good reasons to avoid it here.

7.2.2 Measurement as Process Performed by a Metrological System

Those who imply that only physical properties are measurable can straightforwardly support a position as follows: a property is measurable because it is the input signal of a properly calibrated and operated physical instrument (it is a standpoint assumed, more or less explicitly, in engineering texts on measurement systems, e.g., References 21, 22). The corollary is then derived that it is the quality of the instrument and its operation that guarantees the quality of measurement results, thus justifying the superposition of the—in principle distinct—disciplines of "instrumentation" and "measurement." The tradition of physical sciences supports the plausibility of this position. For example, it is customarily considered that temperature was explicitly identified as a physical quantity and became measurable when thermometers were invented [23]. In this perspective, the conclusion is that the presence of a metrological system for a property constitutes a sufficient condition for its measurability: physical properties are measured because both measurement standards mutually connected via a traceability chain and measuring systems calibrated in reference to such standards are available. On the hand, the concept of metrological system is purely structural, and while the system based on the International System of Units (SI) [24] is only related to physical quantities, nothing in principle prevents the same structure to be realized for non-physical properties. Should this be adopted as a definition, thus making a sufficient condition also a necessary one? Does a (physical or non-physical) property become measurable only when a complete and fully structured metrological system is available for it? Human beings have been measuring long before the very concept of metrological system was even formulated. An atomic clock whose results are metrologically traceable to an SI-compliant definition of a time unit is a way better measuring instrument than the Galileo's heart whose beats he used to count time in his mechanical experiments. Nevertheless, in some general sense also retrospectively Galileo did measure, well before the Meter Convention and the SI. What is the structural reason of the different understanding in the treatment of non-physical properties with respect to physical ones?

7.2.3 Measurement as Process Conveying Quantitative Information

A condition commonly assumed to characterize measurement is that the values it produces are in fact quantitative ones, that is, they represent the measurand as number times measurement unit. Interestingly, this position is equally maintained in physical and non-physical sciences, for example, expressed as "the primary aim of a given theory of measurement is to show in a precise fashion how to pass from qualitative observations to the quantitative assertions needed for more elaborate theoretical stages of science" [25, p. 4] or "if an attribute is not quantitative, then it cannot be measured" [26, p. 287]. This builds on the Euclidean tradition: in *Euclid's Elements* "to measure" is actually a shortcut for "to be (integer) part of," thus explaining the idea that the number in the property value is the ratio of the measurand to the measurement unit (for example, the number 1.23 in the property value 1.23 m representing the length of a given table is interpreted as the ratio of the length of the table to the meter; see Reference [2, 1.20 Note 2]). This also justifies the assumption that only quantities, and not generic properties, are measurable.[6] That a property is indeed a quantity is not sufficient to guarantee that its v-assignment is a measurement: as already mentioned, guesses, and also, for example, technical specifications and forecasts, can lead to quantity values and nevertheless this does not make them measurements. But should the condition for the measurand to be a quantity be accepted as necessary of measurability? In these terms the question is admittedly at least partly conventional. For sure, a long tradition supports an affirmative answer, and assumes that only quantities having a unit are properly measurable. On the other hand, the fact that the eight international organizations which constitute the Joint Committee for Guides in Metrology (JCGM) agreed to include ordinal properties among the measurable entities in the latest edition of the International Vocabulary of Metrology [2, 1.26] shows that there are cogent interests and reasons to widen the scope of measurement science and technology. Furthermore, the very concept of measurement as representation, as interpreted and formalized in the already mentioned scales of measurement [11] and representational theories of measurement [12], suggests a broader vision, in which properties "that have a measurement unit" (as they are identified in Reference 2, peculiarly lacking a specific terms for them) are just a specific, even though very significant, case of measurable entities.[7]

7.2.4 Measurement as Morphic Mapping

In order to assure that correct conclusions about the measurand can be drawn, the data provided by a measurement process should properly represent the information available on the measurand itself. This condition has been formalized in the representational theories of measurement [12] by focusing on the relations possibly observable (in some sense of 'observability') among comparable individual properties, that is, instances of the same general property, and requiring that such relations are preserved by the relations among the property values assigned to such properties. In this context:

- the relations among individual properties are called *empirical*, to emphasize the hypothesis that they hold independently of the fact that a value assignment is performed or not; correspondingly, the set of the considered comparable individual properties together with such relations is called *empirical relational system*[8];

[6]The VIM defines "quantity" as "property of a phenomenon, body, or substance, where the property has a magnitude that can be expressed as a number and a reference" [2, 1.1]. Accordingly, quantities are properties characterized by "having a magnitude," thus including also the so called "ordinal quantities." This complex, and somehow controversial, position is discussed, for example, in Reference 27.

[7]To consider how idiosyncratic is the traditional concept that ties measurement to "properties with unit" it might be enough to remember the ambiguity related to counting, traditionally assumed different from measurement—see the title "Counting and measuring" of the seminal paper [28]—despite that some well-established measuring instruments are in fact counters, for example, Geiger counters and atomic clocks.

[8]Empirical relations are alternatively defined among individual properties or objects as considered with respect to a given general property. Given the assumption—discussed above—that *individual_property = general_property(object)*, we can neglect this distinction here.

- the relations among property values are called *symbolic*, to emphasize that the property values themselves are symbols aimed at conveying information on properties; correspondingly, the set of the property values together with such relations is called *symbolic relational system.*

The delicate concept of empirical relation can be exemplified as follows.

Example

In front of a group of people, their height can be taken into account to compare them on the basis of the empirical relation 'taller than,' as applied to pairs of people, that is, "x is taller than y" is assumed as either true or false. It is a case of a 2-adic (i.e., defined on two arguments, here pairs of persons) relation, as it is the relation "much taller than." This second case shows that whether an empirical relation holds for given objects is not necessarily an unambiguous matter.

Example

The technical novelty of four scientific papers, called here A, B, C, and D, must be assessed. To this aim, 10 expert reviewers were asked to rank these papers and the results they returned are shown in Table 7.1a, each entry representing the percentage of reviewers who assessed that the row paper has a higher technical novelty than the column paper (70% rated A as having a higher novelty than B, and so on). These results can be exploited to inquire whether an empirical relation "higher technical novelty than" holds on the set of the four papers, according to hypothesis that a paper x has a higher technical novelty than a paper y if the review result in the entry (x, y) is at least, for example, 70%, thus considering seven reviews out of 10 as a significant preference. Accordingly, the relation includes the pairs (A,B), (A,D), (C,A), (C,B), and (C,D). In this case the transitive property is satisfied: since C has a higher novelty than A, and A has higher novelty than B and D, C has a higher novelty than B and D. Observe also that neither the pair (B,D) nor the pair (D,B) is in the empirical relation: there is not a clear consensus on the relative novelty of B or D. Let us suppose now that a similar assessment process is arranged for the same four papers, considering their scientific relevance instead of technical novelty, and that the achieved results are those reported in Table 7.1b. In this case the relation includes the pairs (A,B), (A,D), (C,A), and (C,D) and the transitive property is not satisfied: C has a higher relevance than A, and A has

TABLE 7.1 Percentage of reviewers that prefer the row paper to the column paper according to their (a) technical novelty, or (b) scientific relevance

	(a)					(b)			
	A	B	C	D		A	B	C	D
A	–	70	20	100	A	–	80	30	90
B	30	–	10	60	B	20	–	50	40
C	80	90	–	90	C	70	50	–	100
D	0	40	10	–	D	10	60	0	–

higher relevance than B, but C does not have a higher relevance than B. Under the supposition that the considered relation should be instead transitive, it can be concluded that the reviewer's understanding of the property "scientific relevance," as formalized in the given rankings, is still not mature enough to infer the existence of an empirical relation "higher scientific relevance than" from it.

The basic tenet of the representational theories of measurement is that a measurement process must preserve the empirical relations. For instance, if it is agreed that A is taller than B, then a measurement of height must assign to the height of A a value greater than to the height of B. This constraint is formalized in algebraic terms by requiring that for each relation in the empirical relational system there is a relation in the symbolic relational system, and that the properties in the former are mapped to property values in the latter in such a way that if an empirical relation holds among properties then the corresponding symbolic relation holds among the corresponding property values. This constraint, called *representation condition* and mathematically implying that the mapping is a *(homo)morphism*, is essential for measurement since it allows us to draw conclusions about the empirical world on the basis of measurement results.

Example

For a given set of persons, the empirical relation "taller than" can be represented by the symbolic relation ">," so that if $m(x)$ is the height value (e.g., in meters) associated with the height of the person x, the representation condition requires that:

$$\text{if A is taller than B then } m(\text{A}) > m(\text{B})$$

for any pairs of persons A and B.

On the other hand, how can a relation such as, for example, "much taller than" be treated? A possible formal definition is

$$\text{A is much taller than B if } m(\text{A}) > m(\text{B}) + 0.20 \text{ m},$$

while other options typically involve adopting fuzzy set theory. On this basis, the representation condition becomes:

$$\text{If A is much taller than B then } m(\text{A}) > m(\text{B}) + 0.20 \text{ m}.$$

The representational theories of measurement tend to emphasize the role of representational conditions as applied in the direct comparison of objects, while neglecting the crucial fact that measurement is based on the (direct or indirect) comparison of objects under measurement and measurement standards [29]. In this view, representational conditions themselves can be thought of as necessary conditions for measurement, but their sufficiency is hardly justified.

7.2.5 Measurement as Mapping on a Given Reference Scale

A so-called *measurement scale* is formally defined by a morphic mapping from an empirical relational system to a symbolic relational system.[9]

Example

Distances can be measured, for example, in meters or in inches, corresponding to two different measurement scales.

Measurement scales can be classified in types, according to the structural information preserved in the mapping and then conveyed by the property values in the symbolic relational systems. The following five types are usually recognized as the most important ones:

- *Nominal*: Objects can be compared according to the property under consideration, being classified by an empirical relation of property related indistinguishability; no ordering among objects is instead assumed; examples of nominal scales are related to gender (e.g., male or female), geometric shape, and generally to pattern recognition processes, as in the case of graphemes and phonemes;
- *Ordinal*: As for nominal scales, objects can be classified according to the property under consideration but an empirical relation of property-related ordering between objects is also preserved by the morphism; examples of ordinal scales are related to health (e.g., healthy, sick), preference (e.g., completely agree, mostly agree, mostly disagree, completely disagree), perceived quality;
- *Interval*: A property-related distance between ordered objects, and therefore the width of intervals between properties, is also preserved by the morphism; examples of interval scales are related to date, latitude, temperature on Fahrenheit or Celsius scale;
- *Ratio*: A property-related ratio between ordered objects is also preserved by the morphism, together with a zero element representing the lack of the property; physical quantities are typically measured in a ratio scale;
- *Absolute*: A property-related counting of elements in the objects is preserved by the morphism; all properties in the form "number of ..." are typically measured in an absolute scale.

The main motivations behind the formal definition of scale types are that each type

- exhibits structural characteristics in the form of necessary conditions assuring that property values reproduce relations that empirically take place among properties;

[9]In fact, the concept "scale" is used in measurement science with many different meanings. The one presented here is not specifically related to measurement, as it applies more generally to (morphic) representation. A more correct name would have been, then, "representation scale."

TABLE 7.2 Types of measurement scales

Scale type	Defining relations	Admissible representation transformations	Admissible operation	Examples of appropriate statistics
Nominal	Equivalence	One-to-one mapping	$=, \neq$	Mode frequency
Ordinal	Equivalence greater than	Monotonic increasing function	$=, \neq, >, <,$	All the above + median percentile
Interval	Equivalence greater than ratio of intervals	Affine increasing transformation $m' = a \cdot m + b$, $a > 0$	$=, \neq, >, <, +, -$	All the above + arithmetic mean and standard deviation
Ratio	Equivalence greater than ratio of intervals ratio of values	Ratio increasing transformation $m' = a \cdot m$, $a > 0$	$=, \neq, >, <, +, -, *, /$	All the above + geometric mean
Absolute	Counting	Identity	$=, \neq, >, <, +, -, *, /$	All the above

- specifies the conditions that identify all informationally equivalent representations for a property, in terms of the so-called *admissible transformations* of scale.

Example

Length can be represented by property values in meters. However there are equally satisfactory representations in inches, feet, kilometers, or light-years. They can be transformed by simply multiplying property values by a suitable positive constant, that is, in this case multiplication by a positive constant is an admissible transformation.

Given two relational systems X and Y, X is said to be algebraically richer than Y if all relations in Y are contained in X, and X contains at least one relation that is not included in Y. The richer the empirical relational system, the more restrictive the set of representations, and so the more restrictive the class of admissible transformations. The scale types listed above are reported for increasing level of richness. Table 7.2 summarizes the key characteristics of the presented scale types.

Under the condition that information processing on property values should reflect relations in the empirical relational system, a concept of meaningful information emerges. For example, the statement "the (air) temperature in Rome yesterday was twice that in New York" is not meaningful if temperature is measured in Celsius or Fahrenheit degrees, that is, in an interval scale, because multiplication is not an admissible operation in that type. Differently, "the difference in temperature between Rome and New York today is twice what it was yesterday" is meaningful because the ratio of differences is preserved in interval scales.[10] Formally, meaningfulness can be

[10]This notion of meaningfulness is, once again, only related to a necessary condition for measurement, and therefore is not to be confused with the one of trueness. For example, the statement "The length of this object is twice the length of that object" is meaningful even though it might be factually false.

defined as follows: "A statement involving measurement results is said meaningful if its truth value is invariant of admissible transformations for the considered scale type" [30].

7.2.6 Measurement as Process Conveying Objective and Inter-Subjective Information

A complementary option for the characterization of measurement comes from a pragmatic perspective [27]. One may argue about the appropriateness of some measurements and the relevance of some measurement results (a measurement can be undertaken with an unjustified amount of resources or convey useless information), but information brought by a measurement is acknowledged being:

- specifically related to the measurand, and not, in particular, to the opinion of the subjects who operated the process, nor to the state of the surrounding environment: this is a condition of object-relatedness, that is, *objectivity*;
- univocally interpretable in different places and times, whereas an opinion (such as "highly knowledgeable in measurement science") has different meanings in different contexts: this is a condition of subject-transparency, that is, *inter-subjectivity*.

In measurement of physical properties these conditions are guaranteed by the measuring system itself and its appropriate setup and operation. The information conveyed by measurement results is

- *objective* because the measuring instrument operates as a filter, that is, its output depends on the property under measurement and is independent of all other properties (of the object, the environment, the instrument), called here "influence properties";
- *inter-subjective* because the measuring instrument is calibrated against a measurement standard and therefore the results it produces are traceable to that standard, so that different measuring systems in the same metrological system provide comparable results, that is, information that is metrologically traceable to the same reference.

For sure, in practice any measurement result always conveys information not only on the measurand but also, unavoidably, on other properties of the empirical context where the measurement process is performed. Indeed, no physical device is a perfectly sensitive and selective filter. Moreover, also the limited reliability of the adopted measuring instruments calibration and the limited capability of such instruments to maintain a perfectly stable behavior over time prevent the achievement of a complete objectivity and inter-subjectivity of the provided results. As a consequence, instead of having Boolean (i.e., yes-no) states, these two features admit intermediate levels, and a good measurement is the one whose results are considered objective and inter-subjective enough to effectively support its given purposes. Thus, a further ineludible pragmatic aspect of measurement is its being a goal-driven process. The level of objectivity and inter-subjectivity assured by measurement must be then clearly stated, typically in terms of a synthesizing parameter—such as measurement uncertainty—representing the overall quantity of information conveyed by a measurement result. This requirement is clearly asserted by the GUM [14]: "when reporting the result of a measurement, it is obligatory that some quantitative indication of the quality of the result be given so that those who use it can assess its reliability." It is worth noticing that objectivity and inter-subjectivity requirements do not impose any specific constraint on the measurement-scale type used to define the general property under consideration, and that such features are mutually independent, that is, a v-assignment could provide objective but non–inter-subjective results (e.g., when using an uncalibrated measuring system), or vice versa—inter-subjective but non-objective results (e.g., when the result is expressed as a pair "numeral, symbol of measurement unit" but it is just obtained by guess) [27]. Hence both features are essential to characterize the reliability of information delivered by measurement.

7.2.7 The Operative Structure of Measurement

Let us consider, once again, the distinction between properties and property values. If for the sake of simplicity uncertainty is omitted, the relations among such entities are synthesized by three kinds of equalities (see Figure 7.2) [27]:

- $\text{property}_1 =_\alpha \text{property}_2$: for example, the length of this object can be ascertained to be equal to the length of that object, and this may be the outcome of an experimental process in which no property values are involved;
- $\text{property value}_1 =_\beta \text{property value}_2$: for example, $2 \cdot 10^{-2}$ m may be known to be equal to 0.7874 ... in, an assessment which does not require any experimental activity once the involved units have been defined and their relation identified;

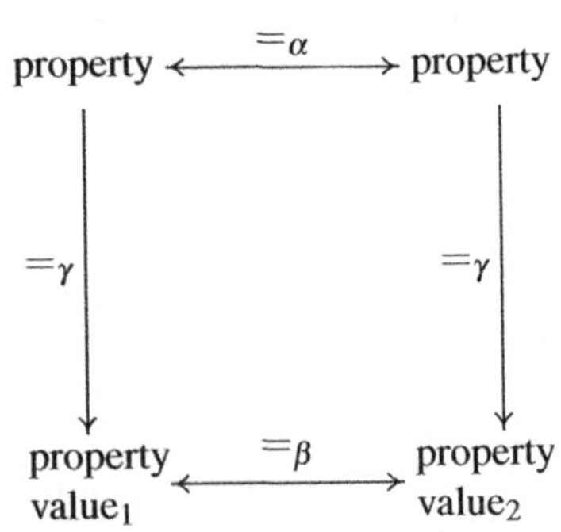

FIGURE 7.2 Equality relations among properties and property values.

- property $=_\gamma$ property value: as discussed above, this can be the outcome of a measurement (of course, the γ–equality is the usual representational relation property = property value).

The simplest structure of measurement, as performed as a direct comparison, is just a specialization of this schema:

1. a measurement standard (or a set of standards) is available that realizes a reference property, to which a property value has been associated through scale construction;
2. such reference property is experimentally ascertained to be α-equal to the measurand;
3. the same property value is then associated with the measurand.

This method is identically applicable, in principle, to both hard and soft measurement.

Example

If a set of scientific papers is available each of them assumed to be a standard for a different level of quality (or a more specific property such as originality of relevance) and with which a different quality value has been associated (step 1), and if by direct comparison a candidate paper can be ascertained to be of the same level of quality as one and only one of the standard papers (step 2), then the quality value of the identified standard paper is chosen for the candidate paper (step 3).

Hence:

- the task of step 1 is to guarantee the inter-subjectivity of the obtained information: the same property value should convey the same meaning (in the example: a given quality value refers to a given standard paper) to different subjects;
- the task of step 2 is to guarantee the objectivity of the obtained information: the α-equality between a reference property and the measurand should be ascertained as specifically related to such properties and not to other ones, including the opinion of the subject who is performing the comparison.

Due to the unavoidably incomplete inter-subjectivity and objectivity, the returned property value is generally affected by uncertainty. Instead of a single measured property value, the measurement result could be a more complex entity, such as a probability distribution over the set of possible property values.

The main drawback of the structure sketched in Figure 7.3 is the requirement for measurement to be operated directly, that is, synchronously, on measurement standards. This is the reason why the usual structure of measurement, as sketched in Figure 7.4, is based on the asynchronous comparison of the object under measurement with the standards, as obtained through a suitably calibrated transducer.

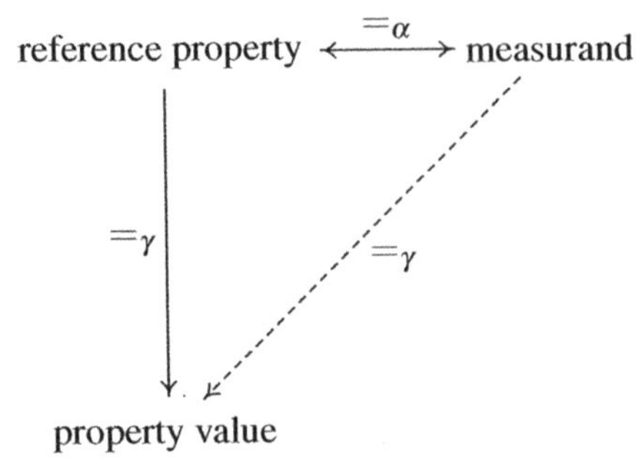

FIGURE 7.3 Structure of measurement by direct, that is, synchronous, comparison.

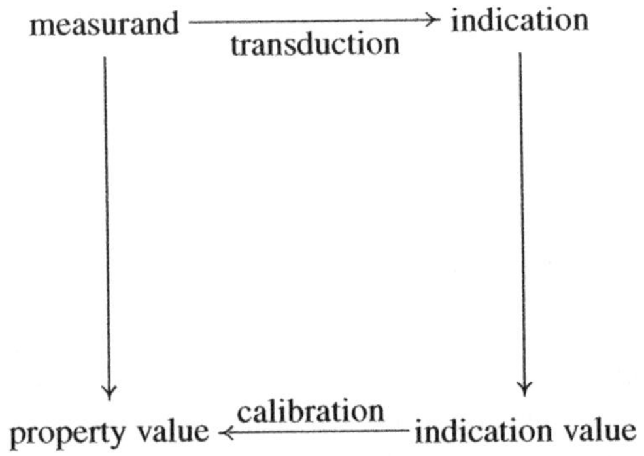

FIGURE 7.4 Structure of measurement by indirect, that is, asynchronous, comparison.

Once again, in principle this method can be applied to both hard and soft measurement.

Example

In a dynamometer the spring transduces the applied force to a length, so that its calibration is required to produce a force value for each length value. Analogously, the quality of a scientific paper is the measurand for a transducer constituted of one or more reviewers, whose work is possibly supported by a form with questions, such as "What is originality of the paper?" and "What is relevance of the paper?", to be answered, for example, with values in a Likert scale 1–5. In this case originality and relevance can be intended as indications[11] (in soft measurement they are usually called "indicators"), and

[11] Are originality and relevance indications for quality, or is quality a multi-dimensional property whose components are originality and relevance? In the first case the emphasis is on the measuring system (an indication is the output property of a measuring instrument), in the second one on the measurand definition / characterization. According to an operational standpoint, the two options actually coincide, provided that a criterion of scalar composition of the two dimensions is available. In either case, the structure can be

the mapping from their values to the sought quality value of the paper under consideration results from the available knowledge of both the general property—the quality of scientific papers—and of the specific behavior of the reviewer operating by means of the questionnaire. Hence, also in this case the critical condition for a measuring system, here including human experts, to be calibrated applies.

7.2.8 A Possible Definition of "Measurement"

Are we ready to give a single, widely agreeable definition of "measurement," broadly encompassing enough to be applicable to both physical and non-physical properties and at the same time specific enough to maintain the distinction between measurement and guess or opinion? In the current evolving situation, where social needs trigger technological trends toward a more and more extensive application of instrumentation, measurement-related concepts have to be systematically revised. Definitions such as the one given by Reference [2, 2.1], according to which measurement is a "process of experimentally obtaining one or more quantity values that can reasonably be attributed to a quantity," are useful means to bridge the tradition and the new scenarios, but they can be hardly considered as the last words on this matter (may really "reasonableness" be accepted as a scientific or technical discriminating criterion?). Grounding on the previous analysis we suggest that objectivity and inter-subjectivity can be adopted as conditions in an operative definition of the concept of reliable information, thus assuming them not only as a necessary condition but also as a sufficient one for characterizing measurement. Hence, the following definition can be drawn, to be considered as a working hypothesis:

> Measurement is a both conceptual and experimental process implementing a morphic property value assignment able to produce information on a predefined property with a specified and provable level of objectivity and inter-subjectivity.

Let us deconstruct this definition.

- Measurement *is a process*, and not the outcome of a process, that can be called instead "measurement result."
- Measurement produces information on a predefined property: the measurement result is referred to a given property intended to be measured, that can be called the "measurand."
- Measurement *includes both conceptual and experimental activities*: neither a purely conceptual process (such as a thought experiment or the analytical solution of a mathematical problem) nor a purely physical process (such as the transformation performed by a transducer) is a measurement, even though they can be part of a measurement. Indeed, conceptual activities are always needed

iterated: relevance might be considered as a "secondary" measurand, an indication of which is / defined in terms of, for example, number of citations, and so on.

to construct a model for the empirical environment where the measurement is performed and, in particular, to define the measurand.

- Measurement *implements a property value assignment*: the aim of measurement is to represent the information that has been acquired experimentally as a property value, that is, an element of a predefined set of reference properties (sometimes called a "scale"), or a more complex structure of property values, such as a subset or a probability distribution of such values.
- Measurement *implements a morphic assignment*: the relations that hold among measurands are mapped into relations among property values, so that the structural information conveyed by the former is maintained by the latter.
- Measurement *is able to produce information with a specified level of objectivity*: the measurement result conveys information on the measurand, and not, for example, on the opinion of the subjects who operated the process, nor on the state of the surrounding environment; measuring instruments are designed, and operated so to guarantee a required level of objectivity through the minimization of the effects of the influence properties.
- Measurement *is able to produce information with a specified level of inter-subjectivity*: the measurement result conveys information that is univocally interpretable in different places and times by different subjects; this can be provided by the calibration of the measuring instruments that guarantees the metrological traceability of measurement results.
- Measurement produces information with a provable level of objectivity and inter-subjectivity: such a level is usually specified as a measurement uncertainty, that is consistently chosen according to the actual information that has been acquired experimentally by using the designed measurement system on the basis of the adopted model for the empirical environment where measurement is performed.

7.2.9 Hard Measurements and Soft Measurements

Physics can be intended as a body of knowledge upon which a network of properties is built, physical laws being the connectors between two or more properties. The existence of such network generates at least two related benefits for measurement:

- from the *foundational* point of view, it provides a tool for the systematic validation of measurement models: were someone in doubt that, for example, a given instrument actually measures force, she/he might measure mass and acceleration and compare the results;
- from the *operative* point of view, it provides a tool for derived measurement: for example, values of force can be obtained without a dynamometer, in function of values of mass and acceleration.

It is then the lack of a network of definitional connections among non-physical properties that makes their v–assignment / measurement a weakly defined one. In soft measurements often a critical issue is indeed the one called *construct significance* or *construct validity*: are we really measuring what we intend to measure? The purpose of this question is actually twofold. On the one hand, it refers to the possibility of definition of general properties: what is length? what is intelligence? Consider, for example, a so-called "intelligence quotient" (IQ) test. Its results can be interpreted as values of a property of the subject who filled the test: but what property?[12] Of course, one can define "intelligence as whatever it is that the tests measure" [31, p. 376] according to an operational approach ("the concept is synonymous with a corresponding set of operations" [32, p. 5]). In this case the knowledge on the pre-metrological meaning for the property, its "underlying model," as it has been called above, is supposedly exploited to design the test, at the same time maintaining the interpretation of the obtained results flexible enough "to discover whether concepts so defined correspond to anything in nature [since] we must always be prepared to find that the concepts correspond to nothing or only partially correspond." [32, p. 5]. But then, recognizing that a complex entity such as intelligence has multiple aspects and it is not plausible that a single test is able to acquire information on all of them, the property should be given a brand new, ad hoc term, say X, and the question immediately arises: why should X be measured? The answer that X values are worth because they are assumed as conveying information on intelligence would lead, of course, to a circular position. Hence, "this definition tells us no more than we knew when we started, and it may tell us less: no two tests measure exactly the same thing, so that one is left with as many definitions of intelligence as there are tests" [31, p. 376]: more refined standpoints are needed and we believe that possible right answers cannot avoid considering the goals motivating the whole measurement process. The second purpose of the question about construct validity is that, once the general property at stake has been somehow characterized, the individual property intended to be measured, that is, the measurand, must be defined in its turn, as a specialization of the corresponding general property: the measurand is the length of this object, or the intelligence of this person, under these and these conditions. In such characterization hard and soft measurements have, interestingly, complementary features. In hard measurement general properties are usually assumed as previously defined, being such knowledge part of physics, and the emphasis is on the measurand definition and then the development of suitable measurement methods, procedures, and instrumentation for specific (ranges of) measurands. Exactly the acknowledgment that physical properties are mutually connected, and therefore that the property of interest generally can depend on some influence properties, drives the endeavor of designing measuring systems of suitable metrological features (sensitivity, selectivity, resolution, repeatability, ...) and then assuring the quality of the information they produce in terms of measurement uncertainty. On the contrary, in soft measurement raw data are usually simple to acquire,[13] and the focus is on how to extract from them information on the measurand. This explains the interest devoted in (some parts of)

[12] "Innumerable tests are available for measuring intelligence, yet no one is quite certain of what intelligence is, or even of just what it is that the available tests are measuring." [31, p. 375].

[13] "Much of what passes for psychological measurement is based upon the counting of frequencies" [26, p. 5], and similarly it can be said in other fields such as, for example, software engineering in terms of number of lines-of-code or number of defects.

soft measurement toward the representational theory of measurement, which gives consistency conditions for property representation in terms of morphic mapping [12] and meaningfulness conditions in terms of scale invariance [33, 34] (representation and invariance are indeed the two basic topics of scale construction [25]). In this perspective, the main distinction between hard and soft measurement becomes a matter of *emphasis*:

- In hard measurement the measurand is the instance of a well-defined general property linked to a network of other general properties which improves the significance of the measurement results and, at the same time, highlights that their objectivity and inter-subjectivity critically depend on properly designed and operated measuring systems; if definitional issues are at stake, they are related to the measurand definition, not to the definition of the general property of which the measurand is an instance.
- In soft measurement, the measurand is the instance of a general property lacking a well-established and widely agreed definition, and unrelated (or just weakly related) to other general properties; this makes the problem of its definition, and therefore of construct validity, a critical one, often worked out in terms of scale construction, that is, identification of consistency and invariance conditions for representation, usually based on the assessment of statistical correlations among the general property and other observable properties (often called "indicators," plausibly to stress their role of expressing one or more aspects of the measurand) by means of techniques such as regression analysis and factor analysis.

7.2.10 Multidimensional Properties

In many situations, the property intended to be measured exhibits multiple relevant components and must be defined and possibly measured in terms of them. This is the case of physical vector quantities, but also in soft measurement multidimensional properties are common.

Example

The quality of four scientific papers, called here A, B, C, and D, must be assessed, and the scientific community recognizes that paper quality involves at least two components: technical novelty and scientific relevance. To this aim two measurement processes are designed and implemented, producing values $m_n(\cdot)$ and $m_r(\cdot)$, respectively, for novelty and relevance, both defined in an ordinal, Likert scale 1–5, 5 being the highest order level. Let us assume that the obtained results are as reported in Table 7.3.

The problem is how to consistently synthesize this information in terms of paper quality. Intuitively, given two papers, x and y, x is ranked superior to (i.e., of higher quality than) y if $m_n(x) \geq m_n(y)$ and $m_r(x) \geq m_r(y)$ with at least one of the two inequalities strictly satisfied. With this rule it follows that papers B and C are not comparable, that is, neither is superior to the other. B is of higher novelty, but of lower relevance than C, so that it would be inappropriate to force an ordering, that could be imposed only

TABLE 7.3 Measurement of quality of scientific papers expressed as technical novelty and scientific relevance

Paper	Technical novelty	Scientific relevance
A	5	4
B	5	2
C	3	4
D	3	2

if we had additional information about the relative priorities of the two considered properties: novelty and relevance. If relevance is assumed to be more important, then C is the preferred choice; if novelty is of higher interest, then B would preferred. Without this further assumption, paper quality does not satisfy the representation condition for ordinal scales.

From the analysis developed in this Section, measurement emerges as a complex, knowledge-based, pragmatic process. For appropriately developing a measurement process, a conceptual framework can be useful: it is the subject of the next section.

7.3 A CONCEPTUAL FRAMEWORK OF THE STRUCTURE OF MEASUREMENT

Once it is acknowledged that measurement is a complex, model-based, goal-driven process, a conceptual framework highlighting its structure and including the representation of its several activities can become a useful tool. Such a framework is presented in Figure 7.5. It is based on the assumption that any empirical property can be, in principle, measured by performing structurally equivalent activities. The framework interprets measurement as a three-level hierarchically structured process constituted by stages, each one composed by activities, which are performed through tasks. Tasks are executed by following a loose temporal sequence (from top to bottom in Figure 7.5), but the systematic presence of feedback paths (highlighted by the bidirectional path on the left side of the diagram) emphasizes the structural complexity of the whole process in which effects may become (delayed) causes. The first stage, *Planning*, is grounded on a priori knowledge, resources, and constraints. On this basis, the following activities are performed.

– *Goal setting*: Measurement is a goal-driven process, aimed at producing effective and reliable information for supporting decisions. Hence, the definition of its purpose provides the foundation to the following activities, while taking into account both available resources and constraints to be satisfied. During this activity, also the object under measurement, the property or properties intended to be measured are identified. In particular, this implies the specification of the minimum amount of information about the measurand required to support effectively the achievement of the selected purpose, typically formalized as the maximum acceptable measurement uncertainty, often called *target uncertainty* [2,27].

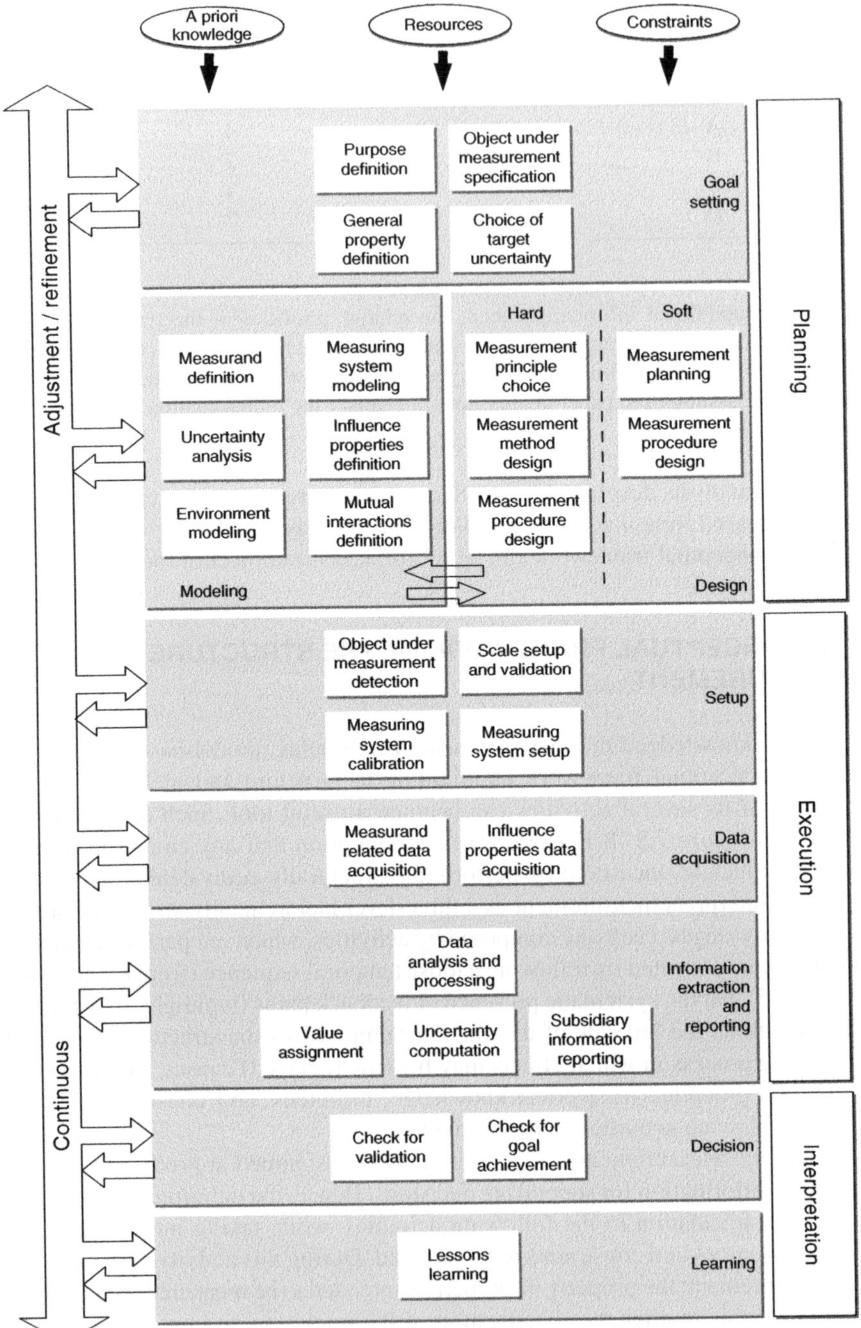

FIGURE 7.5 Diagram synthesizing the conceptual framework of the structure of measurement. The white arrow on the left represents a feedback path, whose presence highlights the complexity of the underlying information-intensive process.

– *Modeling and design*: An operational model of what is intended to be measured, the design of the adopted measurement system and the model of the environment effects on measurement results are the conceptually independent but operatively related issues taken into account. In particular, the measuring system design:

- in soft measurement, establishes the measurement plan and the detailed measurement procedure [35];
- in hard measurement, identifies the measurement principle, method, and procedure [2].

The second stage, *Execution*, requires the following activities.
– *Setup*: The measuring system is arranged and calibrated, so to guarantee that its results are traceable to a standard, and the requirements for an appropriate interaction with the object under measurement are set up. These might require some intervention on the object, aimed at making the property that will be subjected to measurement accessible for interaction.
– *Data acquisition*: the measuring system is put in interaction with the object under measurement and raw measurement data are acquired.
– *Information extraction and reporting*: the raw measurement data are suitably processed to assign a measurement result, generally including one or more property values together with the evaluation of their uncertainty. The obtained information is then presented in a way appropriate for its interpretation.
In the third stage, *Interpretation*, the measurement result is exploited to support fact-based decision-making activities, usually performed to assess the achievement of the measurement goal. The information produced by the execution stage can also be used to validate or improve the measurement process [36].

In the following subsections, these activities are further described and discussed.

7.3.1 Goal Setting

Typically, measurement is aimed at supporting decision-making. Thus, to be useful, the specific goals of measurement must be established in advance on the basis of the available a priori knowledge, while identifying and taking into account the existing resources and constraints. Hence, the key question to be asked before measuring is more of "why do we want to measure?" rather than "what do we want to measure?"

An inverse approach is usually[14] not appropriate in this case because "there are so many possible measures in a complex object that some random selection of properties will not likely turn up anything of value" [37]. Even more sharply, a Murphy's Law tells on this matter: the one property that is not measured is the one that should be. In the scientific literature concerning software measurement several conceptual

[14] Recently the so-called "big data" paradigm seems to foster the emergence of a new approach, sometimes dubbed *datafication*: the main idea is that systems such as social networks and wireless sensor networks allow us to collect huge amounts of data with practically zero marginal costs, so that identifying possible useful information and extracting it is done after data collection.

frameworks have been proposed for the definition of goals and the identification of related properties to be measured, among them the Quality Function Deployment [38], the Goal Question Metric, and the Software Quality Metrics [30] frameworks. Typically, the initial definition of measurement goal is a quite vague, brief and generic statement of intent. Hence, several subsequent refinements and adaptations (often suggested during the following stages of the process) are required before achieving a clear, specific, and operative definition of the information we desire to obtain. We call *purpose* the statement that precisely articulates the desired results. The process of setting measurement purposes is critical for successful measurement design. In order to support a careful purpose definition, the following five facets of information must typically be specified:

- *Object*: The object under measurement (e.g., a physical body like a table, a component of a hardware system, a person, a software system like a word processor, a maintenance process);
- *Purpose* ("why"): The motivation behind the measurement (e.g., to characterize, evaluate, examine, analyze, ...);
- *General property* ("what"): The property of the object we are interested in; it can be a concept of high level of abstraction (e.g., intelligence, effectiveness, reliability, effort, quality, readability, satisfaction);
- *Perspective* ("who"): Viewpoint from which the user of measurement results defines the goal (e.g., relating to properties of the object of interest for the customer);
- *Environment*: The empirical environment where the object is expected to operate (e.g., an indoor/outdoor site, a production process).

Thus, a typical statement expressing a careful purpose definition is like the following one:

{*characterize, evaluate, examine, analyze, ...*} the *general property of object* in order to {*understand, assess, predict, improve, ...*} it from the *perspective* when operating in the *environment*

Example

Estimate the dynamic transfer characteristic of an analog-to-digital converter in order to assess its conformance to given specifications when operating in an acquisition board with given characteristics.

Hence, the definition of the measurement purpose should always drive the identification of both the object under measurement and the measurand. In particular, the general property definition is aimed at recognizing which properties of the object should be measured to obtain the information required to accomplish the measurement purpose.

7.3.1.1 General Property Definition To explain how a candidate general property can be suitably defined, let us consider first the classical introductory example concerning the scale for the Mohs hardness of minerals. The key idea proposed by Mohs for the scale construction was that the hardness of a mineral can be related to its ability to scratch other minerals. He defined a reference measurement scale by selecting a set of 10 minerals chosen in such a way that each one either scratches or is scratched by all other ones, and thus exhibits a different level of hardness, collectively spanning the whole range of mineral hardness encountered in nature. Then he conventionally assigned the numbers from 1 to 10 to them, according to the representation condition that lower numbers are assigned to softer minerals. Each element in the scale operates as a measurement standard, the number associated with it being its hardness value. On the basis of the constructed reference scale and the selected empirical relation among minerals (i.e., scratching), it is then possible to determine the hardness of any piece of material: it is assumed equal to the hardness value of the hardest standard that the given material can scratch, or of the softest standard that can scratch the given material. Clearly, the Mohs scale is ordinal. The above example well illustrates what different steps have to be implemented to attain the definition of a general property. The construction of a reference measurement scale is based on the representational theory of measurement and requires

- to identify the class of objects that instantiate the general property;
- to identify the empirical relations that characterize the general property;
- to define the reference scale, that is, to select a set of objects to be assumed as standards and capable to represent the universe of possible individual properties and to assign property values to them in such a way to preserve the empirical relations among objects;
- to realize the reference scale, that is, to provide a proper implementation of the designed scale;
- to design at least one executable procedure for assigning a property value to elements not included in the reference scale, consistently with empirical relations occurring among individual properties.

The above steps are briefly discussed in the following. As for the definition of the class of objects exhibiting instances of the general property, it is usually advantageous first to consider a restricted set of elements and then, if measurements show to be successful in that set, progressively to extend it. For example, in the measurement of loudness of sound, the following sequence of classes could be logically considered [7]: pure tones at 1000 Hz, pure tones at any frequency, stationary sounds, non-stationary sounds. In principle, a given general property can be defined in different ways,[15] each one leading to a different reference scale. Temperature is a classic example: it can be defined, for

[15] It is always possible to arbitrarily define new properties. However, absolute arbitrariness is limited if the definition of properties is grounded on pragmatic bases, that is, by referring to a measurement goal. In fact, as pointed out in Reference 39 [pp. 18, 28] "the set of derived quantities is open-ended; new ones may be introduced at will in any analysis. [...] A derived quantity does not necessarily represent something

example, on the basis of the order relation "at least as hot as" perceived when touching any pair of objects, or on the variation of the height in a liquid-in-glass thermometer. Different empirical relations could correspond to different scale types, or scales of the same type but with different uncertainties. When dealing with a candidate new general property, the construction of the reference scale is the most critical step, since it strongly influences the subsequent measurement steps. As suggested above, it is frequently advantageous to adopt a progressive strategy. An empirical relation giving rise to an order scale with few standards can be considered first. If measurements performed with such a scale are successful (i.e., measurement data allow a better understanding of the measurand), and it may be useful or necessary to measure in a more specific way, the scale can be improved, by extending its range, improving its resolution, or using standards with lower uncertainty. Also, a completely new and more sophisticated empirical relation leading to a stronger (e.g., interval) scale can be adopted [7] or different empirical characteristics can be considered. For example, instead of scratching, indentation may be chosen as the basic empirical characteristic for hardness definition. As a consequence, different scales are obtained and checking for their agreement may be an effective scale validation strategy. Furthermore, a reference scale may be constructed, not only directly, on the basis of empirical relations involving the considered property, but also indirectly, by deriving it from a known functional relation linking the considered property to other properties whose scales have been previously established. Typically, a property for which a scale is constructed on the basis of the empirical relations defining it is considered to be measured in a *fundamental scale*, while a *derived scale* is one obtained through a functional relation with other measurable properties.

Example

Assuming mass m, length l, and time t as fundamental properties, from the related scales a scale for force F can be derived on the basis of the physical law $F = m \cdot a$, in which a is the acceleration, dimensionally length·time^{-2}, of a rigid body of mass m subjected to the force F.

7.3.2 Modeling

A model is an entity that views an object from a particular perspective and neglects details considered useless. Formal models can be built using different tools (equations, diagrams, mappings, ...) and show how the various object components relate to one another, so that the involved relationships can be analyzed and understood. Models are essential for both measurement planning and interpretation. In particular, modeling activities are aimed at providing a goal-oriented and operational description of all entities in the experimental setting (the object under measurement, the measuring system, the experimental environment, and their mutual interactions), called here

tangible in the same sense as a base quantity, although it may. The square root of time, for example, is a derived quantity because it has the required power-law form, but we cannot point to any physical thing that "is" the square root of time."

the *measurement context*, that have a relevant effect on raw measurement data. As any model provides just a partial description of the object it represents, some potentially useful information could be lost in using it. This occurs also for the various models describing the measurement context, because of the unavoidable approximation of any representation of the empirical world. In particular, properties and interactions that are not identified (because we are not aware they can significantly affect measurement data, so failing in including them in a model) or not modeled (because expected to have, singularly taken, negligible effects on measurement data) limit the amount of information on the measurand conveyed by the measurement result. This incomplete knowledge about the measurement context affects the measurement result, and contributes to the measurement uncertainty. According to this standpoint, measurement uncertainty can be considered as originating in the modeling activity, that is, before the execution stage, which is traditionally considered the main uncertainty source. While, in principle, uncertainty can be considered as unavoidable in measurement, even whether to report it, or instead to assume that it is operatively negligible, is a decision to be made according to pragmatic criteria. On this matter the VIM is clear: "If the measurement uncertainty is considered to be negligible for some purpose, the measurement result may be expressed as a single measured quantity value. In many fields, this is the common way of expressing a measurement result." [2, 2.9 Note 2]. Thus, the evaluation of measurement uncertainty is a crucial task of the modeling activity and it will be extensively analyzed in the following. Unfortunately, identifying all significant sources of measurement uncertainty can be a very hard task, especially when dealing with soft measurements, where measurement uncertainty is often just neglected. Indeed, when considering physical properties, modeling is supported by well-established and universally accepted theories. Opposite, soft measurands are often quite complex and ill-defined properties and validated relations linking them to other properties are often lacking. Anyway, the adopted model must ensure that raw measurement data contain enough information for the achievement of the chosen measurement purpose. In the following, the different tasks involved in modeling activities are discussed with a focus on their effect on the amount of information delivered by measurement result.

7.3.2.1 *Measurand Definition and Definitional Uncertainty* The measurand is an instance of the defined general property. It is the individual property "intended to be measured" [2, 2.3]: This reference to intentions highlights the explicit link between measurement purposes and the modeling stage, primarily aimed at defining the measurand itself. The measurand definition has a crucial role for all measurement activities: In particular, it should contain a description of all the assumptions and conditions (often required to be standardized) under which the measurand has to be considered in order to ensure unambiguous information. It is grounded on the available knowledge on the relations between the measurand and other properties, of the object or the environment, which might affect measurement results and their interpretation. In the measurand definition, different situations may occur depending on the amount of information on the measurand that must be contained in the measurement data. For instance, the effect of the environment properties on raw measurement data can be negligible and therefore none of them needs to be modeled or, at the opposite side, some of them must be modeled and measured in turn. In this latter case a detailed measurand definition is obtained, such as "length of the object at the temperature of ..., atmospheric pressure of ..., ...". The properties of the context unlike the measurand that might significantly, with respect to the target uncertainty, affect the raw measurement data are called *influence properties* [2, 27].[16] They must be identified and included in the context model, together with their mutual interactions and their effect on raw measurement data. Usually, all properties included in the measurement context model are represented in terms of mathematical variables, while their mutual relations are represented by using mathematical relationships among those variables. Either deterministic or non-deterministic (such as stochastic or fuzzy) descriptions can be employed.

Example

In many everyday situations, the assumptions about the conditions defining the measurand are well known and often unexpressed. For example, when evaluating a person's age, an instance of the general property time duration, only years from her/his birthday are counted, even though a different time origin might be adopted, for example, the time of conception.

Example

A definition of "line of code" must provide an operational description of such property and the conditions under which the count must be performed. For example, it is often defined as a non-blank, non-comment line in a code listing. Moreover, the definition must specify whether the measurement should be taken after the first successful compilation or after final system acceptance and whether or not lines of code written in different languages are counted separately.

Definitional uncertainty is the contribution to measurement uncertainty originating from the finite amount of information specified in the measurand definition. It can include the uncertainty due to the effects of environment properties on the measurand. Hence, the definitional uncertainty is a lower bound to measurement uncertainty, so that it makes no sense reducing any other uncertainty contributions at a much lower level than definitional uncertainty itself. Definitional uncertainty is often dominating over other uncertainty contributions in soft measurements.

Example

Let us consider the measurement of the length of a metallic rod. It is known that the size of metallic bodies depends on their temperature, so temperature can be here an influence property. Depending on the chosen target uncertainty, this can be modeled according to two different options:

[16] This definition differs from the one specified in the VIM [2], because of the reference to target uncertainty: influence properties are only those with a "significant" effect on raw measurement data.

- temperature is maintained as a "hidden variable" in the definition of the measurand, whose (definitional) uncertainty should take into account this underdetermination;
- temperature is explicitly included in the definition of the measurand, which is then specified, for example, as the "length at (a reference temperature of) 0°C".

The greater specificity of this second modeling choice virtually allows to obtain measurand values of lower uncertainty, but requires also a system for the control of the temperature or the measurement of temperature and suitable modeling and processing activities for the correction of its effects when the actual temperature differs from the specified reference one.

7.3.2.2 Measuring System Model, Instrumental and Interaction Uncertainty A measuring system is constituted of one or more measuring instruments and often other components are assembled, adapted, and set up so to produce measurement results [2]. The core component of a measuring system is typically a transducer, that establishes a mapping from an input property to an output property, called *indication*. Indication values and measurand values are in a many-to-many relation,[17] usually represented in graphical form as a strip in a calibration diagram, where the width of the strip for a given indication value represents the related instrumental uncertainty. A functional mapping is singled out in this relation as a *calibration curve*, associating each indication value with a measured value [2]. A calibration diagram is obtained through the calibration of the measuring system, assuming that the instrumental measurement uncertainty is large enough compared with the measurement uncertainties associated with the property values provided by the adopted measurement standards. Indeed, the transducing behavior of a measuring instrument is characterized by several parameters such as (the following definitions are adapted from Reference 2)

- *sensitivity*: quotient of the change in the indication and the corresponding change in the property being measured;
- *selectivity*: ability to provide measured property values for the measurand that are independent of other properties in the object under measurement;
- *resolution*: smallest change in a property being measured that causes a perceptible change in the corresponding indication;
- *stability*: ability to maintain metrological properties constant in time.

The performance of the measuring system is affected by these and other (instrumental bias, dead band, response time, bandwidth, ...) parameters, that can be synthesized in terms of:

[17]The relation is many-to-many because a single indication value might correspond to multiple measurand values, but also a single measurand value might be obtained from multiple indication values: this is sign of the presence of uncertainty.

- a location statistic, that Reference 2 calls *measurement truth*, related to the ability to maintain the calibration state and therefore to produce bias-free values;
- a dispersion statistic, that Reference 2 calls *measurement precision*, related to the ability to produce constant values when operating under repeatability conditions.

In their turn, trueness and precision can be synthesized by *measurement accuracy*, conveying the overall metrological information on the measuring system as it is exploited in a given measurement process. By means of calibration, such overall information is finally represented as measurement uncertainty, that can be propagated to measurement result. Two different kinds of uncertainty contributions can be distinguished here:

- *Instrumental uncertainty*, originating in both the model of the measuring system and the model of its interactions with the environment. A lower bound to instrumental uncertainty is the calibration uncertainty, that is the uncertainty obtained in the calibration of the employed instruments. Indeed, any deviation between instrument operative and calibration conditions increases instrumental uncertainty.
- *Interaction uncertainty*, originating in the description of the interaction between the object under measurement and the measuring system and taking into account the fact that the specific interaction might alter the status of the object under measurement, which might differ from nominal conditions.

Observe that the previous uncertainty sources, together with definitional uncertainty, are physiological to any measurement: They are always present and their value can, and should, be estimated during the planning stage and controlled (i.e., kept within predefined bounds) during the execution stage. However, if one or more influence properties are not identified (and so not included in the context model and controlled), the information provided by the measuring system might be wrong. These pathological results should be detected and rejected or, if believed to convey information, they should be explained by improving the empirical context model. This latter circumstance may occur, for example, when discovering new physical phenomena.

7.3.2.3 Hierarchy of Models and Levels of Formal Knowledge In the measurement of physical quantities, the knowledge of well-established theories allows the construction of a hierarchy of models with growing complexity, so enabling better context descriptions and improvements in the significance of measurement results. As an example, consider the goal of ensuring the correct operation of a piston moving inside a cylinder. Information useful for the achievement of this goal can be obtained by measuring the diameter of both the piston and the cylinder and comparing the two results. However, a proper use of the information conveyed by the measurement results requires the construction—maybe implicit and only apparently

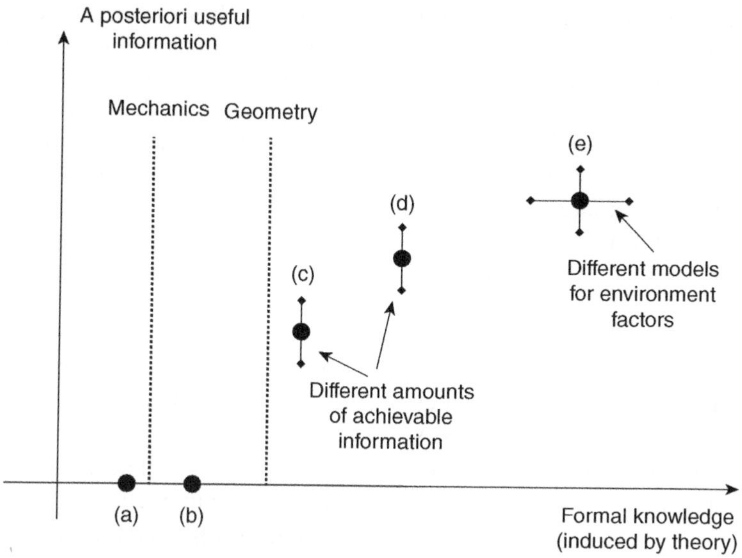

FIGURE 7.6 Qualitative graph describing the relationship between the quantity of a posteriori useful information provided by measurement result and the formal knowledge induced by the underlying theory.

simple—of a model of the two considered empirical objects. To this aim, they can be described using one of the models belonging to the following hierarchical chain, ordered according to increasing level of formal knowledge about the empirical world: (a) *physical object*, (b) *rigid body*, (c) *perfect cylindrical solid*, (d) *cylindrical solid with uncertainty*, (e) *cylindrical solid with uncertainty and sensitivity to environment properties* (e.g., temperature). Measurements based on models (a) and (b) cannot provide information useful to support decisions ensuring the achievement of the defined goal, as shown in the qualitative graph depicted in Figure 7.6, in which the quantity of a -posteriori useful information provided by measurement results is reported versus the formal knowledge induced by the available underlying theory. The concept of diameter applies only to the models (c), (d), and (e), that is, only to models supported by geometry, and measurements based on such models can provide information useful for the achievement of the defined goal. Using models at level (c), the a priori information available should guarantee that deviations of piston and cylinder from a perfectly cylindrical solid are negligible with respect to the predefined target uncertainty. Also the effect of environment temperature variations on the geometric dimensions of the two empirical objects must be ensured to be negligible. Under the above stated constraints, a single measurement of the diameter of the two objects performed with uncertainty less than the target uncertainty can effectively support the decision about the correct operation of the system. Obviously, the achieved information depends on measurement uncertainty, as shown by the vertical segments in Figure 7.6. Similar conclusions can be drawn for models (d) and (e) when one or both the previous hypotheses concerning the deviations from a perfectly cylindrical solid or the effects of the environment temperature variations not necessarily apply and must be verified a posteriori. To this aim, the diameter of both the piston and the cylinder need to be measured repeatedly in different points of their surfaces and, in the case of model (e), for different values of environmental temperature. In that sense, these two levels of knowledge allow the increase in the amount of useful information conveyed by the measurement results, as shown in Figure 7.6.

It is interesting to briefly discuss some characteristics of the above described meta-model structure for the measurement of physical properties [40].

- Properties are not considered as inherent features of the empirical object, but as constituting components of models: models determine properties and any experimental activity aimed at extracting information about the object requires a preliminary description activity.
- Models are goal-driven and they are not inherent features of the modeled object, although models of lower abstraction level are often implicit.
- As level of abstraction increases, the relationship between a model and the modeled object tends to be more and more hidden.
- This hierarchical chain of models is such that lower level knowledge models are embedded, and sometimes hidden (e.g., because implicitly assumed) by higher level models: a higher level model implies all lower level models, so that if a lower level model is changed, the higher level model must be adapted, or even substituted; for instance, in the example above, if the empirical object is not considered as a rigid body, then the concept of its diameter becomes meaningless.
- the co-existence of different (and at least partially independent) theories enables the application of different measurement methods. For instance, the volume of a piston can be derived by measuring its diameter and height (geometrical volume) or by measuring its mass and density (mechanical volume).

Opposite, when dealing with measurements in soft systems, well-established theories involving the measurand are typically not available. As a consequence, a model structure based on different levels of knowledge can rarely be built. Also, once a measurand model has been achieved, it might be not easy to show if a measurement based on that model can provide enough information for the achievement of measurement goal.

7.3.2.4 Uncertainty Analysis and Expression As the result of a measurement, a property should not, generally, be represented by a single measured value. According to the VIM [2], a set of values should be provided that could reasonably be attributed to the measurand, together with any other available relevant information. Probability theory, and in particular the notion of random variable, allows dealing with such a concept in an effective way. In fact, a random variable X is recommended by international organizations [14] to model the output provided by a measurement process and the related probability density function (pdf) represents in a formal way the distribution of possible values attributed to the measurand. In particular—for measurands defined at least on an interval scale and modeled as (almost) Gaussian random variables—the finite amount of information associated with the measurement result can be represented by using the first two moments of the pdf, that is, its expected value and standard deviation. Recently researchers have been proposing to formalize measurement uncertainty also by means of non-probabilistic frameworks, particularly in the context of the mathematical theory of evidence [41], the fuzzy set theory [42, 43], or the theory of random fuzzy variables [44]. In any case, the measurement result is generally provided as a single measured value and a measurement uncertainty, usually obtained, respectively, as estimates of the expected value x and the related standard deviation $u(x)$ of the underlying random variable modeling the output of the measurement process. International guides suggest that these parameters are called *measured value* and *standard measurement uncertainty* respectively [2]. During measurement planning, uncertainty can be evaluated by relying only on a priori available information, for example, instruments specifications, calibration certificates, experience and general knowledge of the measurement designer. The related procedures are called *Type B uncertainty evaluations* and the related standard measurement uncertainty is denoted as $u_B(x)$. Once experimental activities have been completed and acquired data are available, measurement uncertainty can be determined using also Type A evaluations, that is, through statistical techniques applied to indication or measured values. The related standard measurement uncertainty is denoted as $u_A(x)$. Type A and Type B evaluations are generally related to different phenomena: random fluctuations occurring within the measurement context during experimental activities and incomplete knowledge conveyed by the context model respectively. Hence the underlying random variables describing the effects of these phenomena on measurement results can be assumed to be statistically uncorrelated. It follows that an estimate of the standard deviation of the measurement output, called *combined standard measurement uncertainty* $u_C(x)$, can be evaluated by means of the relationship:

$$u_C(x) = \sqrt{u_A^2(x) + u_B^2(x)}.$$

On the other hand, when a measurement result is employed to support decision-making processes, it can be advantageous to represent it as an interval that may be expected to encompass a large enough fraction p of the distribution of values that could be reasonably attributed to the measurand. The fraction p is called *coverage probability* or *level of confidence* [2]. The interval representing the measurement result is usually expressed as $x \pm U(x)$: it is centered around the measured value x, and its half-length $U(x)$ is called *expanded measurement uncertainty*. This parameter is evaluated as the product of the combined standard measurement uncertainty $u_C(x)$ and a suitable *coverage factor* k, that is:

$$U(x) = k u_C(x).$$

The value of the coverage factor k is chosen on the basis of the required coverage probability p and it usually ranges from 2 to 3 [14].

7.3.2.5 Target Uncertainty and Measurement Context Model Complexity

As explained above, to effectively support decisions that promote the goal achievement, the measurement result must convey a minimum quantity of information about the measurand. This is expressed by specifying—during the goal definition activity in the framework of Figure 7.5—an acceptable upper bound $u_T(x)$ to measurement uncertainty, the *target measurement uncertainty*. Thus, modelers must guarantee that the expected measurement uncertainty is lower than the target uncertainty with a sufficiently high confidence level. For sure, context model uncertainty, and as a consequence measurement uncertainty, can always be reduced by including additional levels of detail, for example by increasing the number of influence properties, improving the accuracy of mathematical relationships, or narrowing the ranges of allowed values for influence properties. However, in this way the complexity of the context model increases, which depends on the number of modeled properties, computational complexity of modeled relationships and severity of control requirements for influence properties. The problem is that high complexity models generally impose sophisticated and expensive instrumentation and measurement procedures, which must be defined by high skilled designers and performed by experimenters with a high level of expertise. As a consequence, measurements may become too expensive and their cost may not be justified by the achievement of the intended goal. Moreover, the additional amount of obtained information may exceed requirements, as specified in terms of target uncertainty, thus becoming useless. On the other hand, the amount of information available when using low complexity models may not be sufficient for effectively supporting decision-making activities, in the case the related measurement uncertainty exceeds the specified target uncertainty. It follows that the construction of the most comprehensive context model is not the goal of the modeling activity. Indeed, a suitable context model should pick only the aspects of the context that are relevant for the measurement goal. Among the varieties of possible measurement context descriptions, the selection of a suitable model must consider that

- the amount of information conveyed about the measurand should be sufficient to effectively support decision-making activities, that is measurement uncertainty must be smaller than the target uncertainty with a given confidence level;
- a minimum amount of resources should be allocated, due to both economic or technical constraints.

Observe that the selection of a context model exhibiting the minimum complexity to satisfy the constraint represented by the target uncertainty can be a very hard task, especially for high accuracy and complex measurements.

Example

Consider an apparently simple measurement of mass. The model of the object under measurement and, more generally, the model of the measurement context strongly depends on the chosen value for target uncertainty. For instance, due to economic reasons, target uncertainty when measuring a mass of potatoes or a mass of a gold ring might be very different: of the order of 10 grams or few milligrams respectively. As a consequence, attention paid to the status (and to the model) of these two kinds of objects is quite different: we are not worried about some mud on the surface of potatoes, while we try to avoid the presence of dust on the surface of the gold ring. Also the adopted measuring instruments and procedures are very different.

7.3.2.6 A Procedure for the Selection of a Suitable Measurement Context Model For both hard and soft measurements, the selection of the context model can be performed according to the following iterative procedure, inspired by the Procedure for Uncertainty Management (PUMA) suggested in the International Standard ISO/TS 14253–2 [45].

1. Specify all the a priori information available, the constraints and the available resources.
2. Specify the target uncertainty $u_T(x)$.
3. Identify and model the measurand, the measuring system, and the experimental environment.
4. Using the a priori information available and Type B uncertainty evaluations assess definitional uncertainty $u_{def}(x)$ also by taking into account environment effects.
5. If $u_{def}(x) > u_T(x)$, then
   if allowed by the imposed constraints and the available resources, then return to step 3 and try to improve the measurand model in order to reduce the contribution of the various definitional uncertainty sources by starting from the most relevant ones or return to step 2 in order to increase the target measurement uncertainty; else exit the procedure since the problem cannot be solved.
6. Using the a priori information available and Type B evaluations, identify and assess the contributions of interaction uncertainty $u_{int}(x)$, and instrumental uncertainty $u_{instr}(x)$ to measurement uncertainty.
7. Evaluate overall measurement uncertainty $u_B(x)$ by combining the components of uncertainty budget, $u_{def}(x)$, $u_{int}(x)$, and $u_{instr}(x)$.
8. If $u_B(x) > u_T(x)$, then
   if allowed by the imposed constraints and the available resources, then return to step 3 and try to improve the context model in order to reduce the contribution of the various uncertainty sources by starting from the most relevant ones or return to step 2 in order to increase the target measurement uncertainty; else exit the procedure since the problem cannot be solved.
9. A suitable context model is achieved.

The procedure relies only on a priori available information and Type B uncertainty evaluations. Once experimental activities have been completed, measured values are available and combined standard uncertainty $u_C(x)$ can be evaluated. At that moment, a further comparison between $u_C(x)$ and the target uncertainty $u_T(x)$ should be performed in order to check the validity of the measurement result and, if necessary, improve the context model, redesign the measuring system and perform the experimental activities again.

7.3.2.7 Measurement Significance and Measurement Validity Strictly connected to uncertainty of the model of measurement context (and in particular to definitional uncertainty) is the concept of *measurement significance*. It is related to the query "Are we measuring the right property in the right way?," which refers to whether the measurement result is expected to provide information useful to support decisions for the achievement of the intended purpose. Hence, measurement significance refers to an appropriate description of the measurement context and, more specifically, to a suitable definition of the measurand in reference to the given purpose. It can be improved only through a better understanding of the empirical environment or of the measurand, so that better descriptions can be established. Different from, but strictly connected to, measurement significance is the concept of *measurement validity*. According to References 2 and 46, a measurement process is validated by providing objective evidence that its result allows fulfilling the intended goal with a suitable level of confidence. Hence, the validity query is "Are we measuring the right property, in the right way and with right uncertainty?," and it refers to whether a measurement provides enough information to drive effectively to the achievement of the intended purpose. To be valid a measurement must then be significant and the related uncertainty must be lower than the fixed target uncertainty with a given confidence level. Measurement significance can be confirmed in the interpretation stage, in reference to measurement results, or using theoretical considerations during the design activity. Measurement validation can be accomplished just in the interpretation stage since, by its own definition, it requires fact-based decision-making activities. When dealing with physical properties, the existence of theories and well-established methods for measuring the same property allows their cross-validation, thus ensuring the validity of the implemented measurement process. In many cases, the applied theories have been confirmed for centuries, and any possible contradiction originating in their application would trigger scientific efforts aimed at interpreting the unexplained data. Conversely, when measuring soft properties, often only one way to achieve the measurement result exists. Thus measurement cross-validation cannot be applied and the following two basic methods of proving measurement significance can be employed:

- *Empirical significance*, used to achieve corroborating evidence that measurement results provide information about the specific property intended to be analyzed. For example, standard statistical techniques such as linear regression or correlation analysis can corroborate the assumption that measuring a given property can provide information about another non-measured property. Empirical corroboration, although necessary, is in general not sufficient to confirm measurement significance.

- *Theoretical significance*, which allows confirmation that the structure of the designed process does not violate any necessary characteristic to be properly considered a measurement.

Some reasons limiting measurement significance are listed in the following.

Construct significance refers to the limits of the operational definition representing a measurand and it is related to the query: "Are we actually measuring what we want to measure?" When measurands are of high level of abstraction, it may be very difficult to recognize whether a certain definition for a measurand is significant or not. It may be even more difficult to improve it or to create a new and more significant definition.

Example

Different intelligence tests may be considered. However, in many situations, the test results may not be useful to predict the problem solving capabilities of a person [16,47].

Content significance refers to the degree to which the operational definition of a measuring covers the range of different aspects included in it [48].

Example

A test designed to assess software engineering skills must not be limited to the evaluation of programming capabilities only, but should contain also an assessment of process management capabilities.

Example

If our aim is checking the correct operation of a piston moving inside a cylinder, a single measurement of their respective diameters can be of little significance and could lead to wrong decisions. Indeed, a single measurement cannot ensure the circular profile of the two objects.

Consistency significance refers to a structural constraint imposed by the representational theory. This requires the measurement scale adopted for the general property to be a homomorphism, that is a mapping in which empirical relations between individual properties are preserved in abstract relations between property values [12, Chapter 1]. This representation condition allows us to express in symbolic terms the empirical relations occurring between individual properties. Hence, consistency provides a necessary condition for the adopted definition to actually capture the aspects of the measurand we are looking for, so preserving the behavior exhibited by individual properties in the empirical world. Especially when dealing with soft measurements, a specific care is reserved to the symbolic relational system, while the empirical relational system is often analyzed too quickly. Data acquisition procedures are defined and implemented without a proper investigation about the relations among empirical properties intended to be measured. As a result, it might happen that the actual meaning of acquired data is not clear.

Processing consistency significance refers to the relationship that links two or more properties when indirect measurements are considered. A basic issue in this case is whether the values returned by the relationship actually represent information on a property of the object under measurement. In fact, unlike physical measurements, when dealing with soft properties, directly measured properties can often be acquired with high accuracy and using simple activities (e.g., number of lines of code of a piece of software), but the relationships linking them with the measurand are often unknown.

Example

Let us suppose that a measurement of the diameter d of a cylindrical object is performed. The expression $\pi \cdot d$ can be evaluated, but only the knowledge of the theory known as geometry guarantees that the computed value is empirically meaningful and conveys information about the cylinder circumference.

Example

Let us assume that we are able to directly measure the mass m and the acceleration a of a physical object. Then the product $F = m \cdot a$ can be evaluated and, thanks to a well-known physical law, the obtained result conveys information on another property, the applied force F. But, if we have no a priori knowledge about the empirical significance of the result obtained by processing the directly measured property values, what can we state? For example, what about the result of the relationship $x = m + a$? Has it empirical meaning? Because of the knowledge of the classical physical theory we can state that no physical property corresponding to x can exist. This is a basic result of dimensional analysis, requiring properties to have the same physical dimension to be added.

The theory of scale types [11] identifies the scale constraints for an operator to be applied, thus leading to consistency significance of relationships involving properties values [33]. For instance, purely ordinal values cannot be added or doubled. Thus, given the statement "the temperature of 40°C is double as compared to the temperature of 20°C" the inference that the first temperature is twice the second one is not valid, because these properties are evaluated in an interval scale, in which ratios are not invariant. Scale constraints are necessary but not sufficient conditions for the significance of the performed processing, and usually some other constraints must be satisfied to guarantee the significance of a candidate property. As a theory cannot be verified but only corroborated or falsified [49], a significant or a valid measurement is one that we are unable to show it is not significant or valid yet.

7.3.2.8 Overall Targets of Modeling The modeling stage is globally aimed at identifying the conditions according to which the measurement process can be reliably replicated, and its results reliably interpreted, in different contexts. These overall conditions are described:

- from the point of view of the process, in terms of *measurement reproducibility*, that is, the ability to replicate measurements on the same or similar objects under a set of conditions including different locations, operators, and measuring instruments, that is, by changing different features of the experimental context in which the measurement results are achieved;
- from the point of view of the results, in terms of *metrological traceability*,[18] that is, the possibility for the measurement results to be related to a reference through a documented unbroken chain of calibrations.

Measurement reproducibility and metrological traceability are then the ultimate targets for a process that, as emphasized in Section 7.2.6, is structurally aimed at conveying objective and inter-subjective information.

7.3.3 Design

Although representational theories maintain that measurement could be characterized even only by the measurement scale constructed for the defined general property, the use of a measuring system should be considered as a mandatory condition, as typically assumed in hard measurement. This may be assumed as a general paradigm also in the case of soft measurements, provided that a proper definition of what a measuring system is, holding in both situations, is adopted. A measuring system can be defined as "a set of one or more measuring instruments and often other devices, including any reagent and supply, assembled and adapted to give information used to generate measured quantity values within specified intervals for quantities of specified kinds" [14].

This definition is general enough not to exclude soft measurements, but at the same time it requires that the employed measuring system, however implemented, be subjected to basic metrological conditions, so ensuring measurement traceability and a proper control of instrumental uncertainty.

When measuring physical properties, the design activity implies establishing [2]:

- a *measurement principle*, that is, a physical phenomenon serving as a basis of measurement;
- a *measurement method*, that is, a generic description of the logical organization of operations required for a proper application of the adopted measurement principles; methods are qualified in various ways, such as substitution methods, differential methods, null methods, direct and indirect methods;
- a *measurement procedure*, that is, a detailed and documented description of operations an operator is required to perform to execute measurement in an effective and efficient way according to the chosen measurement method.

[18]The term "traceability" is sometimes used to mean "logistic traceability," that is a description of the context in which a measurement result was obtained and more generally a product was delivered (e.g., date and time, environment status, employed instruments, ...).

When dealing with soft measurements no general theory involving the measurand usually exists. As a consequence, the measuring system is usually designed by first establishing a generic description of the operations to be performed, often called *measurement plan*, and then a detailed measurement procedure [35]. When dealing with either physical or soft properties, procedures can be very different. However, they should normally include

- a formal definition (and possible textual description) of the properties to be measured, of both the object under measurement and the empirical environment, to support the established goals;
- ranges of values of the properties to be measured;
- when measuring physical properties, a description of the adopted principles and methods;
- a detailed description of the equipment used for acquiring measurement data and of its metrological characteristics;
- the role of the persons responsible for experimental activities, including their competence and training experience (it is often important that experimenters understand the whole procedure: why the measurement is needed, how it is going to be used, and how their action contributes to the overall validity of the delivered information);
- the period of time when data acquisition must be performed (or the frequency of replicated acquisitions);
- a detailed description of the activities to be performed to acquire the required information about the properties of interest;
- a detailed description of the processing activities to be performed on the acquired data to achieve the desired measurement result including information needed for the evaluation of measurement uncertainty;
- the definition of the reporting format for measurement results (e.g., single measurements, trend charts or control diagrams, only measurement results, or added detailed analysis texts), reporting timing (when and how often the measurement results are provided), and reporting dissemination policies (which restriction on information access are applied);
- information needed to the subsequent interpretation stage;
- the interpretation purpose (conformance assessment, measurement validation, measurement improvement, ...) together with the adopted decision criteria and thresholds.

Any measurement procedure must satisfy various constraints, including time and costs limitations, expected maximum uncertainty of the achieved results, and experimenter experience.

Example

Consider the measurement of the temperature of a body. Well-known physical laws allow the identification and modeling of the measurand, the environment properties, and their mutual interactions. The chosen instruments can acquire information about the measurand using different physical phenomena (e.g., thermal expansion of the considered body, thermal expansion of another body used as a transducer, thermoelectric effect, and so on). Once the measurement principles and instruments have been selected, the measurement method is chosen (e.g., the transducer is put in contact with the considered body), and a measurement procedure is defined in which each activity is described with sufficient detail.

Example

The resistance of a resistor can be measured either directly or indirectly using a priori knowledge provided by well-known physical laws and measurement methods. Definition of the measurand is then driven by this knowledge (e.g., the dependence of resistance on both the environment temperature and the applied current).

When measuring soft properties, the two distinct concepts of measurement procedure and measurand definition often collapse as a whole. Indeed, in soft measurements, a property is often defined by providing a complete and detailed procedure for measuring it, in accordance with the Bridgman's operational definition of properties. Thus, any other procedure would measure, in principle, a different property [32].

Example

Functional size of a software product is defined by providing a detailed procedure to measure it.

Usually, clarity and self-consistency of a measurement procedure are checked by means of peer review. However, any measurement procedure should be properly tested before its actual use for the implementation of the execution stage: it must be applied during a trial period, in order to reveal possible flaws to be corrected. This is usually achieved by validating the acquired data, that is, by checking them for completeness and involved uncertainty.

7.3.4 Execution: Setup, Data Acquisition, Information Extraction and Reporting

The setup activity is a set of tasks aimed at the proper implementation and arrangement of the designed measuring systems. In fact, before acquiring information about the measurand, the designed reference scale has to be properly implemented and the designed measuring system must be correctly assembled and calibrated in order to guarantee the traceability of measurement result. As already mentioned, the outcome of calibration may be represented as a calibration diagram, that is, the "graphical expression of the relation between indication value and corresponding measurement result," which also provides information on instrumental measurement uncertainty [2]. In order to acquire the information on the measurand correctly, the measuring system must be rightly arranged with respect to both the object under measurement and the surrounding environment. An example of this may result in turning on in advance instruments and apparatuses to guarantee the achievement of thermal steady-state status, or in preparing or cleaning containers according to predefined procedures before acquiring data regarding substances and materials. Similarly, when dealing with soft measurements, forms to support a structured data collection are arranged, test items are calibrated and the environment where the interviews or the test will be conducted is set up. After preliminary setup activities, data acquisition processes are performed according to the defined measurement procedure, and raw measurement data are obtained. Some major features of the acquired data, related to their usefulness to the subsequent information extraction and interpretation activities, are

- *sufficiency*, related to the extent to which the information associated to acquired data (e.g., influence properties, acquisition time interval, adopted measurement method) suffices for the extraction of information required to support the achievement of measurement goal;
- *correctness*, related to whether the experimental activities are actually performed according to the measurement procedure;
- *logistic traceability*, associating measurement data to the time period when they were acquired.

From raw measurement data, the object under measurement must first be detected (e.g., the presence of a given signal must be established from noisy data or a pattern must be identified in the acquired image). The information extraction and reporting activity is then performed. This is a set of tasks aimed at extracting information about the measurand from measurement data, addressing how measurement result should be presented to provide the information required to answer the questions pertaining to the measurement goal. Indeed, raw measurement data may directly provide information about the measurand only in the so-called direct measurements. Conversely, when no trivial measurements are concerned, measurement result is often obtained as the output of a functional relationship involving acquired data and processing activities are needed, as described in the defined measurement context model. This occurs not only for the so-called *indirect* (or *derived*) measurements, but also for direct measurements when the effect of influence properties is significant and must be taken into account [14]. During this crucial activity, also measurement uncertainty has to be computed. It is worth emphasizing that in measurement result presentation digits corresponding to orders of magnitude lower than measurement uncertainty should not be reported. According to the VIM [2], the measurement result is a "set of quantity values being attributed to a measurand, together with any other available relevant information." Thus, when reporting the result of a measurement the following information should be provided:

- a property value, that is, the measured value;
- the amount of information provided, by means of, for example, the combined standard uncertainty or an interval of values and the related coverage probability or, as in the most general option, of a pdf;

- the measurement scale, stated, for example, by specifying a measurement unit;
- the limits of validity of the information provided, in reference to the considered influence properties and their allowed ranges of variations;
- additional subsidiary information needed (or useful) for the reproduction of the measurement result.

Example

The result of the measurement of a resistor resistance could be expressed as $R = 1.234 \pm 0.003\ \Omega$, with a coverage factor equal to 2 (corresponding to a probability of about 95% when considering a normal distribution), and specifying that the environmental property taken into account is the temperature T and its allowed range is $0°\text{C} \leq T \leq 60°\text{C}$.

7.3.5 Interpretation

The interpretation stage can be implemented either online or offline. It is always aimed at exploiting the result of measurement to establish whether or not the intended purpose is met. This purpose can be, for example, *conformance assessment, measurement validation*, or *measurement improvement*. To ensure that interpretation is clearly focused on the intended purpose and does not fluctuate depending on measurement result, it is crucial to define the interpretation plan during the design activity, that is, before experimental activities are performed. In particular, the decision criteria and thresholds used to interpret the measurement results should be clearly defined. Due to measurement uncertainty, the decision result is an uncertain event and there is a risk of wrong decisions. For example, when the conformance or non-conformance of a product or service has to be decided with the support of measurement results, the decision has to be assessed under specified consumer and producer risks [50,51]. As a consequence, information concerning measurement uncertainty must be taken into account in the decision rule. This issue is tackled in an International Standard related to geometrical property measurements [52], but its basic assumptions can be adopted in many decision-making activities involving hard or soft measurements: when a fact-based decision about conformance has to be taken, those who want to prove conformance or non-conformance are responsible for the effects of measurement uncertainty on decision results. The interpretation stage can also be addressed at recognizing the lessons learned during measurement planning or execution and at preserving the acquired knowledge for future use. This may result in new procedures or best practices, thus addressing decision-making activities about possible improvements in both the measurement process and the use of measurement results. These feedback activities may lead to update the previous measurement activities, as shown in Figure 7.5. For example, through interpretation activities the excessive sensitivity of an instrument to irradiated electromagnetic fields may be discovered, thus leading to decide of screening the sources, or adopting an instrument with better immunity. Sometimes, even measurement goals may be changed as a result of the interpretation stage, for example leading to the refinement of the whole process so to obtain better information with the same amount of devoted resources.

7.4 AN APPLICATION OF THE MEASUREMENT STRUCTURE FRAMEWORK: ASSESSING VERSUS MEASURING RESEARCH QUALITY

7.4.1 Motivations for Research Quality Measurement

Knowledge development and scientific research are unanimously recognized as main drivers for the socioeconomic innovation and growth of nations. Intellectual assets, enabling innovation through the advancement of the state of the art in practical application fields, are rapidly becoming the keys to value creation and increase in national gross domestic products [53]. Consistently with this development model, most advanced nations have been investing significant portions of public funds to finance research activities. To ensure the effectiveness of expenditure, nations are increasingly resorting to research quality assessment programs, aimed at ranking research institutions and funds recipients. While the goal is clear in its intentions, the assessment process has proved to be hard to be implemented. The United Kingdom has had a leading role in the management of these processes and it is argued that quality assessments are the reasons why UK-based research is top ranked worldwide and nationally better funded than used to be when assessment programs were not performed. This policy and best practice has spread in other nations as well, with Europe playing a leading role in the usage of somewhat complex and documented quality assessment models. The topic is controversial, because of the necessity to balance indicator-based and peer review-based evaluations and because of the behavioral consequences induced by v-assignment processes. However, the lack of public trust in the self-government of scientists and the need to introduce methods that are objective and inter-subjective enough to allow an appropriate ranking of researchers or institutions on the basis of their merit are a prevailing attitude within many public administration sectors. Result-oriented management styles, based on verifiable objectives and on proofs of efficiency and effectiveness in the expenditure of public money done by scientists and researchers, are additional motivations. The literature on this topic is huge and has investigated the many societal, technical, and practical implications of research assessment exercises, but has largely ignored the preliminary structuring activities needed in any process aiming to be a measurement. This is evident also in the terminology used in this domain: many avoid the term *measurement* when it comes to research quality, to underline the overall unstructured nature of the associated information gathering processes. Terms such as "assessment," "evaluation," and "appraisal" become substitutes, plausibly used to emphasize the limited reliability of the information conveyed by the process outcomes. So, to which extent is research quality measurable? Both assessment and measurement are v-assignment processes about the same observable property, but that provide different quality of information, as it will be clarified in the following. This section is aimed at using the conceptual framework introduced in the previous section to describe and compare both research quality assessment and measurement, identifying the differences between them and highlighting which features are applicable to measurement only. This example intends to stress the usefulness of a framework, the one we have proposed in particular, to address and model the process of performing measurements. A framework eases discussions on a topic, by setting the base for a common language and for shared concepts [63]. It favors interpretation of evidence and assists researchers in attributing a meaning of subsequent findings. Moreover, casting research quality assessment as a measurement process helps understanding the potential weaknesses of existing approaches and provides elements for discussions and improvements.

7.4.2 Measurement Goal Definition

As shown in Figure 7.5, in the first stage of measurement its purpose has to be defined. Various options apply in the case of research quality: information is collected, processed and analyzed to fund projects, to promote or hire people, to verify the appropriate usage of resources, to rank organizations, to assess the societal impact of research outcomes, to explore reputational value of institutions. On the basis of this information, retrospective studies are performed to find technological antecedents of scientific breakthroughs and to assess the influence of scientific results over long periods of past times [54, 55, 64]. The goals often depend on the parties expressing the needs: while governments are often interested in measuring for accountability and funding reasons, local organizations such as research institutions may also be interested in assessing for promotion or hiring purposes. The clearness in the goal definition is usually more limited here than when setting goals in the case of hard measurements, where requirements and specifications can often be declared using well-established communication styles and symbolisms, strongly reducing risks of misinterpretation and leading to more mature inter-subjective definitions.

Example

One of the clearest statements regarding the purpose of a research quality measurement process is that of the UK Research Excellence Framework (REF) [56]: "The primary purpose of the REF is to produce assessment outcomes for each submission made by institutions:

- The funding bodies intend to use the assessment outcomes to inform the selective allocation of their research funding to higher education institutions, with effect from 2015–16.
- The assessment provides accountability for public investment in research and produces evidence of the benefits of this investment.
- The assessment outcomes provide benchmarking information and establish reputational yardsticks."

7.4.2.1 *Object Identification and General Property Identification* The planning stage, as shown in Figure 7.5, requires the description of the class of objects that manifest the property of interest and the identification of the specific object under measurement. The description should be clear and complete enough to allow deciding whether a given object belongs to the class [62].

Example

Possible objects are research papers in a given subject category, published in a specified set of journals, within a specified time interval. Other entities, such as research projects in a given research field, individual researchers, research teams, or research institutions, can also be considered.

Of course, different general properties apply to objects of different type.

Example

The content quality of a scientific paper sent for publication in a journal is a different property from the quality of research done by a research group.

Regardless of the general property of interest, the single research item represents the base unit upon which any research assessment exercise is based. A research item represents a codified form of knowledge and can take several forms, in relation to communications habits applicable in different research domains, such as: scientific journal paper, book, book chapter, scientific letter, seminar, conference talk, poster, software product, prototype, patent. Assessment procedures may consider a single item or composite sets of items depending on the subject whose research outcome is under assessment.

Example

Research items might be: papers published in a journal over a given period of time; journal papers published by a given publisher; scientific outcomes of a single researcher, of a group of researchers affiliated to the same research institution, working in a certain research domain (disciplinary field) or in the same territory (region, nation, ...).

In assessing a research plan (e.g., when distributing funds on a competitive basis) results not only are unknown beforehand, but also nobody can assure them to be achieved at the end of the research process. In this case, the proposal is assessed also considering results previously obtained by the proposer or by his collaborators.

7.4.2.2 Environment Identification As it occurs for any measurement process, when assessing research quality all properties of the environment possibly affecting v-assignment results must be identified.

Example

Cultural aspects can influence v-assignment results. For example, different patterns of research publishing styles are found in western and Asian countries: while many western countries are characterized by the *guilt culture*, in many Asian nations the *shame culture* is prevailing and senior researchers keep out from the research competition, not to risk the scientific reputation in front of the society in which they live [57]. This leads to younger researchers being authors of research papers and senior scientists acting as reviewers or anyhow behaving such that they do not become publicly accountable.

Example

When highly ranked scientists and their scientific products are evaluated, there is the risk of a Matthew effect ("the rich get richer and the poor get poorer") in science [58]. In this case, it might be possible that assessors are influenced by their prejudices, and make biased judgments, not exclusively based on evidences, but also, or even primarily, on the scientific prestige of the assessed person.

Example

Historical events may influence v-assignment results too. It is known that the number of citations of publications by Marx and Lenin dropped significantly, after the events that lead to the destruction of the Berlin Wall [59].

Example

The identification of the environment related to a journal paper requires analysis of the journal reputation, of the prestige of publishing authors, and of the editor-in-chief.

Various measurement methods can be applied to perform v-assignments of research quality properties, with peer-review processes and bibliometric data analyses playing a major role. Description of these methods is done in the design stage. However, as emphasized by the presence of feedback paths in the diagram of Figure 7.5, the choice of the measurement method influences the identification and modeling stages, and vice versa. As for the peer review process, its outcomes depend on competency and prejudices of reviewers. In addition they can be subjected to distortions due, for example, to opportunistic behaviors induced by conflicts of interest. Individual characteristics of the reviewed researchers, such as sex, personal reputation, affiliation to a religious institution, or the mother tongue, may influence the assessment outcome. Also the type of research outcomes (e.g., patents) may influence the assessment output. For these reasons, often synthetic quantitative indicators are required, that can act as proxies of research quality having more objective properties. Of fundamental importance for the outcome is clearly the context in which the assessment takes place. Which incentives drive the reviewer to act superficially or with great attention? Who selects reviewers? Is the reference scientific community large enough to guarantee against opportunistic behaviors? As far as bibliometric data are concerned, it is often assumed that citation indexes are less influenced by subjectivity or opportunistic behaviors than peer review processes. However, citations are used not only to recognize the importance of a research outcome but also to stigmatize low quality methods or inconsistencies in the results. High quality outcomes, obtained in a narrow field or related to frontier problems, may also have low influence because outside the disciplinary mainstream, to which higher influence research items usually belong. Moreover, also citations may be affected by opportunistic behaviors. Finally, the quality of a citation index depends on the quality of the database used to evaluate the indicator, which in turn depends on the criteria used for inclusion (or exclusion) in the archive of the elementary information and on its accuracy, for example, coverage degree of the considered disciplinary area.

7.4.2.3 *Setting of Target Uncertainty* When assessing research quality, uncertainty is seldom taken into account and therefore target uncertainty is not pre-set. This issue may be referred to, for example, the number of levels chosen as v-assignment results: while it might be considered reasonable to assume three or five different levels to represent the quality of a scientific publication, it is plausible that 10 levels correspond to a too much resolute scale with respect to the uncertainty of assessments. Not addressing explicitly the issue of uncertainty evaluation is a major difference between assessment and measurement.

7.4.3 Modeling

In the modeling stage the general features of the measurement process, established in the previous steps, are operatively characterized, in particular by choosing and defining one or more observable properties of the considered object. The construction of a model for the measurement context may be a complex activity, because of the consequences implied by the decisions driven by the measurement results: funding and promotion opportunities, scientific reputation and other outcomes, may all be affected by the adopted model. Hence, all relevant stakeholders (funding agencies, researchers, and the society at large) should be involved so to avoid underestimating pertinent viewpoints and to converge to an agreed model able to properly grasp all significant aspects of the property undergoing v-assignments.

7.4.3.1 *General Property Definition* In the case of research quality, the most critical task of the modeling stage is probably that in which the general property of interest is defined. In fact, in order to ensure that v-assignments provide meaningful information it is essential that the underlying concepts, as commonly understood, are clarified prior to operatively describing them. Some frameworks assume that the first step is the definition of research and then research quality [56], where *quality* is commonly intended here as the *degree to which a set of interrelated characteristics fulfills requirements* [46, 61]. Such broad concepts ask for additional, context-dependent details to be defined in a quantifiable way, but they are so multifaceted and even fuzzy that some of their determinants can hardly be captured by any set of operatively defined properties. In fact, often the selected features focus only on specific aspects of the general property of interest, but do not convey information on it as a whole. The quality of research has many distinct views that could be identified, modeled and assessed separately. The model should describe the concept "research quality" as intended in the context where the information will be used, so to limit its definitional uncertainty. Typically, a distinction is made between:

- *conformance*, intended as coherence of a research result with the standards usually adopted by the reference community of scientists (national or international) over a specified period of time. Standards may specify the properties to be considered that, for a bibliographic product, include: originality, relevance (or *importance*) for the development of knowledge in a certain research field, methodological rigor, clarity, writing style, robustness of data, and evidences on which conclusions are drawn;
- *influence* of research on the scientific community, intended as effects produced by the object under assessment on the development of knowledge in the given field. There is a slight difference between *relevance* for the development of science, intended as potential influence and effective *influence*, typically assessed by means of the number of citations over a given time period. For example, a research result may not be appreciated or may be appreciated only in a long period of time, because of limitedness of communication mechanisms, difficulties in fully appreciating the consequences of the contribution, costs or cultural issues. Also, conformance of a scientific product is (usually) a necessary but not a sufficient condition to guarantee its influence. For example, if a paper appears in a qualified scientific journal, it can be assumed that it satisfies the quality standards adopted in that period by the scientific community but not necessarily it will have an influence;
- *impact* of research, intended as societal relevance of the research, that is, the society returns on investments, on the basis of priorities defined by the society itself. While conformance and influence refer to scientific aspects only, impact includes also the research capability of solving societal problems, e.g., by underpinning innovation or by promoting occupational and economic returns. Hence, assessment of impact can be difficult to perform, because it requires the adoption of complex and uncertain cause–effect models. Moreover, impact v-assignments should be carried out independently from conformance and influence v-assignments, to avoid biases in the process results.

Example

The mentioned UK REF research assessment exercise is based on the evaluation of an overall quality profile, obtained by the combination of 3 sub-profiles: *outputs*, *impact*, and *environment* [56]. Specific definitions are given for each of the concepts captured by the sub-profiles in different knowledge domains (e.g., scientific and humanistic). To include the influential effects of the various research environments, 36 Units of Assessment (e.g., clinical medicine, mathematical sciences) are considered.

Example

If scientific reputation is the concept to be captured, the selected quality research aspects to be evaluated can be defined in terms of properties such as scientific production and acknowledgments from scientific community.

Observe also that within modeling, the task of *uncertainty analysis* is generally not performed. Thus, the disciplined approach provided by the framework proposed in this chapter highlights the need for an increase in the understanding of uncertainty-generating phenomena in research quality assessments.

7.4.4 Design

7.4.4.1 V-assignment Design In the following we consider only v-assignments of the research quality dimensions that are performed internally to the scientific community, that is, conformance and influence. As for the assessment of conformance, it is essentially a two-step process:

- the identification of the many properties of the object undergoing assessment that may impact conformance decisions;
- comparison between these elements and the standards widespread in the reference scientific community.

Differently from what applies in hard measurement, the prevailing research standards, that is, the reference scale, are not inter-subjectively defined and formalized. Their knowledge only pertains to members of the specific scientific community, who also have the capability of identifying the relevant elements of a research product and of comparing them to standards: these are competencies acquired in a long process that scientists activated with other goals in mind: the development of their own research. As a consequence, assessments regarding conformance must be carried out only by members of the scientific community (the peers), using reviewers' panels. Editorial peer reviews are largely based on the information used in bibliometric analyses of research product quality: an article published in a high reputation journal, after the peer review phase, is automatically considered a research product, independently of its content. From this point of view, the peer review process is equivalent to a quality assurance process. This reflects an undisputed faith in the scientific community, considered completely autonomous and self-regulating. The presence of conflicts of interest and the social importance of research push toward even more systematic and transparent assessments, so to reduce biases. The involvement of persons that virtually act as measuring systems poses fundamental issues about measurement objectivity and inter-subjectivity. Thus, some criteria should be followed in the selection of reviewers' panels, for example:

- involving reviewers with expertise in the considered subject: seniority in the role and scientific reputation is often associated to similar reactions (e.g., the judgment) to the same stimuli (e.g., the product to be assessed) because of the harmonizing effect of individual experience in the profession and of the extensive exposition to international scientific contexts;
- avoiding conflict of interest and preventing interactions among reviewers to ensure unbiased and independent reviews.

Also, the panel composition should be representative of the scientific community (e.g., an international panel should be selected). The number of reviewers forming the panel is critical for the inter-subjectivity of the assessment results and must be great enough to ensure statistical stability: five or more reviewers could be needed. Monitoring of convergence when new results are acquired may suggest the proper number of reviewers in the panel. Conversely, influence of a research product is assessed by using citation indexes that are considered good approximations (proxies) of this quality property. Sometimes, other types of recognitions are used such as awards, invited talks and presentations. Details concerning data acquisition processes may play a significant role and their comprehensive analysis is always required.

Example

Once decided that citations coming from scientific international journals must be counted, the journal list is compiled and counting is subsequently performed. However, details are of major importance: which journals to include in the list? What to do with self-citations? How to attribute correctly products to authors?

When performing a retrospective assessment (e.g., in a departmental research quality assessment) two choices are possible:

- ad hoc peer review;
- processing of information available as outcome of other peer review processes done for different goals (e.g., bibliometric analyses).

Both approaches, sometimes improperly called "qualitative" and "quantitative" respectively, have strengths and weaknesses. Deciding which procedure is to be used depends on issues such as costs or context (e.g., when privacy of results matters). In fact:

- peer review processes are expensive, both because of direct costs and in terms of time not dedicated to research, but provide high resolution judgments. Thus, peer review proves its validity when assessing single projects or the research outcomes of a single researcher;
- indicator-based processes are less expensive and are preferable when complex sets of products, researchers or institutions are assessed.

In general, procedures to be used for acquiring indicator values are straightforward and the focus is on the way the involved general properties should be related to the acquired values by definitional relations. However, indicator-based only assessments are usually discouraged since they cannot fully grasp the original abstract concept associated with the considered object property [60]. Also, the correct usage of indicators requires suitable technical skills to guarantee the absence of abuses. Frequently a single-value composite property defined as combination of various indicators is used for ranking purposes. However, if the relative importance of the different aspect of research quality is not known, as it often occurs, the usage of single numbers to capture the multifaceted aspects of research quality should be avoided [60]. Because of the intrinsic limitations of each approach, joint peer review-based and indicator-based v-assignments provide often the best assessment result.

7.4.4.2 Reference Scale Design

V-assignments related to research quality properties are based on scales that are mostly ordinal both in prospective and retrospective research quality assessments. Because of unavoidable (definitional, instrumental, and interaction) uncertainties, only a limited number of scale elements can guarantee a sufficiently high trust in v-assignment results. The characteristics of each element of the given scale should be clearly specified during the design phase. When peer review assessments are concerned, an evaluation form should be predefined that may guide reviewers in structuring and standardizing their assessment activities and results.

Example

A structured and standardized form for supporting the review assessment could contain questions about the different aspects of quality (e.g., "What is the originality of this paper?", "What is the relevance of this paper?"), to be answered, for instance, with values in a Likert scale 1–5.

Example

The dimensions on which a judgment must be expressed according to the UK REF are output, impact, and environment, in accordance with instructions based on national and international standards. Judgments are expressed using 1 to 4 stars (1 star=national level research, 2 stars=international level research, 3 stars=international excellence, 4 stars=world-leading results). Products that are not assessed do not comply with national standards or with the definition of research.

The chosen scale is often self-validating. In fact, reviewers selected for peer review usually share the same viewpoints, having comparable scientific backgrounds, analogous experiences and similar roles in their institutions. At the same time, harmonizing sessions can be performed beforehand, to favor similar assessment behaviors. Whether peer review- or indicator-based processes have to be performed or not, in the end of the design stage, clear operative procedures should be provided to acquire the needed information. Also, the design stage must include policies adopted to cope with unreliable or not-responding reviewers, as well as with technical difficulties arising when interfacing with required databases.

7.4.5 Execution

During this stage, suitable data processing procedures and decision rules might be applied to compose contrasting reviews, to decide about rankings when comments do not match or to exclude inconsistent results. Overall, data collection tasks may involve a large set of tools and stakeholders. Information extraction and reporting is the activity in which all collected information is processed and prepared for public disclosure.

Example

The results of a research assessment process can be expressed by means of an assessment report that must justify the final assessment of research quality. If a nation-wide assessment is performed, a report is written to highlight the strengths and weaknesses of the overall assessment procedure. In both cases this activity may serve for improvement purposes.

At present, there is no evidence of any assessment program providing estimates of result uncertainties. Nevertheless, uncertain outputs may be due to limited construct significance, to uncertain inputs, or to the variability associated to peer review judgments.

Example

Is the true number of citations a knowable quantity? In principle, if a very narrow and specific definition is provided for such a quantity, the answer is positive. However, if a broad sense is given to this concept, as it is customary in actual research quality assessment exercises, uncertainties in the properties characterizing the elements stored in databases used for future analysis performed by assessors and constraints in the available resources may make such "true value" an elusive concept, similar to what happens in the domain of hard measurements. In fact, both commercial and freely available databases are continuously updated by adding new records (e.g., references to journal papers or books). Thus, the date of extraction of the database information becomes certainly relevant, but the doubt still remains whether citing documents have all been included or not, so that the issue of incomplete coverage is important. Moreover, citation databases collect information from different sources and follow criteria for inclusion or exclusion in the selection of citing documents, according to strategies that can also be driven by broader market needs. This happens, for instance, when the database manager is a privately owned company. For instance, freely available sources of citation information, such as Google Scholar, include *gray literature* as possible citation sources, a choice not always acceptable for effective v-assignments to research quality properties.

7.4.6 Interpretation

The last stage of a measurement process is that of results interpretation, that includes analyzes aimed at verifying the achievement of the specified goals, possibly together with the validation or the improvement of the whole process. When dealing with research quality v-assignment, this stage may be characterized by a long latency, because of the nature of research. According to outcomes of this last stage, assessment programs are modified over time to account for lessons learned and for changes in the measurement environment. It is also worth noticing that mature assessment programs are based on pilot studies carried out to test procedures and to validate processes.

Example

The effectiveness of a peer review process in guaranteeing the quality of published products in a journal can be estimated by determining the number of published work that do not comply with disciplinary standards (Type I error) and the number of rejected high quality contributions (Type II error). Type I errors can be reduced by adopting a higher rejection rate. Accordingly, Type II errors will increase, and vice versa. The optimal rejection rate depends on how easily the scientific community is capable to recognize Type I and Type II errors. This is easier in natural sciences and in mathematics, where journals adopt a relatively low rejection rate, in comparison to human and social sciences, where in fact journals usually have a higher rejection rate.

Research quality assessments become policy-set processes: researchers are also driven by opportunistic behaviors adopted so to fit definitions and criteria. Therefore, as it happens also in the domain of hard measurements, observing the system may induce undesired side effects and system modifications.

Example

If only journal papers count in a certain research quality assessment framework, researchers are discouraged to present their research results at scientific congresses. This might lead to less participation at scientific gatherings and reduce, in turn, their fundamental role to favor discussion and reasoning about research outcomes.

7.5 CONCLUSIONS

Since measurement activities are performed in an increasingly larger number of knowledge domains, the quest for a modern interpretation of this concept appears to be nontrivial. In this chapter we made the effort to propose and describe a measurement framework that can be used to analyze the major aspects associated to hard and soft measurements. The framework provides a discipline for planning and executing a measurement and for interpreting experimental outcomes. The application of this framework to the assessment of research quality has evidenced the need for such disciplined reasoning when issues of measurability and metrological model identifiability become evident. This example has highlighted also that information producing processes, such as measurement and assessment, (both cases of the generic process that we have called v-assignment), usually have different features. Assessment differs from measurement in at least three aspects: uncertainty not explicitly evaluated, reference scale implicitly defined, comparison with standards performed in a subjective, or only partially inter-subjective, way. The usage of the measurement framework eases reasoning about the various steps, both formal and informal, that could be followed to enhance the objectivity and the inter-subjectivity of any v-assignment and of its results.

FURTHER READING

On the subject of measurement fundamentals many authors have expressed their views from different perspectives. Interesting material is covered in:

Finkelstein, L., Gratten, K. (eds), *Concise Encyclopedia of Measurement and Instrumentation*, Pergamon, Oxford (1994).

Krantz, D., Luce, R.D. Suppes, P., Tversky, A. *Foundations of Measurement*, Academic Press (1971, 1989, 1990).

Kyburg, Jr., H.E. *Theory and Measurement*, Cambridge University Press (1984).

Narens, L. *Abstract Measurement Theory*, MIT Press (1985).

Rosen, R. *Fundamentals of Measurement and Representation of Natural Systems*, North–Holland, Amsterdam (1978).

Rossi, G.B. *Measurement and Probability: A Probabilistic Theory of Measurement with Applications*, Springer (2014).

The concept of measurement was analyzed in the context of several disciplines other than the hard sciences:

Boumans, M. *Measurement in Economics—A Handbook*, Academic Press (2007).

Fenton, N.E., Pfleeger, S.L. *Software Metrics: A Rigorous and Practical Approach*, PWS Publishing (1996).

Kan, S.H. *Metrics and Models in Software Quality Engineering*, 2nd edition, Pearson Education (2003).

Michell, J. *Measurement in Psychology—Critical History of a Methodological Concept*, Cambridge University Press (1999).

In relation to research quality assessment consider the books:

Bailin, A., Grafstein, A. *The Critical Assessment of Research*, Chandos Publishing (2010).

Vinkler, P. *The Evaluation of Research by Scientometrics Indicators*, Chandos Publishing (2010).

Sources of additional information are represented by the journals *Scientometrics*, published by Elsevier that debates issues related to science and scientific research and by the open-access journal *Research Trends*.

REFERENCES

1. Chang, H., Cartwright, N. Measurement, in *The Routledge Companion to Philosophy of Science*, S. Psillos, M.Curd (eds), Routledge (2008), pp. 367–375.
2. JGCM 200:2012, *International Vocabulary of Metrology—Basic and General Concepts and Associated Terms (VIM)*, Joint Committee for Guides in Metrology (2012) (2008 edition with minor corrections), Available at http://www.bipm.org/en/publications/guides/vim.html.

3. Mari, L. A quest for the definition of measurement, *Measurement*, **46**, 2889–2895 (2013).
4. Gould, S.J. *The Mismeasure of Man*, Norton (1981).
5. De Morgan, A. *The Connexion of Number and Magnitude: An Attempt to Explain the Fifth Book of Euclid*, Taylor and Walton (1836).
6. Heath, T.L. *Archimedes: Works*, Cambridge University Press (1897).
7. Rossi, G.B. Measurability, *Measurement*, **40**, 545–562 (2007).
8. Luce, R.D., Tukey, J.W. Simultaneous conjoint measurement: a new scale type of fundamental measurement, *Journal of Mathematical Psychology*, **1**, 1–27 (1964).
9. Andrich, M. *Rasch Models for Measurement*, Sage (1988).
10. Mari, L., Wilson, M. An introduction to the Rasch measurement approach for metrologists, *Measurement*, **51**, 315–327 (2014). doi:10.1016/j.measurement.2014.02.014.
11. Stevens, S.S. On the theory of scales of measurement, *Science*, **103**, **2684**, 677–680 (1946).
12. Krantz, D., Luce, R.D., Suppes, P., Tversky A. *Foundations of Measurement*, Academic Press (1971, 1989, 1990).
13. Finkelstein, L. Fundamental concepts of measurement: definitions and scales, *Measurement and Control*, **8**, 105–111 (1975).
14. JCGM 100:2008, *Evaluation of Measurement Data—Guide to the Expression of Uncertainty in Measurement (GUM, originally published in 1993)*, Joint Committee for Guides in Metrology (2008), Available at http://www.bipm.org/en/publications/guides/gum.html.
15. Mari, L., Carbone, P., Petri, D. Measurement fundamentals: a pragmatic view, *IEEE Transactions on Instrumentation and Measurement*, **61** (8), 2107–2115 (2012).
16. Finkelstein, L. Widely, strongly and weakly defined measurement, *Measurement*, **34**, 39–48 (2003).
17. Finkelstein, L. Problems of measurement in soft systems, *Measurement*, **38**, 267–274 (2005).
18. Finkelstein, L. Widely-defined measurement: an analysis of challenges, *Measurement*, **42**, 1270–1277 (2009).
19. BIPM. *Evolving Needs for Metrology in Trade, Industry and Society and the Role of the BIPM*, International Bureau of Weights and Measures (2007).
20. Michell, J. *Measurement in Psychology—Critical History of a Methodological Concept*, Cambridge University Press (1999).
21. Bentley, J.P. *Principles of Measurement Systems*, Pearson (2005).
22. Morris, A.S. *Measurement and Instrumentation Principles*, Butterworth (2001).
23. Sherry, D. Thermoscopes, Thermometers, and the Foundations of Measurement, *Studies in History and Philosophy of Science*, **42**, 509–524 (2011).
24. BIPM. *The International System of Units (SI)*, 8th edition, International Bureau of Weights and Measures (2006), Available at http://www.bipm.org/en/publications/si-brochure.
25. Suppes, P. *Representation and Invariance of Scientific Structures*, CSLI Publications (2002).
26. Michell, J. The logic of measurement: a realist overview, *Measurement*, **38**, 285–294 (2005).
27. Mari, L., Giordani, A. Quantity and quantity value, *Metrologia*, **49**, 756–764 (2012).
28. von Helmholtz, H. Zahlen und Messen erkenntnis–theoretisch betrachtet, *Philosophische Aufsutze Edward Seller gewidmet*, Leipzig, (1887); reprinted in Gesammelte Abhandl, 3,

pp. 356–391 (1895); English translation by C.L. Bryan, *Counting and measuring*, Van Nostrand (1930).

29. Mari, L.Beyond the representational viewpoint: a new formalization of measurement, *Measurement*, **27**(2), 71–84 (2000).
30. Fenton, N., Pleeger, S.L. *Software Metrics*, 2nd edition, Pws Pub Co (1996).
31. Sternberg, R.J. Intelligence, in *The Oxford Companion to the Mind*, R. L. Gregory (ed), Oxford University Press (1987).
32. Bridgman, P.W. *The Logic of Modern Physics*, MacMillan (1927).
33. Roberts, F. *Measurement Theory: With Applications to Decision Making, Utility, and the Social Sciences*, Cambridge University Press (1984).
34. Narens, L. *Theories of Meaningfulness*, Routledge (2002).
35. ISO/IEC 25040:2011 Systems and Software Engineering—Systems and Software Quality Requirements and Evaluation (SQuaRE)—Evaluation Process, ISO Geneva (2011). Systems and Software Engineering—Systems and Software Quality Requirements and Evaluation (SQuaRE)—Evaluation Process.
36. Carbone, P., Buglione, L., Mari, L., Petri, D. A comparison between foundations of metrology and software measurement, *IEEE Transactions on Instrumentation and Measurement*, **57** (2), 235–241.
37. Humphrey, W.S. *Managing the Software Process*, Addison–Wesley (1989).
38. Akao, Y. *Quality Function Deployment: Integrating Customer Requirements into Product Design*, Taylor & Francis (2004).
39. Sonin, A.A. *The Physical Basis of Dimensional Analysis* (2001) Available at http://web.mit.edu/2.25/www/pdf/DA_unified.pdf
40. Mari, L., Lazzarotti, V., Manzini, R. Measurement in soft systems: Epistemological framework and a case study, *Measurement*, **42**, 241–253 (2009).
41. Ferrero, A., Salicone, S. Fully comprehensive mathematical approach to the expression of uncertainty in measurement, *IEEE Transactions on Instrumentation and Measurement*, **55** (3) 706–712 (2006).
42. Mauris, G., Berrah, L., Foulloy, L., Haurat, A., Fuzzy handling of measurement errors in instrumentation, *IEEE Transactions on Instrumentation and Measurement*, vol. **49** (1), 89–93 (January 2000).
43. Mauris, G., Lasserre, V., Foulloy, L., Fuzzy modeling of measurement data acquired from physical sensors, *IEEE Transactions on Instrumentation and Measurement*, vol. **49** (6), 1201–1205 (December 2000).
44. Ferrero, A., Salicone, S., The random-fuzzy variables: a new approach to the expression of uncertainty in measurement, *IEEE Transactions on Instrumentation and Measurement*, **53** (5), 1370–1377 (October 2004).
45. ISO 14253–2:2013: Geometrical Product Specifications (GPS)–Inspection by Measurement of Workpieces and Measuring Equipment–Part 2: Guide to the Estimation of Uncertainty in GPS Measurement, in Calibration of Measuring Equipment and in Product Verification, ISO, Geneva (2013).
46. EN ISO 9000:2005, Quality Management Systems. Fundamentals and Vocabulary, ISO Geneva (2005).
47. Heim, A.W. Intelligence: Its assessment, in *The Oxford Companion to the Mind*, R.I. Gregory (ed), Oxford, UK, Oxford University Press (1987), pp. 379–381.

48. Kan, H. *Metrics and Models in Software Quality Engineering*, Reading, MA, Addison–Wesley (2002).
49. Popper, K. *The Logic of Scientific Discovery*, Hutchinson Co. Publisher Ltd. London (1980).
50. JGCM 106:2012, *Evaluation of Measurement Data—The Role of Measurement Uncertainty in Conformity Assessment*, Joint Committee for Guides in Metrology (2012), Available at http://www.bipm.org/utils/common/documents/jcgm/JCGM_106_2012_E.pdf
51. *Evaluation of Measurement Data—The Role of Measurement Uncertainty in Conformity Assessment*, International Organization for Legal Metrology (2012), Available at http://www.oiml.org/en/files/pdf_g/g001-106-e12.pdf.
52. ISO 14253–1:2013 Geometrical Product Specifications (GPS)—Inspection by Measurement of Workpieces and Measuring Equipment—Part 1: Decision Rules for Proving Conformity or Nonconformity with Specifications, ISO, Geneva (2013).
53. *Innovation and Growth Rationale for an Innovation Strategy*—Organization for Economic Co–operation and Development (2007). Available at http://www.oecd.org
54. Sherwin, C.W., Isenson, R.S. Project hindsight. A Defense Department study of the utility of research, *Science*, 156, 1571–1577 (1967).
55. Illinois Institute of Technology. *Technology in Retrospect and Critical Events in Science (The TRACES Study)*, Washington, National Science Foundation (1968).
56. Research Excellence Framework REF (2014), Available at http://www.ref.ac.uk
57. Goldfinch, S., Yamamoto, K. *Prometheus Assessed?: Research Measurement, Peer Review, and Citation Analysis*, Woodhead Publishing Limited (2012).
58. Merton, R.K. The matthew effect in science, *Science*, **159**, 3810 (January 5, 1968).
59. Moed, H.F. *The Multi–Dimensional Research Assessment Matrix*, Seminar Research Evaluation in Practice, National Geographic Society, Washington DC (17 October 2012), Avaliable at www.researchtrends.com
60. Research Quality Framework: Assessing the quality and impact of research in Australia (March 2005). Available at http://www.csu.edu.au/__data/assets/pdf_file/0019/51472/rqf_issuespaper.pdf
61. Garvin, D.A. What does product quality mean?, *Sloan Management Revalue System*, **26**, 25–43 (1984).
62. REF2014. Assessment framework and guidance on submissions (July 2011), Available at http://www.ref.ac.uk/pubs/2011-02/
63. Smyth, R. Exploring the usefulness of a conceptual framework as a research tool: a researcher's reflections, *Issues In Educational Research*, **14** (2), 167–180 (2004), Available at http://www.iier.org.au/iier14/smyth.html.
64. Tidd, J., Trewhella, M.J. Organizational and technological antecedents for knowledge acquisition and learning, *R&D Management*, **27** (4), 359–375 (October 1997).

EXERCISES

7.1 The temperature of a body is measured in different days. On one day, measurement provides 20°C . On a second day the measured temperature is 40°C . Answer the following questions:

1. Identify the general property, the individual property and the measured value in the single temperature measurement.
2. A statement is issued that the temperature on the second day is twice as high as the temperature of the first day. Is this statement meaningful? Explain.
3. A change of scale is performed for the two measurements, that are now expressed using the absolute temperature scale (in K). Then, the ratio of the two values so obtained is calculated. Does this ratio have a metrological meaning? Explain.
4. On different days other two temperatures are collected: 40°C and 50°C . Is it reasonable to say that the difference between the two original temperatures is twice as large as that of the two additional measurement values? Explain.

7.2 Professors at a university want to measure the knowledge acquired by their students in a given class. Accordingly, they ask the students to answer questions on a given subject. The report is then read and a grade expressed as an integer value between 1 and 10 is assigned to the author. Answer the following questions:

1. Is this process a measurement?
2. How would you design this process to increase its inter-subjectivity?
3. How would you evaluate the uncertainty to be associated with the assigned value?
4. Comment on the validity of this process to determine amount of knowledge possessed by the students.

7.3 Wine judging at competitions has strong economic implications and consequences. In this example a panel of four experts judges 30 wines and compiles a ranking of the first five on the basis of pre-set criteria. Same wines may score much differently in different competitions: they may win in one competition and fail to classify in an another one. Also, judges may not show consistent behavior: same wine judged by the same judge in different competitions may not get the same score. Questions:

1. Is this process a measurement?
2. How would you improve the repeatability of judges and repeatability of measurements?
3. How would you design the context to assure intersubjectivity of results?
4. How would you evaluate assessment uncertainty?

7.4 Measurement of service quality may be based on the so-called RATER model. According to this model the "service quality dimensions" are five: reliability, assurance, tangibles, empathy, responsiveness. Each dimensions is then characterized by sub-dimensions.

The model can be used to design a questionnaire to be supplied to customers. For each dimension, several questions are defined in the questionnaire (e.g., under empathy: does the staff have good relationships with the customer?). Customers are then asked to classify the performance of the service provider by answering such questions using an ordinal scale. Obtained grades are represented using integer numbers and may then be composed to provide a single number, indicating the overall service quality.

Answer the following questions:

1. Is this a measurement process?

2. Do you expect repeatable and reproducible measurement results?

3. How could you use results of this assessment to improve the process?

4. Could you evaluate measurement uncertainty?

7.5 Attempts to measure pain tolerances and thresholds were done in the past by inflicting pain inducing mechanisms, such as skin burnings, to volunteers. Think of a process that, starting from the identification of the pain inducing mechanism and taking into account personal tolerances and volunteer emotional state, can be thought of as a pain measurement process.

PART II

APPLICATIONS

CHAPTER 8

SYSTEM IDENTIFICATION

GERD VANDERSTEEN
Dept. ELEC, Vrije Universiteit Brussel, Brussels, Belgium

8.1 INTRODUCTION

System identification is the general term of extracting parameter and model information from gathered data. Such data-driven modeling is widespread in a lot of scientific domains including engineering, natural science, and even econometrics. This chapter focuses on the basics of identification techniques for the instrumentation and measurement society, including the predictive capabilities of the models, the physical interpretation of the parameters, and a quality stamp through the computation of parameter uncertainties and model validation. All parameter estimation techniques, their uncertainty, and the model validation will all be based on statistics.

The aim of this chapter is to give an educative introduction in system identification using a simple example which covers a lot of system identification aspects and positions these aspects in an instrumentation and measurement framework.

The examples are worked out in the same systematic way to stress the different choices in identification. Furthermore, the complexity will gradually increase, each time highlighting an important aspect of identification.

After completing the example, some initial directions are given to state some typical problems and to help the reader in finding some starting literature.

8.2 A FIRST EXAMPLE: THE RESISTIVE DIVIDER

The strength of system identification in instrumentation and measurements will be illustrated using a voltage divider circuit, as shown in Figure 8.1, where a voltage generator, V_r, with a finite output impedance R_g stimulates the series connection of two impedances, R_1 and R_2. The mid-point between the two impedances will be considered as the output voltage V_y, while V_u represents the input voltage.

Modern Measurements: Fundamentals and Applications, First Edition.
Edited by Alessandro Ferrero, Dario Petri, Paolo Carbone, and Marcantonio Catelani.

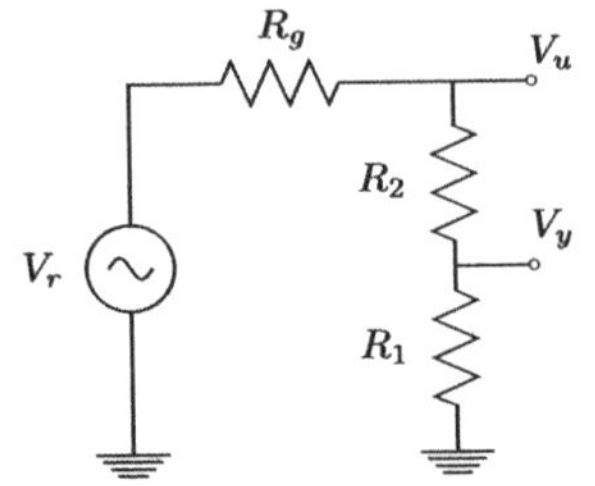

FIGURE 8.1 Schematic representation of the resistive divider.

To simplify the notation in the rest of this chapter, it is chosen to represent the input, output, and reference quantities (V_u, V_y, and V_r) with, respectively, u, y, and r, and to express all signals in Volts. The measured quantities equal $u = u_0 + n_u$ and $y = y_0 + n_y$ where the subscript $_0$ represents the exact (possibly unknown) quantities, and where $n_.$ represent the additive noise.

To start, we can say that the *underlying system* is a resistive divider with two linear resistors, R_1 and R_2, in series. The (exact) input and output voltages, u_0 and y_0 then equal

$$u_0 = \frac{R_1}{R_1 + R_2 + R_g} r = \beta_u r$$

$$y_0 = \frac{R_1 + R_2}{R_1 + R_2 + R_g} r = \beta_y r,$$

which are both described using a linear, static relationship with r. The measurements will be perturbed with measurement errors $u = u_0 + n_u$ and $y = y_0 + n_y$. In this context, it will be assumed that the measurement errors are additive and zero-mean. A non-zero mean deviation occurs when the voltmeter has a constant offset, which should be removed using a calibration procedure. Furthermore, it is assumed that the noise introduced by both voltmeters are uncorrelated.

The *system model* links the output to the input in a linear way: namely $y_0 = \alpha_0 u_0$ with

$$\alpha_0 = \frac{R_1}{R_1 + R_2} = \frac{\beta_u}{\beta_y}.$$

For now, it is assumed that there is no prior knowledge on the model parameter α. It is important to note that the model is linear in the model parameter α, that is, $y = \alpha u$. An alternative system model is the parametrization using R_1 and R_2. The model is then nonlinear in R_1 for a given R_2.

The *excitation* applied to the system, r, is assumed to be non-zero. A zero-valued excitation will result in $r = u_0 = y_0 = 0$, making it impossible to determine α. Within identification, one says that the excitation needs to be persistent.

The *noise model* captures the statistical behavior of the perturbations on the measurements. For now, it is assumed that the additive noise is

- zero-mean Gaussian distributed, $E[n_u] = E[n_y] = 0\,\text{V}$,
- with a constant and known variance σ_u^2 and σ_y^2,
- independent over input and output $\sigma_{uy} = 0\,\text{V}$, and
- independent over the different measured samples.

8.3 A FIRST TRIAL OF ESTIMATORS

To motivate the use of identification techniques, consider some (seemingly straightforward) methods to estimate the parameter α given the measured u and y in the above framework:

Simple averaging:

$$\alpha_{AVG} = \frac{1}{N} \sum_{i=1}^{N} \frac{y(i)}{u(i)}. \tag{8.1}$$

Classical least squares solutions (which assumes that $u(i)$ is measured without noise)

$$\alpha_{LS} = \frac{\sum_{i=1}^{N} y(i)u(i)}{\sum_{i=1}^{N} u^2(i)}. \tag{8.2}$$

Errors-in-variables solution which assumes noise on both input $u(i)$ and output $y(i)$

$$\alpha_{EV} = \frac{\sum_{i=1}^{N} y(i)}{\sum_{i=1}^{N} u(i)}. \tag{8.3}$$

Instrumental variable method which uses the (known) reference voltage $r(i)$

$$\alpha_{IV} = \frac{\sum_{i=1}^{N} y(i)r(i)}{\sum_{i=1}^{N} u(i)r(i)}. \tag{8.4}$$

Consider N measurements of the divider circuit with $R_1 = R_2 = 1\,\text{k}\Omega$, $R_g = 10\,\Omega$. Hence, $\alpha_0 = 0.5$. Furthermore, assume that the generator sweeps its DC value r from 0.25 V to 1.25 V in an equidistant way, and that the additive noise has a standard deviation of $\sigma_u = \sigma_y = 0.1$ V. Figure 8.2 shows the results obtained with the considered estimators for increasing number of samples N.

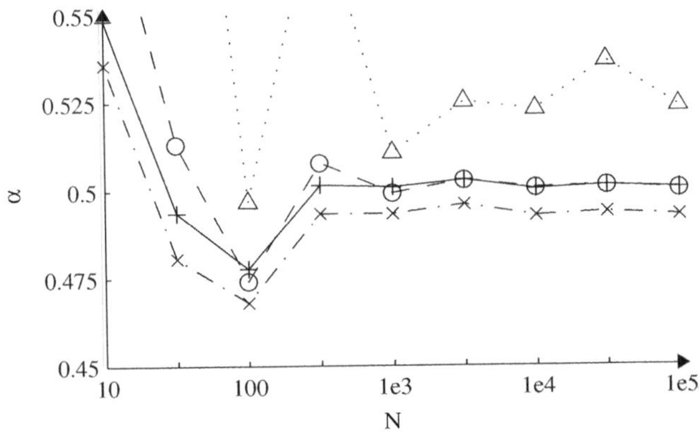

FIGURE 8.2 Comparison of the different estimators for an increasing number of data samples: "△": α_{AVG}; "×": α_{LS}; "o": α_{EV}; "+": α_{IV}.

The straight averaging α_{AVG} fails completely. This is due to the fact that the noise realizations on the input $u(i)$ make that some of the contributions $y(i)/u(i)$ tend to large values.

The least-squares solution α_{LS} results in a systematic underestimate of α. This bias originates from the uncertainty on the input u since it produces an extra term in the denominator of (8.2)

$$E\left[\frac{1}{N}\sum_{i=1}^{N} u^2(i)\right] = \frac{1}{N}\sum_{i=1}^{N} u_0^2(i) + \sigma_u^2.$$

Both the errors-in-variables estimator α_{EV} and the instrumental variable estimators α_{IV} lead to an unbiased estimate. In both cases, one can prove that their numerator and denominator both converge asymptotically to the expected value.

The instrumental variable estimators α_{IV} uses the known reference signal $r(i)$ to transform the errors-in-variables problem into two output error problems with $r(i)$ the (noiseless) input and $u(i)$, $y(i)$ the (noisy) outputs, that is, $\alpha_{IV} = \beta_{y,LS}/\beta_{u,LS}$ with

$$\beta_{u,LS} = \frac{\sum_{i=1}^{N} u(i)r(i)}{\sum_{i=1}^{N} r^2(i)}$$

$$\beta_{y,LS} = \frac{\sum_{i=1}^{N} y(i)r(i)}{\sum_{i=1}^{N} r^2(i)}.$$

8.4 FROM TRIAL-AND-ERROR TO A GENERAL FRAMEWORK

The previous section clearly shows that a systematic approach is necessary for two reasons: first, one can not take it for granted that all estimators provide correct (e.g.

asymptotically unbiased) estimates; second, no parameter uncertainty or model validation is available. This section will use the maximum-likelihood (ML) framework to show that

- the errors-in-variables estimator (8.3) is indeed an ML estimator;
- the parameter uncertainty can be estimated in a simple way;
- the model validation can be done using a statistical analysis of the (unmodeled) residuals;
- the assumption that the noise model is known (i.e., σ_u and σ_y are known) can be relaxed.

This section uses the same resistive divider example as in the previous section with $N = 100$. Three variations on the divider example are considered to demonstrate the identification capabilities:

Case 1: No modeling errors: No errors in the system model. The noise model, with $\sigma_u = \sigma_y = 0.02\,\text{V}$, is known.

Case 2: System model errors: The only difference with case 1 is that there is a constant offset of 0.05 V on the input variable u.

Case 3: Noise model errors: The only difference with case 1 is that the data are generated with the noise model $\sigma_u = 0.05\,\text{V}$ and $\sigma_y = 0\,\text{V}$, while the estimator has the (incorrect) noise model with $\sigma_u = \sigma_y = 0.02\,\text{V}$.

The probability density functions (pdf) are in all cases extracted using a Monte Carlo simulation with 10^6 independent experiments.

8.4.1 Setting up the Estimator

Since both input u and output y are perturbed with additive Gaussian noise, it is a natural choice to determine the ML estimator in an Errors-in-Variables framework [1]. The ML estimator is (for the given setting) obtained by minimizing the following ML cost function

$$V_{ML}(\alpha) = \sum_{i=1}^{N} \frac{(\alpha u(i) - y(i))^2}{\alpha^2 \sigma_u^2 + \sigma_y^2} \tag{8.5}$$

with respect to the parameter α:

$$\alpha_{ML} = \arg\min_{\alpha} V_{ML}(\alpha). \tag{8.6}$$

Note that the minimizer of (8.5) can be computed explicitly and equals (8.3).

The cost (8.5) can also be written in matrix notation as a least squares problem

$$V_{ML}(\alpha) = e^t(\alpha)e(\alpha) \tag{8.7}$$

with e^t the vector transpose and $e(\alpha)$ the error column vector where the i-th element equals

$$e_i(\alpha) = \frac{\alpha u(i) - y(i)}{\sqrt{\alpha^2 \sigma_u^2 + \sigma_y^2}}. \tag{8.8}$$

Note that elements of the error vector evaluated in the exact parameters $e_i(\alpha_0)$ are independent over i and distributed according to a zero-mean normal distribution with a unity variance, $N(0, 1)$.

Finding the minimum of the cost leads to a least-squares optimization where the error vector depends in a nonlinear way on the model parameters. Classical nonlinear least squares optimization methods can be used to determine the minimum of this cost [2, 3]. These optimization methods are in general iterative methods. When the error vector is linear in the parameters, then the global linear least squares minimum can be determined non-iteratively, without the need of starting values. If the error vector is linear in a subset of the parameters, then it is possible to lower the computational complexity by eliminating these linear parameters from the cost function [4].

8.4.2 Uncertainty on the Estimates

Computing the uncertainty can be done through the Jacobian matrix, which equals the first derivative of the error vector with respect to the parameters

$$J_i(\alpha) = \frac{\partial e_i(\alpha)}{\partial \alpha}. \tag{8.9}$$

The general framework provides an efficient way to compute the (co)variance matrix on the estimates, namely

$$\sigma_\alpha^2 = [J^t(\alpha) J(\alpha)]^{-1}. \tag{8.10}$$

This only requires the computation of the error vector and its Jacobian. Figure 8.3 illustrates the excellent agreement between the estimated α_{ML} from a Monte Carlo simulation and the normal distribution predicted by the theory.

If the user is interested in R_1 for a given R_2, that is,

$$R_1 = \frac{\alpha R_2}{1 - \alpha} = f(\alpha), \tag{8.11}$$

then it is possible to use of the propagation of uncertainty to determine $\sigma_{R_1}^2$ from σ_α^2 using

$$\sigma_{R_1}^2 = \sigma_\alpha^2 \left| \frac{\partial f(\alpha)}{\partial \alpha} \right|^2. \tag{8.12}$$

This is demonstrated nicely in Figure 8.4.

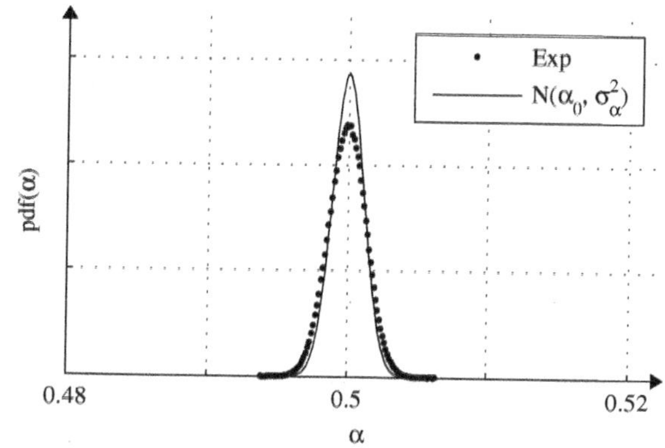

FIGURE 8.3 The pdf of α_{ML} in the absence of modeling errors (case 1) determined using a Monte Carlo run with 10^6 samples (•), compared to the normal distributions $N(\alpha_{ML}, \sigma_\alpha^2)$ (–).

8.4.3 Model Validation

The aim of model validation is to ensure the use of the correct system model structure and order, and the correct noise model. This is important in order avoid both under- and over modeling.

Model validation can be done on the data used during estimation, that is, on the so-called estimation dataset. Model validation on the estimation dataset studies the statistical properties of the residuals after identification. This can be done using the (known) statistics of

- the cost function (cost function validation) and/or
- the cross-correlation test of the different residuals (whiteness test of the residuals).

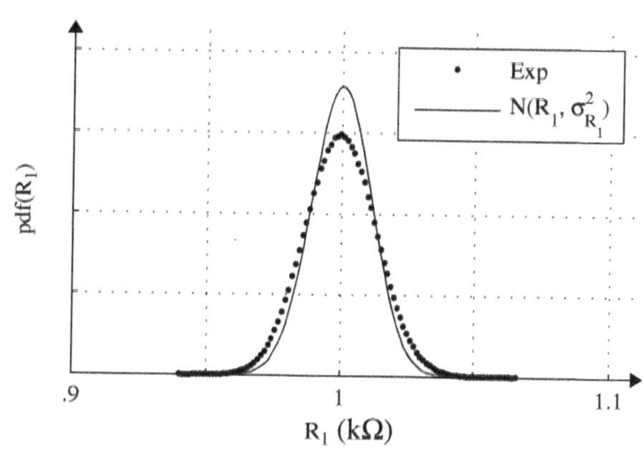

FIGURE 8.4 The pdf of R_1 in the absence of modeling errors (case 1) determined using a Monte Carlo run with 10^6 samples (•), compared to the normal distributions $N(1\ \mathrm{k}\Omega, \sigma_{R_1}^2)$ (–).

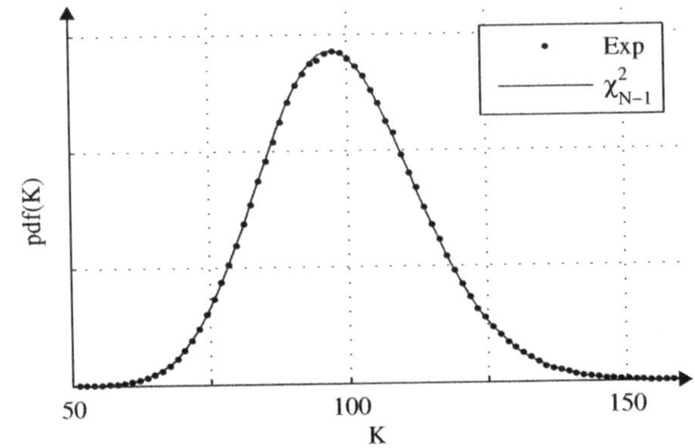

FIGURE 8.5 The pdf of the maximum likelihood cost V in the absence of modeling errors (case 1), compared to the χ^2 distribution with $N-1=99$ degrees of freedom (–).

Model validation can also be performed on data that was not used during estimation, that is, on the so-called validation dataset. The use of this unseen data has a major advantage when it comes to model validation. If the model has too much flexibility (over-modeling), then the estimator might use this flexibility to model the measured noise realization. This can easily be detected using a validation dataset since the (independent) noise realization of the validation dataset cannot be explained by the over-fitted model.

8.4.3.1 *Cost Function Validation* The model validation can be done through the (unmodeled) residuals by evaluating the error vector in the ML estimate $e(\alpha_{ML})$ [5, 6]. If the data from the underlying system match the system model, and if the noise model is correct (case 1), then the residuals will be uncorrelated (whiteness test). In addition, it can be shown that the cost evaluated in the estimates, $K(\alpha_{ML})$, has a χ^2_{N-1} distribution with $N-1$ degrees of freedom (i.e., the number of samples minus the number of free parameters). Figure 8.5 shows the excellent agreement of the theoretical χ^2_{99} distribution and the probability density function of the cost $V(\alpha_{ML})$ in the absence of modeling errors (case 1). Hence, $E[V(\alpha_{ML})] = N-1 = 99$ and $\sigma^2_{\alpha_{ML}} = 2(N-1)$.

If the system model and/or the noise model are incorrect, then the ML cost will no longer converge toward the predicted χ^2 distribution. This can be seen in Figure 8.6 and in the case of system model errors (case 2). Similar results can be found in the case of noise model errors (case 3). The theoretical mean and standard deviation of the ML cost, and the simulation results using 10^6 independent samples are shown in Table 8.1.

8.4.3.2 *Whiteness Test on the Residuals* The residuals on the estimates, e_i, as a function of the reference value, r_i, can be used to study the presence of model errors. Figure 8.7 shows the residual errors in the presence of a system model error, and clearly shows that there is an unmodeled trend in the residuals.

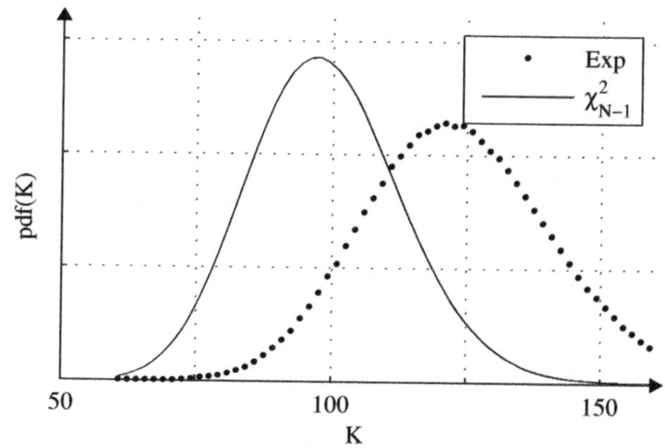

FIGURE 8.6 The pdf of the maximum likelihood cost V in the presence of modeling errors (case 2), compared to the χ^2 distribution with $N - 1 = 99$ degrees of freedom (–).

TABLE 8.1 Statistics of the cost function for the different cases

	$E[K]$	σ_K^2	$E[\alpha_{ML}]$	σ_α
Theoretical value	99.0	198.0	0.5	0.0028
No model error (case 1)	99.2	198.7	0.5000	0.0028
System model errors (case 2)	116.2	276.2	0.4686	0.0026
Noise model errors (case 3)	123.9	311.9	0.5000	0.0028

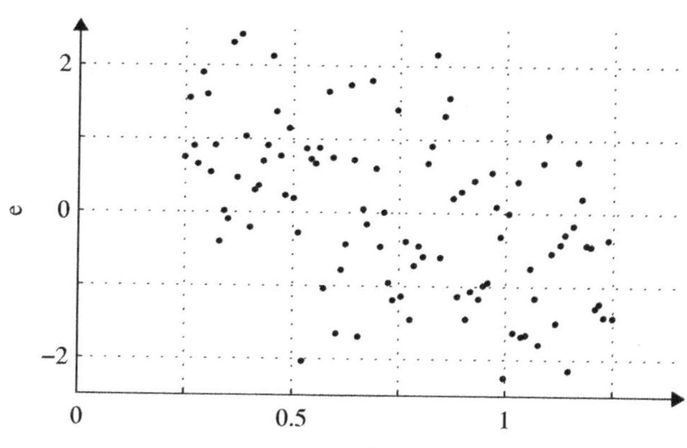

FIGURE 8.7 The residuals as function of the reference r in the presence of modeling errors (case 2).

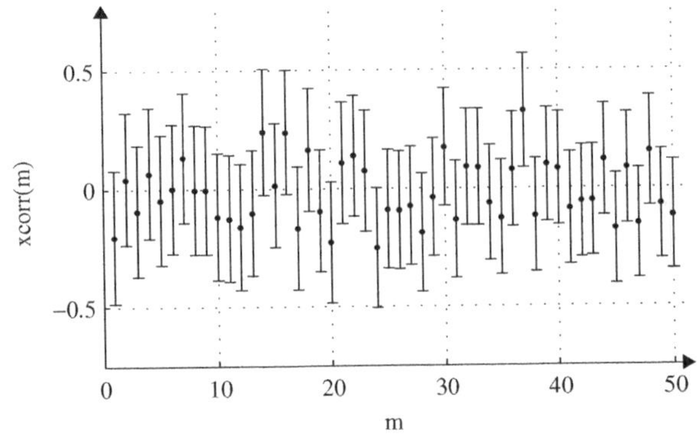

FIGURE 8.8 The $xcorr(m)$ together with its 2σ bounds when no modeling errors are present (case 1).

The systematic trend in the residuals can be quantified in a statistical way using for example, a whiteness test of the residuals. In this case, we first ordered the different experiments for increasing r_i. A whiteness test of the residuals then uses the cross-correlation of the residuals e_i to determine whether these residuals are uncorrelated over the index i

$$xcorr(m) = \frac{1}{N-m} \sum_{i=1}^{N-m} e_i e_{i+m}. \tag{8.13}$$

Note that $xcorr(0)$ corresponds to the cost function divided by N, whose statistics are already studied during the cost function validation.

If the residuals are zero-mean, unity-variant, normally distributed, $N(0, 1)$, and independent over i, then the variance of (8.13) equals $1/(N - m)$. This enables the user to determine uncertainty bounds on these correlations and to perform a hypothesis test for example, by determining if more than 5% of the cross-correlated values of (8.13) are outside the 95% interval of $\pm 2\sigma$. Figure 8.8 shows the whiteness test in the absence of modeling errors (case 1). Figure 8.9 demonstrates the use of the whiteness test when modeling errors are present (case 2).

8.4.4 Extracting the Noise Model

The previous sections assumed that the (additive) noise model is known. In practice, one needs to extract the noise model from measurements. This can be done in different ways.

Separate noise measurements are performed to measure the noise model in the absence of an excitation. This demands an extra measurement (and hence measurement time) and potentially leads to a noise model which is different from the noise model in the presence of the excitation. This difference can be induced by changing the range/operating point of the measurement equipment between the measurements with and without excitation.

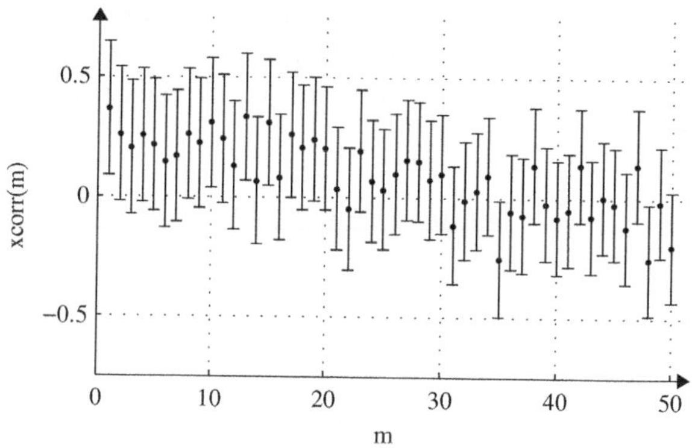

FIGURE 8.9 The *xcorr(m)* together with its 2σ bounds when if system model errors are present (case 2). Note (1) the trend in the cross-correlation, and (2) more that 5% of the values are outside the 2σ bounds.

Statistical analysis over repeated experiments applies the same excitation (reference) signal M times and enables the extraction of a non-parametric noise model using a statistical analysis over the different (independent) realizations [1, 7]. The (co)variances in the estimators are then replaced by these sample (co)variances. The main advantage of this method is that the identification techniques can easily be adapted to this sample (co)variance if the number of samples is sufficiently large ($M \geq 7$). The only drawback is that it must be possible to perform repeated measurements.

Estimating a parametric noise model together with the plant model does not require repeated measurements and is often used within the control community. The main drawback of this method is the need of a double model selection problem: both the model structure for the system and the noise model are required.

Estimating a non-parametric noise model from non-repeated measurements became recently possible using the so-called Local Polynomial (LP) method. This LP method estimates a local (smooth) model and extracts a noise model in this local domain. This method has the advantage that a noise model selection is avoided. In a latter step, this non-parametric noise model can be used within an identification framework.

Within the example of the resistive divider, consider the analysis over repeated experiments. For each reference, $r(i)$, M independent realizations of the input and the output are measured. The sample mean

$$\hat{x} = \frac{1}{M} \sum_{k=1}^{M} x_k$$

and the sample variance (co)variance

$$\hat{\sigma}^2_{\hat{x}\hat{y}} = \frac{1}{M-1} \sum_{k=1}^{M} \left(x_k - \hat{x}\right) \left(y_k - \hat{y}\right)$$

can then be used to replace the measured inputs and output, and the (co)variance matrix of the noise model. This estimator is known as the sample maximum likelihood (SML) estimator [1, 7], since it uses the sample covariance information as input. Hence,

$$V_{SML}(\alpha) = \hat{e}^t(\alpha)\hat{e}(\alpha) \tag{8.14}$$

with

$$\hat{e}_i(\alpha) = \frac{\alpha\hat{u}(i) - \hat{y}(i)}{\sqrt{\alpha^2\hat{\sigma}_{\hat{u}}^2 + \hat{\sigma}_{\hat{y}}^2}}. \tag{8.15}$$

All the previously discussed techniques and properties (estimation of the uncertainty, interpretation of the cost function, whiteness test of the residuals) are still available for the SML estimator if $M \geq 7$. Only some scaling factors, which depend on M, need to be taken into account. In the absence of modeling errors, it can be shown that

- the efficiency loss of the SML with respect to the ML equals

$$\sigma^2_{\alpha_{SML}} = \frac{M-2}{M-3}\sigma^2_{\alpha_{ML}} \tag{8.16}$$

  and hence is limited to 12% or less on the standard deviation for $M \geq 7$.
- the uncertainty on the estimates can be computed using

$$\sigma^2_{\alpha_{SML}} = \frac{M-1}{M-3}\left[\hat{J}^t(\alpha_{SML})\hat{J}(\alpha_{SML})\right]^{-1} \tag{8.17}$$

  with

$$\hat{J}_i(\alpha) = \frac{\partial\hat{e}_i(\alpha)}{\partial\alpha}. \tag{8.18}$$

- The SML cost converges asymptotically toward a distribution with mean value

$$E[V_{SML}] = \frac{M-1}{M-2}(N-1) \tag{8.19}$$

  and variance

$$\sigma^2_{V_{SML}} = \frac{(M-1)^3}{(M-2)^2(M-3)}\sigma^2_{V_{ML}}. \tag{8.20}$$

  This distribution can be used for model validation.

8.5 PRACTICAL IDENTIFICATION FRAMEWORK FOR INSTRUMENTATION AND MEASUREMENTS

The aim of this section is to guide the reader to standard literature and published examples. It is, by no means, the intention to provide a complete overview of identification in instrumentation and measurement, but to provide guidance to specific identification (sub)problems.

General introductions to identification can easily be found in the literature. The work of Ljung [8] starts from a control point of view and describes both theoretical and practical applications of (mainly) time-domain identification in a prediction error framework (PEM). The work of Pintelon [1] is more focused on (mainly) frequency-domain identification techniques within an errors-in-variable framework. Complementary to [1], there exists an exercise oriented book [5] that provides an excellent exercise-based introduction into identification.

This section contains a short overview of identification in instrumentation and measurements concerning dynamic linear time-invariant (LTI) Systems, nonlinear systems, sine fitting, and calibration/compensation techniques.

8.5.1 Dynamic Linear Time-Invariant (LTI) Systems

The measurement and modeling of linear dynamic systems has a lot of practical applications and is well understood. The main advantage of linear time-invariant systems is that they can be represented (exactly) either in the time domain (using the impulse response), or in the frequency domain (using the Frequency Response Function, or FRF) [1, 8].

The measurement of the Frequency Response Function (FRF) using a network analyzer often starts with sampling both the input and the output signal in the time domain. If the input signal is periodic (with periodicity T), it is known that the response of an LTI system is also a periodic signal (with the same periodicity T). The amplitude and phase changes introduced by the LTI system depend on the applied frequency.

8.5.1.1 *From Single-Sine to Multisine Measurements* A first way to measure the FRF is by applying the individual sinewaves of the frequency under test and to measure the response after the transients died out. However, the user then losses a lot of measurement time, proportionally with the number of frequency lines. This was one of the main motivations to introduce periodic excitation signals that contain various frequency components. Since such a periodic signal contains various sinewaves with common base frequency $f_0 = 1/T$, it is possible to write these signals as

$$u(t) = \sum_{k=1}^{F} A_k \sin(2\pi k f_0 t + \phi_k) \tag{8.21}$$

The main advantage is that it is then possible to measure the FRF at F frequencies, while one only needs to wait for 1 transient die out. Note that, if all phases are equal to zero ($\phi_k = 0$) the signal will have large peak values. Therefore, it is important to choose the phases ϕ_k appropriately to reduce the so-called crest factor (CF) of the signal (also known as peak-to-average-power ratio):

$$CF = \frac{peak(u(t))}{RMS(u(t))} \tag{8.22}$$

8.5.1.2 Different Kind of Multisine Excitations exist since they enable the separation of the steady-state behavior of the system, the transient response of the system, and the additive noise when measuring different successive periods of the same multisine excitation. In addition, some multisine excitations enable to determine the level of nonlinear distortion and even to separate the distortion in its even and odd-order nonlinear components.

Different periodic excitation signals are possible depending on the number of sines F, the amplitudes A_k and the phases ϕ_k [1]:

Pseudo Random Binary Signals (PRBSs) are popular binary signals which can easily be generated using simple hardware. Their auto-correlation function is "dirac-like". Although a PRBS looks like a noisy signal (when considering the auto-correlation function), it is important to note that a PRBS is in essence a periodic signal. Hence, the correct choice of the clock rate and the length of the PRBS makes it possible to generate a large number of sinewaves in the appropriate frequency band.

Log-tone multisines are often used when measuring an FRF over several decades in frequency. Log-multisines have the advantage that the number of frequencies for each decade remains (approximately) the same, and hence better distribute the power over the different frequency bands. Full multisines will overstress the higher frequency bands in a parametric modeling phase since the higher frequency bands contain more spectral lines.

Schroeder multisines are a first step to reduce the peak value of the multisine in a systematic way by choosing $\phi_k = -k(k-1)\pi/F$ for a constant A_k. The resulting crest factor is typically 1.7. The advantage of this signal is that it can easily be generated with a (relatively) small crest factor. However, the fact that ϕ_k is not random over k makes it less suited for nonlinear detection.

Crest factor optimized multisines optimize the signal in order to have a minimal crest factor. This is important in the case where one needs to restrict the input (or the output) range to avoid nonlinear distortion (e.g., saturation) or to stay within operational limits (e.g., the maximal applied signal imposed by legal regulation).

(Full) Random phase multisine excite all harmonics $1 \ldots F$, but the phases ϕ_k used are chosen randomly from a uniform distribution over $[0, 2\pi)$ and this independently over k. In most cases, a flat power spectrum is used (i.e., A_k is constant). Random phase multisines have the particular property that the probability density function (pdf) of $u(t)$ approaches a Gaussian distribution. Full random phase multisines also make it possible to determine the level of non-linear distortion using multiple experiments with different phase realizations. However, they are unable to separate even and odd nonlinear distortions.

Odd random phase multisines only excite the odd frequency bins, hence $A_{2k} = 0$. This excitation signal has two main advantages over a full multisine. First, it enables the detection of the even non-linear distortions at the even frequency bins, just by looking at the power spectra. Second, the even non-linear distortions are not falling on top of the excited frequency bins and hence are not perturbing the FRF measurement. The only drawback is that the frequency resolution of the FRF is halved.

Random odd random phase multisines take a set of randomly selected excited bins from an odd random phase multisine and put their input amplitude to zero. The main advantage of these detection lines at odd frequency bins is the immediate detection of the level of odd nonlinear distortion at these odd frequency bins. The generation of such spectral pure excitation puts, however, a stringent constraint on the signal generator used.

8.5.1.3 Transient and Leakage are considered as two main issues when using DFTs or FFTs to convert the time domain samples into the frequency domain. There exists a wide range of methods to solve this problem.

Classical windowing techniques multiply the time domain data with a (predefined) window prior to performing an FFT. This reduces the spectral leakage (introduced, for example, by the transient of the system or the noise) at the cost of widening the spectral line. A Hann window will, for example, spread a sinewave at a single bin over three bins.

(Cyclic) Overlap processes different (overlapping) subrecords of a data sequence using classical windowing techniques, followed by a statistical analysis of the different subrecords. This makes it possible to determine both the spectra, the FRFs, and their uncertainty on these values [9, 10].

Local Polynomial methods separate the (linear) steady state FRF, the transient, and the nonlinear distortions from the additive noise [9, 11, 12]. This is done using a least-squares estimation of a frequency-domain polynomial model in a small frequency band for both the transfer function of the system and the transient response. The main advantages of this methods are (1) their high performance with respect to transient suppression, (2) their ability to detect the nonlinear distortion, and (3) their use in parametric identification. The output of the methods (i.e., nonparametric FRF and its uncertainty) can be used directly within frequency domain identification techniques to extract a parametric model. The main constraint on this method is that the input spectra must be non-smooth in the spectral domain. This enables the use of noise excitations and random multisines, but rules out the use of, for example, step responses.

8.5.2 From Linear to Nonlinear Systems

The class of nonlinear systems is defined as all systems that don't have a linear behavior. Hence, it is not surprising that there are a huge number of possible model structures and applications.

8.5.2.1 Volterra Theory extends the concepts of linear dynamic systems toward weakly nonlinear dynamic systems [13,14]. Volterra models consider that the classical (linear) impulse response is the first-order term in the Volterra series development. The higher order terms of the series development around the operating point are then the higher-order Volterra kernels. If the nonlinear system is static, then the Volterra series boils down to a Taylor series expansion.

The main advantage of the Volterra theory is that it provides a theoretical framework to compute the Volterra series of composite (nonlinear) system (cascading, feedback systems). Volterra theory also allows the computation of nonlinear pre- or post-compensation of nonlinear distortion. The main issue with Volterra theory is that it only works for weakly nonlinear systems: Volterra series are series expansions that, like other series expansions, have a limited convergence range.

The higher order kernels are in general multidimensional, time-dependent functions. The increase of the number of dimensions and the combinatorial explosion of the number of parameters to measure makes it hard to identify the different kernels. This explains why various special purpose excitations are proposed to determine the Volterra kernels from measurements [15, 16].

8.5.2.2 Block-Oriented Models describe the system as an interconnection of dynamic LTI systems in combination with static nonlinear functions. Classical examples are Wiener and Hammerstein systems, or combinations of these systems. The block-oriented models have the advantage that they do not suffer from convergence problems like in the Volterra case. However, determining/breaking up the different contributions of the different sub-systems using input and output data only is not an easy mission [17, 18].

8.5.2.3 Best-Linear-Approximation (BLA) models are recently introduced to characterize nonlinear systems with a best (in least squares sense) approximation of the nonlinear dynamic system by an LTI system, together with an additive output noise source. This additive noise source represents both the additive measurement noise, and the nonlinear distortions that are not captured by the (best) linear dynamic system.

The BLA assumes that the input is excited with Gaussian-like signals. Since random phase multisines also belong to the class of Gaussian signals, it is possible to use them to extract the BLA.

8.5.3 Sine Fitting

The measurement/modeling of sinewave(s) has various applications ranging from sinewave testing for digitizers [19], the measurement of oscillators, toward the monitoring of an AC power grid.

Consider the simplest case of a single sinewave first

$$y(t) = A\sin(2\pi ft + \phi) + n(t). \tag{8.23}$$

If $n(t)$ is independent over t, and if the frequency f is known, then it is more opportune to write this sine-fitting problem as

$$y(t) = A\sin(2\pi ft) + B\cos(2\pi ft) + n(t) \tag{8.24}$$

since this leads to a linear least-squares problem in A and B. The main advantage of rewriting the problem to become linear in the parameters is that all the linear parameters can be eliminated explicitly [4,20]. Hence, no starting values are required for these linear parameters. It can be concluded that the model structure (8.24) is a better representation when it comes to identification in a least-squares sense.

8.5.3.1 *Multi Harmonic Sine Fitting* considers a sum of harmonically related sinewaves

$$y(t) = A_0 + \sum_{k=1}^{K}[A_k\sin(2\pi kft) + B_k\cos(2\pi kft)] + n(t). \tag{8.25}$$

References 4 and 20 develop a Maximum Likelihood estimators using

- the elimination of the amplitude variables A_k and B_k in the nonlinear optimization process,
- the cost-function interpretation for the model validation, and
- the model order selection problem to select the number of harmonics, K.

The example nicely illustrates that, although the initial problem has $2K + 2$ unknowns (A_0, A_k, B_k, and f), the final nonlinear optimization problem only needs to be solved with respect to a single (nonlinear) parameter f.

8.5.3.2 *Time-Base Distortion* during high speed sinewave measurements introduce uncertainty on the time instance of the nth measured sample, t_n

$$y_n = A_0 + \sum_{k=1}^{K}[A_k\sin(2\pi kft_n) + B_k\cos(2\pi kft_n)] + n(t_n). \tag{8.26}$$

In this case, one needs to identify not only the properties of the sinewaves, but also the (unknown) time instances t_n. Different strategies can be used, depending on whether the user assumes that t_n is represented as a function of n (e.g., a sum of harmonic functions [21,22]) or if each t_n is an independent value (also known as the nonparametric modeling of the timebase [23]). Referred works use a similar strategy as followed for the resistive divider: it starts with the system model and noise model assumptions, sets up the ML cost to be minimized, extracts the uncertainty on the parameters, and validates/selects the model using the statistical properties of the residuals and the cost function.

8.5.3.3 ***Noise Characterization*** is an application of identification techniques onto stochastic processes. An example is the measurement of the time-jitter which is present during the digitization of high speed signals [24]. The use of a system identification approach enables the use of ML-based techniques and to extract not only the jitter parameters, but also to determine the uncertainty on the estimates and to validate the model [25, 26].

8.5.4 Calibration and Compensation Techniques

The proposed identification techniques are applicable not only to the extraction of user-requested quantities, but also to calibration and compensation of the (imperfect) measurement equipment.

8.5.4.1 ***Spectral-Clear Power Spectra*** are demanded in various measurement applications. This is particularly important if the spectral purity of the signal is used to measure, for example, the nonlinear distortion introduced by the device [27, 28]. The presence (or absence) of a certain frequency component can be either measured using the spectral analysis of a time domain waveform, or using a spectrum analyzer. In all cases, it is important that the user can determine whether or not the measured spectral line is (or is not) above the detectable noise level. Statistics-based methods once more give the tools to do so [28].

8.5.4.2 ***Identification-Based Calibration Techniques*** enable two improvements with respect to classical calibration techniques. First, the uncertainties on the determined error model (and its coefficients) make it possible to know not only the value, but also the quality of the measurements [29]. Large uncertainty bounds can indicate that the considered set of standards is not informative enough and that the user should look to additional standards. Second, the modeling of the calibration standards enable novel calibration techniques that are otherwise unfeasible from a practical point of view [30].

8.6 CONCLUSIONS

This chapter essentially has two contributions. The first contribution is a high-level introduction to identification, illustrated using the basic measurement example: a resistive divider. Even though this example is extremely simple, namely a static and linear system, it nicely illustrates all the steps and advantages of an identification approach.

The second contribution of this chapter is a set of guidelines and references that illustrate the introduced identification techniques into instrumentation and measurement problems.

It is the author's hope that this chapter convinced the reader that identification methods are valuable within the instrumentation and measurement context and that he/she is able to apply/develop identification methods for their problems at hand.

FURTHER READING

The papers and books that to be considered for future reading depends on the interest of the user and the application at hand. Most of the future reading is integrated within different subsections and can be summarized as follows:

General introduction to identification methods can easily be studied reading Reference 1 with the focus on frequency domain methods, and Reference 8 with the focus on time-domain methods. The former book is accompanied with an exercise book [5]. The latter book focusses more on control-related modeling issues.

General optimization methods are described in References 2 and 3, while the mathematical explanation of why and how to eliminate the linear parameters in the error vector of the least-squares problem is explained in [4].

FRF measurements of linear time-invariant systems are described in chapter 2 of Reference 1. This chapter also includes the impact of the choice of the excitation signals. More advanced, state-of-the-art methods can be found in References 9, 11, and 12.

Volterra theory for nonlinear systems is nicely introduced in Reference 13.

Sinewave fitting, including model selection in an identification framework, can easily be understood when studying Reference 20.

REFERENCES

1. Pintelon, R., Schoukens, J. *System Identification: A Frequency Domain Approach,* 2nd Edition, Wiley-IEEE Press (2012).
2. Fletcher, R. *Practical Methods of Optimization, Volume 1: Unconstrained Optimization*, John Wiley & Sons, New York (1980).
3. Fletcher, R. *Practical Methods of Optimization, Volume 2: Constrained Optimization*, John Wiley & Sons, New York (1980).
4. Cadzow, J. Signal processing via least squares error modeling, *IEEE ASSP Magazine*, **7**, 12–31 (October 1990).
5. Schoukens, J., Pintelon, R., Rolain, Y. *Mastering System Identification in 100 Exercises*, Wiley-IEEE Press (2012).
6. Pintelon, R., Schoukens, J., Vandersteen, G. Model selection through a statistical analysis of the global minimum of a weighted nonlinear least squares cost function, *IEEE Transactions on Signal Processing*, **45** (3), 686–693 (February 1997).

7. Schoukens, J., Pintelon, R., Vandersteen, G., Guillaume, P. Frequency-domain system identification using non-parametric noise models estimated from a small number of data sets, *Automatica*, **33** (6), 1073–1086 (June 1997).
8. Ljung, L. *System Identification (2nd ed.): Theory for the User*, Prentice Hall PTR, Upper Saddle River, NJ, USA (1999).
9. Pintelon, R., Barbe, K., Vandersteen, G., Schoukens, J. Improved (non)parametric identification of dynamic systems excited by periodic signals, *Mechanical Systems and Signal Processing*, **25** (7), 2683–2704 (October 2011).
10. Barbe, K., Schoukens, J., Pintelon, R. The use of nonparametric noise models extracted from overlapping subrecords for system identification, *IEEE Transactions on Signal Processing*, **59** (10), 4635–4647 (October 2011).
11. Pintelon, R., Schoukens, J., Vandersteen, G., Barbe, K. Estimation of nonparametric noise and FRF models for multivariable systems – part I: theory, *Mechanical Systems and Signal Processing*, **24** (3), 573–595 (April 2010).
12. Pintelon, R., Schoukens, J., Vandersteen, G., Barbe, K. Estimation of nonparametric noise and FRF models for multivariable systems-part II: extensions, applications, *Mechanical Systems and Signal Processing*, **24** (3), 596–616 (April 2010).
13. Schetzen, M. *The Volterra and Wiener Theories of Nonlinear Systems*, John Wiley & Sons, New York (1980).
14. Wambacq, P., Sansen, W. *Distortion Analysis of Analog Integrated Circuits*, The Springer International Series in Engineering and Computer Science. Springer (1998). Available at http://books.google.be/books?id=aOyxwdroH2cC
15. Evans, C., Rees, D., Jones, L., Michael, W. Periodic signals for measuring nonlinear Volterra kernels, *IEEE Transactions on Instrumentation and Measurement*, **45** (2), 362–371 (April 1996).
16. Boyd,S., Tang, Y., Chua, L. Measuring Volterra kernels, *IEEE Transactions on Circuits and Systems*, **30** (8), 571–577 (1983).
17. Schoukens, J., Nemeth, J., Crama, P., Rolain, Y., Pintelon, R. Fast approximate identification of nonlinear systems, *Automatica*, **39** (7), 1267–1274 (July 2003).
18. Schoukens, M., Rolain, Y. Parametric identification of parallel wiener systems, *IEEE Transactions on Instrumentation and Measurement*, **61** (10), 2825–2832 (October 2012).
19. I. S. 1057-1994. *IEEE Standard for Digitizing Waveform Recorders*, Std. (1994).
20. Pintelon, R., Schoukens, J. An improved sine-wave fitting procedure for characterizing data acquisition channels, *IEEE Transactions on Instrumentation and Measurement*, **45** (2) 588–593 (April 1996).
21. Schoukens, J., Pintelon, R., Vandersteen, G. A sinewave fitting procedure for characterizing data acquisition channels in the presence of time base distortion and time jitter, *IEEE Transactions on Instrumentation and Measurement*, **46** (4), 1005–1010 (August 1997).
22. De Ridder, F., Pintelon, R., Schoukens, J., Verheyden, A. Reduction of the gibbs phenomenon applied on nonharmonic time base distortions, *IEEE Transactions on Instrumentation and Measurement*, **54** (3), 1118–1125 (June 2005).
23. Vandersteen, G., Rolain, y., Schoukens, J. An identification technique for data acquisition characterization in the presence of nonlinear distortions and time base distortions, *IEEE Transactions on Instrumentation and Measurement*, **50** (5), 1355–1363 (October 2001).

24. Souders, T., Flach, D., Hagwood, C., Yang, G. The effects of timing jitter in sampling systems, *IEEE Transactions on Instrumentation and Measurement*, **39** (1), 80 –85 (February 1990).

25. Vandersteen, G., Pintelon, R. Maximum likelihood estimator for jitter noise models, *IEEE Transactions on Instrumentation and Measurement*, **49** (6), 1282–1284 (December 2000).

26. Verbeyst, F., Rolain, Y., Schoukens, J., Pintelon, R. System identification approach applied to jitter estimation, 23rd IEEE Instrumentation and Measurement Technology Conference, Sorrento, Italy (April 24–27, 2006), pp. 1752–1757.

27. Rabijns, D., Van Moer, W., Vandersteen, G. Spectrally pure excitation signals: only a dream? *IEEE Transactions on Instrumentation and Measurement*, **53** (5), 1433–1440 (October 2004).

28. Rabijns, D., Vandersteen, G., Van Moer, W. An automatic detection scheme for periodic signals based on spectrum analyzer measurements, *IEEE Transactions on Instrumentation and Measurement*, **53** (3), 847–853 (June 2004).

29. Van hamme, H., Vanden Bossche, M. Flexible vector network analyzer calibration with accuray bounds using an 8-term or a 16-term error-correction model, *IEEE Transactions on Microwave Theory and Techniques*, **42** (6), 976–987 (June 1994).

30. Gommé, L., Schoukens, J., Rolain, Y., Van Moer, W. On the use of a crystal detector for a phase calibration of the large signal network analyzer, *Measurement Science & Technology*, **19** (8) (August 2008).

EXERCISES

8.1 Section 8.2 showed that estimators (8.2)–(8.4) have different asymptotic properties when the number of samples tends to infinity. Assume that both input and output are perturbed with additive, zero-mean Gaussian distributed noise as assumed in Section 8.1. Determine the asymptotic value for the number of samples tending to infinity ($N \to \infty$) for the least-squares and the errors-in-variables estimator, respectively (8.2) and (8.3), using the fact that the mean value converges asymptotically toward its expected value.

8.2 Section 8.3 builds up a simple estimator from scratch. Implement all the steps proposed in this section and verify the simulation results with the graphs presented.

8.3 Perform a Monte Carlo analysis of estimators (8.2)–(8.3) and verify the asymptotic results with the results from Exercise 8.2.

8.4 Consider the least-squares problem of a sinewave with additive output noise (see (8.23) and (8.24)) for a given frequency f. Determine the two maximum-likelihood estimators that correspond to these two model equations.

8.5 Starting from the maximum-likelihood cost functions of Exercise 4, prove that it is possible to explicitly eliminate the linear parameter A for (8.23).

CHAPTER 9

RELIABILITY MEASUREMENTS

MARCANTONIO CATELANI
Department of Information Engineering, University of Florence, Italy

9.1 INTRODUCTION

Nowadays, in many fields of application, it is fundamental to guarantee high reliability performance for components and systems. This is a topic in transportation, avionics, automotive, electronics, telecommunications, mechanics, and so on. The knowledge of reliability allows to establish the time interval in which the item (a component, an equipment, or a system) is capable of performing its function, under specified conditions of use. So, it is evident how this information represents the starting point for more complex evaluations concerning availability, maintainability, and safety that is, in technical terms, the RAMS (reliability, availability, maintainability and safety) requirements. After a brief description of the term "Quality" and its evolution over time, this chapter focuses on reliability measurements. To this aim, the fundamental terms with the corresponding definitions according to international standards are proposed; a classification of failures and faults is also given. Then, the basic concepts for the evaluation of both the reliability models and the failure rate plot are introduced, as well as the fundamental statistical parameters that, traditionally, are used in this context: mean time to failure (MTTF), mean time between failures (MTBF) and mean time to restore (MTTR). In this chapter a brief presentation of the most important reliability tests that can be implemented in laboratory for the evaluation of components degradation is also proposed. Being a complex system constituted by a set of sub-systems, in the second part of this chapter the models of system reliability are presented with details and examples for series, parallel and *koon* structures. The knowledge of such models represents a fundamental aid for engineers and technicians in the design phase in order to evaluate reliability growth with the use of redundant configurations. In many contexts knowledge of the system reliability has to be implemented with availability evaluation. To this aim,

Modern Measurements: Fundamentals and Applications, First Edition.
Edited by Alessandro Ferrero, Dario Petri, Paolo Carbone, and Marcantonio Catelani.

many techniques can be found in literature. Among these, FMEA (failure modes and effects analysis), FMECA (failure modes, effects and criticality analysis), FTA (fault tree analysis), ETA (event tree analysis), HAZOP (hazard and operability analysis), Markov analysis, and others, can be considered for the design review of equipment and complex systems. A brief description of FMECA and FTA is proposed as the conclusion of this chapter.

9.2 BRIEF REMARKS ON THE CONCEPT OF QUALITY

The concept of *Quality* has undergone important and significant conceptual transformations over time. Until the 1950s, in many industries and the manufacturing area, the word "product quality" was associated with "compliance". In fact, it was assumed as "quality product," a product whose characteristics—denoted as functional parameters—when measured, appeared to be subjected to requirements set out in the design phase, that is, the technical specifications. In other words, it was often considered that the terms *quality* and *compliance* are synonyms, this last nowadays widely used as "conformity to specifications." In the present scenario, where the increase of technological developments is often combined with the changing needs of society, the concept of quality assumes a much more extensive significance. At the base is the concept of *product* as result of a *manufacturing process* in which a series of activities transforms raw materials, technologies, and resources into the output: the product. So, the current concept of "quality" cannot be attributable only to the concept of "compliance", as in the past. Rather, a more complex monitoring activity is necessary in all the phases of the manufacturing process by means of input/output measurements as well as measurements during the process. The aim is to check all its phases and the meeting of characteristics with all the product requirement as established by designers. On the basis of such considerations, the actual definition of *quality* corresponds to the degree to which a set of inherent characteristics of an item fulfills requirements, being the item, a system or sub-system, an equipment, a component, and so on. For a given item, the characteristics could be intrinsic or assigned, qualitative or quantitative; it can be also of physical nature (electric, mechanic, etc.) or time-dependent (reliability, availability, and maintainability), and other. To this aim it is fundamental to translate each characteristic into a measurable quantity in order to check the fulfillment of requirements; in other terms, it is necessary to verify the compliance and the ability of the item to maintain its performance in the time.

9.3 RELIABILITY, FAILURE AND FAULT: BASIC CONCEPTS AND DEFINITIONS

In this paragraph we emphasize the concepts and definitions of *reliability*, *failure* and *fault* with a brief, also if not exhaustive, classification for these last. Given a generic item – component, device, sub-system, functional unit, equipment or system that can be considered individually and that carries out one or more functions – the reliability represents a time-dependent feature that can be expressed both in qualitative and quantitative terms.

In qualitative terms [1], it represents the ability of the item to perform the required function (or a set of functions) without failure for a given time interval, under given conditions. From the point of view of measurable performance, it is fundamental to consider the quantitative definition of reliability which is based on the concept of probability. Always in reference [1], assuming a time interval and the item capable of carrying out its required function at the beginning of such an interval, **reliability** corresponds to the "*probability that the item is capable of performing its required function in the established time interval, under established conditions.*"

On the basis of this last definition, reliability can be evaluated through analytical models (law of reliability) that allow to establish a particular value at a fixed time, or measured and estimated through statistical parameters (reliability parameters) as the MTTF, MTBF, and so on. Some concepts concerning the reliability models and reliability parameters are given in Section 9.4.1.

As said in the introduction, the reliability measurements represent a fundamental topic in many fields of application: transportation, avionics, automotive, electronics, telecommunications, and mechanics represent some examples. But it is important to remember that, in the above-mentioned contexts, the reliability information represents the starting point for more complex evaluations of RAMS requirements concerning performance of availability, maintainability, safety and risk assessment.

According to standards [1], the basic concepts and definitions are given in the following.

- **Availability**: the ability of an item to be in a state to perform a required function under given conditions at a given instant of time or over a given time interval, assuming that the required external resources are provided.
- **Maintainability**: the ability of an item under given condition of use to be retained in, or restored to, a state in which it can perform a required function, when maintenance is performed under given conditions and using stated procedures and resources.
- **Maintenance support performance**: the ability of a maintenance organization, under given conditions, to provide upon demand the resources required to maintain an item, under a given maintenance policy.
- **Dependability**: the collective term used to describe the availability performance and its influencing factors: reliability, maintainability, and maintenance support performance.

In some applications it is fundamental to define the concept of **safety** as *the freedom from unacceptable risk of harm*. Combining this concept with those defined previously, it is clear the acronym RAMS introduced above.

Safety is connected to the evaluation of **risk**, meaning the probable rate of occurrence of an hazard causing harm and the degree of severity of the harm.

Combining the RAMS analysis with a risk assessment it is possible to evaluate hazardous deviations with respect to the normal functionality for an industrial plant (offshore oil and gas platforms, chemical plants, nuclear plants, etc.), a production process or a complex system (airbag systems in automobile, railway signalling systems, parking brakes, etc.). For these applications a safety system is often required in order to prevent critical situations and protect people and environment for dangerous events. In this chapter the RAMS performances related to an equipment or a complex system are not considered in detail and more information can be found in References 2, 3, 4, and 5. Recalling the definition of reliability, the capability of the item to perform the required function ceases when a failure occurs. On the basis of this consideration a **failure** is an event that results in a **fault** state of the item.

It is important to remember that the failures are classified according to different criteria. An important classification can be made in function of the causes responsible for their occurrence [1, 4]:

- *misuse failures*, due to the application of stresses during use which exceed the stated capabilities of the item;
- *primary failure*, failure not caused either directly or indirectly by a failure or a fault of another item;
- *induced or secondary failures*, generated by the failure of another device;
- *early life failure*, due to intrinsic weaknesses in construction of the element and whose causes are normally identifiable during the manufacturing process and which are manifested during the initial use of the device;
- *random failure,* for uncontrollable factors which occur during the "useful life" of the component and with a probability independent of time;
- *wear-out failure*, generated by the cumulative degradation caused by the stresses imposed in use. The probability of occurrence of a wear-out failure typically increases with the accumulated operating time, number of operations, and/or stress applications.
- *ageing failure*, a failure whose probability of occurrence increases with the time due to cumulative deterioration. The ageing process is a physical or chemical phenomenon that involves changes in characteristics of the material with time, in some circumstance related to interaction with its environment.

A different classification concerns the severity of consequences generated by the failure, that is:

- *critical failure* that can cause, with a high probability, damage to persons or material unacceptable to other parts of the system;
- *failure of primary importance*, that, although different than those mentioned previously, can reduce the functionality of a system;
- *failure of secondary importance*, a failure which does not reduce the functionality of a system;

Another classification can be made considering the impact of the failure at system level:

- *total failure*, when variations in the characteristics of the element are such to completely compromise the function of the system;
- *partial failure*, when the variations of one or more characteristics of the element do not compromise its complete functioning;
- *intermittent failure*, characterized by a succession, generally casual, of periods of normal operations with periods of fault, without any maintenance operations carried out on the device.

When the status of system operation and/or fault has to be checked in order to undertake any maintenance tasks, other terms related to detection activities can be defined such as:

- *fault detection*, the event by which the presence of a fault becomes apparent;
- *fault localization*, the action to identify the faulty sub-item at the appropriate indenture level for maintenance;
- *fault diagnosis*, the action to identify and characterise faults;
- *fault correction*, the action to correct faults, after fault localization.

Other definitions and classifications can be found in References 1 and 4 and will be discussed in more detail later, while some symbols and abbreviations used in the text for non-repaired and repaired item are summarized in the following glossary.

- A_o, asymptotic availability
- $A(t)$, instantaneous availability (availability function)
- $f(t)$, probability density function of the time to failure
- $F(t)$, cumulative distribution function of the time to failure
- MTBF, mean operating time between failures
- MTTF, mean time to failure
- MTTR, Mean time to restoration
- $R(t)$, reliability function or instantaneous reliability, that is, the probability of survival until time t
- $R(t_1, t_2)$, reliability for the time interval (t_1, t_2)
- $R(t, t + x|t)$, conditional reliability for the time interval $(t, t + x)$, assuming that the item survived to time t
- TTF_i, observed time to failure of item i
- t, continuous random variable modeling the time to failure TTF
- λ, constant failure rate, that is, the reciprocal of MTTF when time to failure (TTF) is exponentially distributed
- $\lambda(t)$, instantaneous failure rate

- μ, constant repair rate, that is, the reciprocal of the mean of the corrective maintenance time (CMT) when CMT is exponentially distributed
- $\mu(t)$, instantaneous repair rate

9.4 RELIABILITY THEORY

Referring to the definitions given in Section 9.3, the reliability evaluation allows to determine the analytical behavior over time of the reliability function and the corresponding parameters. Such analysis can be implemented by means of an experimental approach based on laboratory tests conducted on the item. In this case, if a set of measurements is collected and the observed failure data are available from the tests, different reliability functions (also denoted as reliability models) can be defined and evaluated as well as the related parameters. In addition, assuming the time to failure as a positive and continuous random variable, the experimental approach can be supported by the probabilistic framework.

9.4.1 Reliability Models and Measures Related to Time to Failure

In general terms, we can consider a *test* [1] as a procedure or a set of operations implemented to determine or verify one or more characteristic of an item. A *laboratory test* [1], in particular, is a test made under prescribed and controlled conditions that may, or may not, simulate and reproduce the field conditions. To this aim different types of stress can be considered in laboratory such as temperature, humidity, vibration, and so on. A classification of environmental stress is proposed in the Standard IEC 60068-1: Environmental testing, Part.1-General and Guidance [6] as shown in Table 9.1.

The tests can be more carefully detailed in function of the particular type of stress. For example, test U for the durability of terminals and devices integrally assembled with a component can concern traction (Ua1), compression (Ua2), bending (Ub), torsion (Uc), and the torque factor measurement.

Without going into details of the type of stress and the corresponding level applied to the item, a laboratory test carried out to determine or verify a reliability performance can be represented, in a simplified form, as in Figure 9.1 where two different levels of the same stress are identified. More complex sequences may provide, during the test run, the application of different values of the same stress (e.g., step-stress test), the simultaneous application of different types of stress (e.g., temperature-humidity combined test), the test runs with the item in service, etc.. Referring to the profile plotted in Figure 9.1, a sequence of k-different cycles with time length Δt_i, $i = 1, \dots, k$ and only two levels of stress—S_o, assumed as reference value and S_p, up stress level—are taken into consideration.

Independently from the type of stress, the selection of levels and duration, a set of activities have to be implemented according to the following sequence [4].

Phase 1: preliminary adjustment—It is performed on the item being tested in order to eliminate the effects of its preceding states or conditions.

TABLE 9.1 Classification of environmental tests [6]

Test	Environmental stress
A	Cold
B	Dry heat
C	Heat with high humidity (continuous)
D	Heat with high humidity (not continuous, cyclic)
E	Mechanical impacts (bumps and jerks)
F	Vibrations (sinusoidal, random occasional)
G	Constant acceleration
J	Mold
K	Corrosive atmosphere (e.g. salty fog)
L	Powder and sand
M	Atmospheric pressure (high and low)
N	Temperature changes
Q	Hermetic sealing (for liquids and gas)
R	Water (rain, dripping)
S	Radiation (solar, excluding electromagnetic radiation)
T	Welding
U	Sturdiness of terminals (of components)

Phase 2: controls and initial measurements—The aim of this step, normally identified with δ_0 in Figure 9.1, is to ascertain that all components to be tested are functioning correctly. We assume this phase and this stress level S_0 as the reference condition for measurements on component under test.

Phase 3: treatment—During this phase the items are subjected to the stress profile according to standards or determined by other experimental criteria. An example could be the application of a temperature level for a certain time period, identified with Δt in Figure 9.1, using an oven or the application of heat with high humidity in a climate controlled chamber.

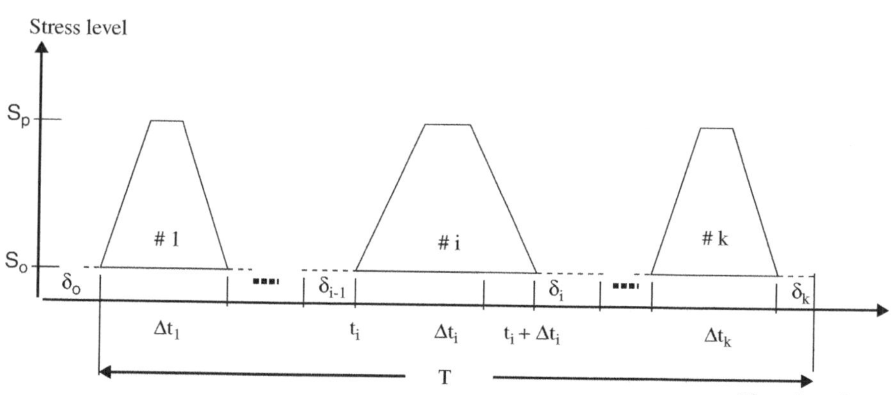

FIGURE 9.1 Generic test sequence.

Phase 4: readjustment—After the stress is applied, it is necessary to restore items to reference conditions of phase 2 and verify the level of degradation or the occurrence of failure. This can be made in the time interval from the end of the i-th cycle and the beginning of the successive $(i + 1)$-th.

The test profile shown in Figure 9.1 and implemented in laboratory allows to determine the reliability functions and measures related to time to failure. To this aim a sample of n items (sampling units) taken from a production process is considered for the test. Referring to the profile shown in Figure 9.1, n represents the number of items that are operational at the time $t = 0$ and verified in δ_0. Denoting as $n_S(t)$ the items which are still operational (survivors) at the instant t and $n_F(t)$ the number of failures observed in t, we can deduce

$$n = n_S(t) + n_F(t) \tag{9.1}$$

with $n_S(0) = n$ and $n_F(T) = n$, if T represents the cumulative test time and all the items are faulty in $t = T$. Considering the contributions of the above-mentioned phases, the cumulative test time, normally in hours, can be expressed as

$$T = \sum_{i=1}^{k} \Delta t_i + \sum_{j=1}^{k} \delta_j = t_k + \Delta t_k + \delta_k. \tag{9.2}$$

The time to failure by the test on the basis of the selected stress profile allows to determine some relevant experimental reliability functions. These, identified in the following using the upperscript "*", are the *experimental frequency of failure* f^*, the *experimental cumulative function of failure* F^*, the *experimental reliability* R^* and the *experimental failure rate* λ^*. Starting from the test results and the discrete-time experimental functions, by introducing the random variable *time to failure* $\boldsymbol{t}$ the corresponding theoretical models can be obtained. Then, the reliability law is expressed as a continuous-time relationship between the failure rate of the item and the probability that such item maintains its performance over time.

9.4.1.1 *Probability Density Function of the Time to Failure* A fundamental function representing the temporal distribution of failures is the probability density function of the time to failure. Considering Equation (9.1) and monitoring the number of items that fails in each cycle of Figure 9.1, it can be evaluated starting from the experimental frequency of failure.

$$f_i^* = \frac{n_S(t_i) - n_S(t_i + \Delta t_i)}{n \Delta t_i} \qquad i = 1, \ldots k, \tag{9.3}$$

where t_i is the starting time of the i-th test cycle:

$$t_i = \sum_{r=1}^{i-1} \Delta t_r + \sum_{j=0}^{i-1} \delta_j \quad t_1 = \delta_o \quad 1 < i \le k. \tag{9.4}$$

From Equation (9.3), if $\Delta t \rightarrow 0$, it follows the estimated value of the failure frequency function as

$$\hat{f}(t) = \lim_{\Delta t \to 0} \frac{n_S(t) - n_S(t + \Delta t)}{n\Delta t} = \frac{1}{n}\frac{dn_F(t)}{dt} \tag{9.5}$$

with $\hat{f}(0) = 0$, assuming all items without failure at the beginning of the test. The frequency density of failure is often expressed in hours. Introducing now the continuous random variable time to failure $\mathbf{t}$ the corresponding *probability density function (pdf)* of the time to failure can be obtained as

$$f(t) = \lim_{\Delta t \to 0} \frac{\text{Prob}\,\{t \le \mathbf{t} \le t + \Delta t\}}{\Delta t}. \tag{9.6}$$

Consequently, the probability that the item fails in the time interval $[t_1, t_2]$ is evaluated as

$$\text{Prob}(t_1 \le \mathbf{t} \le t_2) = \int_{t_1}^{t_2} f(t)dt. \tag{9.7}$$

9.4.1.2 *Cumulative Function of Failure* The trend of failures in time can be described also by the experimental cumulative function of failure F^*:

$$F_i^* = \sum_{r=1}^{i} f_r^* \Delta t_r = \frac{n_F(t_i + \Delta t_i)}{n} = 1 - \frac{n_S(t_i + \Delta t_i)}{n} \quad 1 < i \le k \tag{9.8}$$

with $F_k^* = 1$. It is easy to observe from Equation (9.8) that F^* is a dimensionless quantity. Assuming also in this case $\Delta t \rightarrow 0$, the corresponding estimate is

$$\hat{F}(t) = \frac{n_F(t)}{n}, \tag{9.9}$$

where $\hat{F}(0) = 0$ and $\hat{F}(T) = 1$. By introducing the time to failure $\mathbf{t}$ the cumulative function of failure is obtained

$$F(t) = P(\mathbf{t} \le t) = \int_0^t f(t)dt. \tag{9.10}$$

If a value of t is established Equation (9.10) represents the probability that the item fails in a time less than, or equal to, t. In other terms, the cumulative function of failure represents the probability that the item stops performing the required function in the interval $[0, t]$. For this reason $F(t)$ is denoted also as *Unreliability function*.

9.4.1.3 *Reliability law* Similar to the experimental cumulative function of failure, the experimental reliability is obtained as the ratio:

$$R_i^* = \frac{n_S(t_i + \Delta t_i)}{n} = 1 - F_i^* \tag{9.11}$$

and considering $\Delta t \to 0$, the estimate value:

$$\hat{R}(t) = \frac{n_S(t)}{n} = 1 - \hat{F}(t) \tag{9.12}$$

with $\hat{R}(0) = 1$; $\hat{R}(T) = 0$ and dimensionless. Introducing now the random variable $\mathbf{t}$

$$R(t) = P(\mathbf{t} > t) = \int_t^{+\infty} f(t)dt = 1 - F(t). \tag{9.13}$$

For an established value of t, Equation (9.13) represents the probability that the item fails after t. In other terms, according to the definition given in Section 9.3, Equation (9.13) identifies the survival function of the item, that is the probability that such item will operate failure-free in the time interval $[0, t]$. As a consequence of Equation (9.13) the probability that the item fails during the time interval (t_1, t_2), $0 \leq t_1 < t_2$, is given by

$$R(t_1) - R(t_2) = P(t_1 \leq \mathbf{t} < t_2) = \int_{t_1}^{t_2} f(t)dt.$$

9.4.1.4 *Instantaneous Failure Rate Function* The failure rate over the interval $[t_i, t_i + \Delta t_i]$ is defined as the ratio of the number of failures occurring in Δt_i to the number of survivors $n_S(t_i)$ at the beginning time t_i, divided by the length Δt_i of the i-th interval.

$$\lambda_i^* = \frac{n_S(t_i) - n_S(t_i + \Delta t_i)}{n_S(t_i)\Delta t_i} = \frac{n_F(t_i + \Delta t_i) - n_F(t_i)}{n_S(t_i)\Delta t_i} = \frac{f_i^*}{R_i^*} \quad i = 1, \ldots .k \tag{9.14}$$

expressed in failures (or fractional of failures) per hours^{-1}. The corresponding theoretical function (called instantaneous failure rate $\lambda(t)$), can be evaluated assuming that the event "failure of the item" occurs in the time interval $(t, t + \Delta t]$, assuming the item functioning in t. It is expressed by the formula:

$$\begin{aligned} \lambda(t) &= \lim_{\Delta t \to 0} \frac{\text{Prob}\,\{(t < \mathbf{t} \leq t + \Delta t) | (\mathbf{t} > t)\}}{\Delta t} \\ &= \frac{1}{\text{Prob}\,\{\mathbf{t} > t\}} \lim_{\Delta t \to 0} \frac{\text{Prob}\,\{t < \mathbf{t} \leq t + \Delta t\}}{\Delta t}, \end{aligned} \tag{9.15}$$

which is in accordance with the definition given in [1], that is, *the failure rate is the limit, if it exists, of the ratio of the conditional probability that the failure of a non-repairable item occurs within time interval* $(t, t + \Delta t]$, *to* Δt, *when t tends to zero, given that failure has not occurred within time interval* $(0, t)$.

Recalling Equations (9.6) and (9.13) the failure rate can be rewritten as

$$\lambda(t) = \lim_{\Delta t \to 0} \frac{1}{\Delta t} \frac{R(t) - R(t + \Delta t)}{R(t)} = \frac{f(t)}{R(t)} \tag{9.16a}$$

and

$$\lambda(t) = -\frac{1}{R(t)} \frac{dR(t)}{dt} = -\frac{d}{dt} \ln R(t) \tag{9.16b}$$

consequently

$$R(t) = \exp\left\{-\int_0^t \lambda(\varsigma) d\varsigma\right\}. \tag{9.17}$$

Equation (9.17) represents the *fundamental law of reliability*, that is the relationship between the failure rate λ of the item and the probability that such item maintains its performance over time. The reliability model allows to deduce some important considerations. According to the definition given in Section 9.3, it appears evident that the reliability of an item assumes the maximum value, equal to 1, at the instant $t = 0$, that is at the beginning of the test sequence of Figure 9.1. In addition, the reliability decreases over time in function of the failure rate and tends to zero for $t \to \infty$. The reliability functions above are summarized in Table 9.2.

TABLE 9.2 Reliability functions

	$f(t)$	$F(t)$	$R(t)$	$\lambda(t)$
$f(t)$	=	$\frac{d}{dt}F(t)$	$-\frac{d}{dt}R(t)$	$\lambda(t) \cdot \exp\left(-\int_0^t \lambda(\varsigma)d\varsigma\right)$
$F(t)$	$\int_0^t f(\varsigma)d\varsigma$	=	$1 - R(t)$	$1 - \exp\left(-\int_0^t \lambda(\varsigma)d\varsigma\right)$
$R(t)$	$\int_t^\infty f(\varsigma)d\varsigma$	$1 - F(t)$	=	$\exp\left(-\int_0^t \lambda(\varsigma)d\varsigma\right)$
$\lambda(t)$	$\frac{f(t)}{\int_t^\infty f(\varsigma)d\varsigma}$	$\frac{dF(t)/dt}{1 - F(t)}$	$-\frac{d}{dt}\ln R(t)$	=

9.4.2 Life Distributions

The knowledge of the probability density function of the time to failure represents an important aspect for the evaluation of the life of a component or an equipment. Equations (9.10), (9.16a), and (9.17) can be modified in function of the pdf of the time to failure. In the following, the life distribution is derived for two widely used expressions of failure time pdfs.

9.4.2.1 *Exponential Distribution* If the probability density function $f(t)$ is defined as:

$$f(t) = \begin{cases} \lambda e^{-\lambda t} & \text{for } t \geq 0 \text{ and } \lambda > 0 \\ 0 & \text{otherwise} \end{cases} \tag{9.18}$$

t is said an *exponential random variable* with parameter λ, which is constant over time. To specify that an exponential random variable is exponentially distributed, the notation $\mathbf{t} \sim exp(\lambda)$ can be used. From Equation (9.10) the related cumulative distribution function becomes

$$F(t) = P(\mathbf{t} \leq t) = \int_0^t \lambda \exp(-\lambda \varsigma) d\varsigma = 1 - \exp(-\lambda t) \tag{9.19}$$

and the reliability law is

$$R(t) = P(\mathbf{t} > t) = \int_t^{+\infty} \lambda \exp(-\lambda \varsigma) d\varsigma = \exp(-\lambda t) \tag{9.20}$$

as shown in Figures 9.2b and 9.2c, respectively. Equation (9.20) represents the simplest mathematical model for the reliability of a non-repaired item. It is easy to verify that the reliability decreases exponentially over time with a decaying rate that is related to the failure rate λ of the item, as shown in Figure 9.2c. The exponential distribution $\mathbf{t} \sim exp(\lambda)$ is characterized by the following statistical properties [7]: mean or expected value $E(t) = 1/\lambda$, variance $Var(t) = 1/\lambda^2$, median $m(t) = \ln 2/\lambda$.

9.4.2.2 *Weibull Distribution* Assuming the probability density function $f(t)$ as

$$f(t) = \begin{cases} \dfrac{\beta (t)^{\beta-1}}{\alpha^{\beta}} \exp\left\{-\left(\frac{t}{\alpha}\right)^{\beta}\right\} & \text{for } t \geq 0 \\ 0 & \text{otherwise} \end{cases} \tag{9.21}$$

the time to failure is said to be Weibull distributed (i.e., $\mathbf{t} \sim Weibull(\alpha, \beta)$), where α (> 0) and β (> 0) denote the scale parameter and the shape parameter, respectively. From Equation (9.10) the cumulative distribution function is given by

$$F(t) = P(\mathbf{t} \leq t) = \int_0^t f(t) dt = \begin{cases} 1 - \exp\left\{-\left(\frac{t}{\alpha}\right)^{\beta}\right\} & \text{for } t \geq 0 \\ 0 & \text{otherwise} \end{cases} \tag{9.22}$$

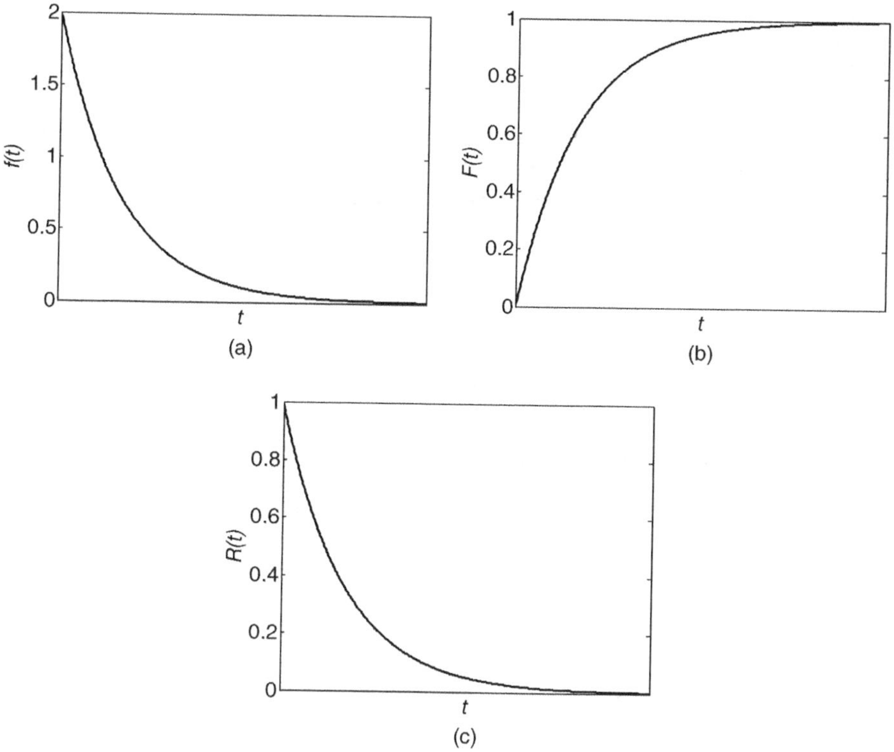

FIGURE 9.2 Probability density function (a), cumulative distribution function (b) and reliability function (c) for a constant failure rate λ.

and, from Equation (9.17), the reliability law is

$$R(t) = P(\mathbf{t} > t) = \exp\left(-\frac{t}{\alpha}\right)^{\beta}. \tag{9.23}$$

By comparing Equation (9.18) with Equation (9.21) it can be observed that the Weibull distribution coincides with the exponential distribution when the shape parameter $\beta = 1$. In this case we also obtain $\alpha = 1/\lambda$. By increasing β, other important distributions can be identified as the Rayleigh ($\beta = 2$) and an approximation of the Gaussian distribution when $\beta = 4$ or higher. In Figure 9.3 the plots of the probability density function, the cumulative distribution and the reliability model for the scale parameter $\alpha = 1$ and different values of the shape parameter β are represented.

Considering Equation (9.16a) the failure rate related to $\boldsymbol{t} \sim \mathit{Weibull}(\alpha, \beta)$ can be written as

$$\lambda(t) = \frac{f(t)}{R(t)} = \frac{\beta(t)^{\beta-1}}{\alpha^{\beta}}. \tag{9.24}$$

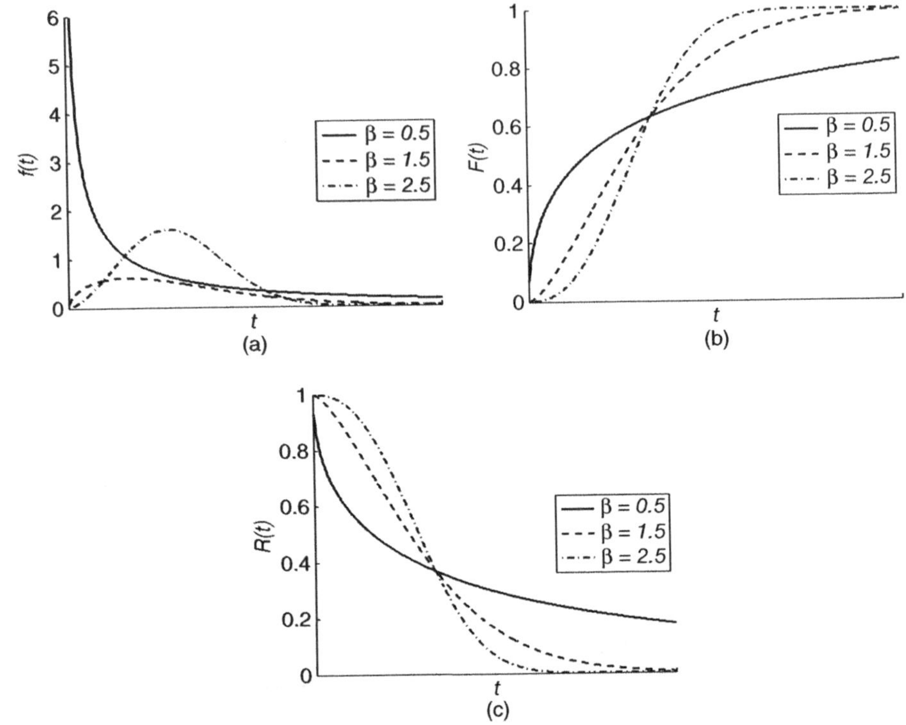

FIGURE 9.3 Probability density function (a), cumulative distribution function (b) and reliability function (c) for different values of the shape parameter β, with $\alpha = 1$.

It is easy to see that the failure rate is constant over time if $\beta = 1$, it decreases with $0 < \beta < 1$ and, vice versa, it increases for $\beta > 1$; the failure rate is linearly increasing over time in the particular case of $\beta = 2$. Figure 9.4 shows the behavior of (9.24) for different values of β and $\alpha = 1$.

The mean $E(t)$ and variance $Var(t)$ for the random variable $t \sim Weibull(\alpha, \beta)$ are given by

$$E(t) = \frac{\alpha}{\beta}\Gamma\left(\frac{1}{\beta}\right); \; Var(t) = E(t)^2\left[2\beta\frac{\Gamma\left(\frac{2}{\beta}\right)}{\Gamma^2\left(\frac{1}{\beta}\right)} - 1\right], \tag{9.25}$$

where $\Gamma(\cdot)$ denotes the Gamma function.

9.4.3 Reliability Parameters

When it is necessary to quantify the reliability performance of an item in established working conditions some statistical parameters may be used in addition to the reliability model of Equation (9.17). Such parameters, classified for non-repaired or

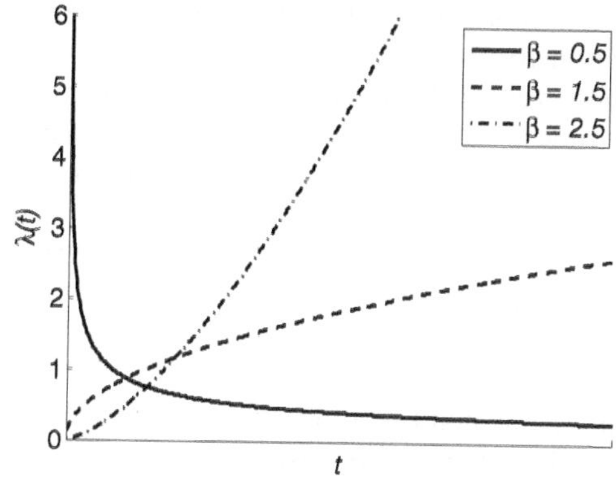

FIGURE 9.4 Plots of the failure rate $\lambda(t)$ for different values of the shape parameter β, with $\alpha = 1$.

repaired item, represent a measure of the reliability features. They can be experimentally evaluated by measuring the time to failure TTF using laboratory tests or defined in probabilistic terms by considering the time to failure t as a continuous random variable. If observed failure data are available for n non-repaired items, that is an item that at any instant of time may be either in an up-state or in a down-state, an estimate of the MTTF is given by

$$\hat{MTTF} = \frac{\sum_{i=1}^{n} TTF_i}{n}, \tag{9.26}$$

where TTF_i denotes the observed time to failure of the item i. In probabilistic terms $MTTF$ corresponds to the expected value of t, that is

$$MTTF = E(t) = \int_0^{+\infty} tf(t)dt = \int_0^{+\infty} R(t)dt. \tag{9.27}$$

From Equation (9.27) if the time to failure is exponentially distributed with constant failure rate

$$MTTF = \frac{1}{\lambda}.$$

Considering n repaired items, that is items that can be restored after a failure, the observed MTTF is

$$\hat{MTTF} = \frac{\text{total operating time}}{k_F} = \frac{\sum_{i=1}^{n} (\text{operating time})_i}{k_F}, \tag{9.28}$$

where the *total operating time* denotes the cumulated operating time of all n items taken in to consideration for the test, during a given time interval; k_F is the total number of failures observed in the interval and (*operating time*) i represents the cumulated operating time of the i-th item in such period. For repaired items an important reliability measure, often used for a system or equipment, is given by the MTBF

$$MTBF = \int_0^{+\infty} tf(t)dt = \int_0^{+\infty} R(t)dt. \tag{9.29}$$

It is important to underline that Equation (9.27) allows to calculate the mean time to failure for a non-repaired item as, for instance, a component of a more complex system. In this case $R(t)$ and λ represent the reliability law and the failure rate of the component, respectively. A different meaning is assumed by Equation (9.29). When the component fails and the system moves from the up state to the down state, the faulty component has to be substituted in order to restore the functionality of the system in the up state. In this case $R(t)$ denotes the reliability law of the system and can be expressed in function of the failure rates of the constituting sub-system and components as described in Section 9.5. Recalling the definition of availability $A(t)$ given in Section 9.3, for repaired items, another fundamental parameter, denoted as asymptotic availability, can be introduced as the limit for $t \to \infty$ of $A(t)$.

$$A_O = \frac{MTTF}{MTTF + MTTR}, \tag{9.30}$$

where MTTR represents the mean time to restoration that is the mean time necessary to restore the functionality of the item. MTTF, MTBF and MTTR are normally expressed in hours.

9.4.4 The Bath-Tube Curve

Recalling Equation (9.24) and considering the plots shown in Figure 9.4 a qualitative representation of the failure rate behavior over time can be depicted as shown in Figure 9.5. In this curve, usually denoted as *bath-tub curve* [4] for its typical shape, three different parts can be identified in an item life cycle.

In the first part the failure rate, often high in the initial phase, decreases rapidly in early item life. This phenomenon is known as "infant mortality" and it is due to undiscovered defects or failures of the item. Such failures, classified in the standard [1] as *early life failures*, are due to intrinsic weaknesses in construction of the element whose causes, which are manifested during the initial use of the device, are normally identifiable in variations of the manufacturing process, marginal tolerance of the parts, insufficient or inadequate test and controls, variations in calibration process. Typically, early life failures occur in the first year of the item use and can be identified by means of screening techniques. Among these, the burn-in test is frequently implemented in ICT field where a particular level of temperature is

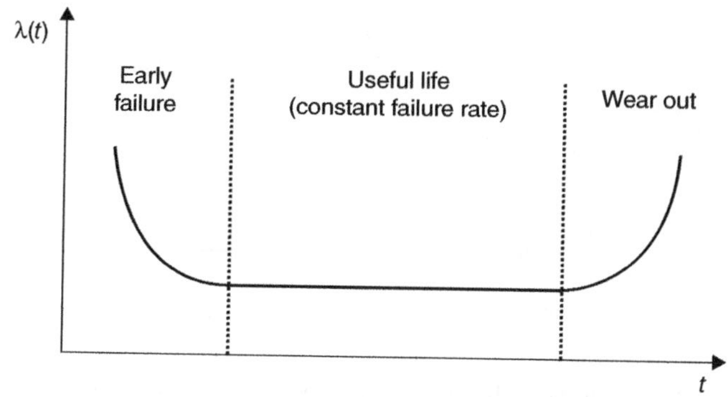

FIGURE 9.5 General representation of the failure rate: the "bath-tub" curve [4].

applied at the device under test in order to induce failures. The second part of the bath-tube curve in Figure 9.5 represents the *useful life period.* During this interval the failure rate can be considered approximately constant over time and the failures are classified as *random failures.* It is important to underline that the range of the useful life strictly depends on technology: an example is given by electronics with respect to mechanics. As a consequence, the assumption of constant failure rate leading to the exponential distribution expressed by Equation (9.18) holds only in the useful life period of the bath-tub curve. In this part of an item life cycle the failure rate and the probability of occurrence of random failures can be decreased by derating techniques. The last part of the bath-tub plot of Figure 9.5 represents the *wear-out period.* In this situation the item is characterized by the ageing process and delivered for use. Early life failure, useful life and wear-out period can be represented using a random variable $t \sim Weibull(\alpha, \beta)$ whose distribution has been described in Section 9.4.2.

9.5 SYSTEM RELIABILITY ASSESSMENT

After the presentation of the theoretical approach to reliability in Section 9.4, it is interesting to know how the reliability performance of a system, or an equipment, can be evaluated on the basis of the reliability performance of the constituting elements. In general terms, a system is a set of elements, components or sub-systems connected among themselves with the aim of guaranteeing functional requirements [1]. Thus, it results that information concerning system reliability can be obtained only if the system functional configuration is known, as well as the reliability requirements of its elements. A graphical representation of the systems structure in terms of sub-systems and/or components is represented by the reliability block diagram (RBD). This approach allows system successful functioning to be related to the ways its elements are mutually interacting. At first, a hierarchical level decomposition has to be performed as depicted in Figure 9.6.

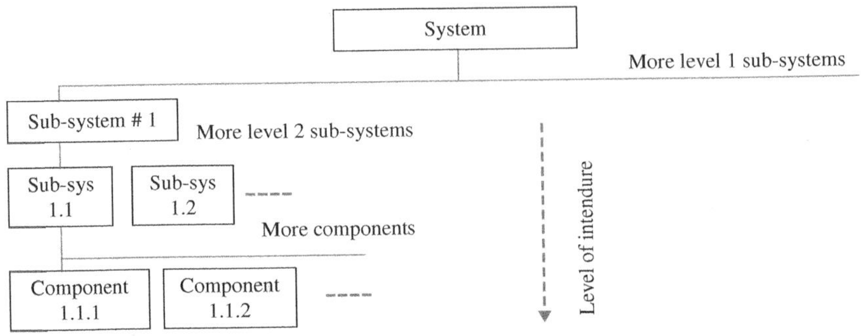

FIGURE 9.6 Hierarchical decomposition of the system.

After decomposition, the inter-relationship among sub-systems and components have to be identified and expressed by means of elementary models such as series, parallel, k-out-of-n, stand-by, and their combination. The system reliability is then obtained as a mathematical combination of the sub-system and/or component's reliability on the basis of the RBD configuration. Considerning a system constituted by n items E_i $(i = 1, \ldots n)$, assuming e_i $(\overline{e}_i)$ the event "*item functioning without failure (item in failure)*" in the interval $(0, t]$, the quantities $P(e_i) = P(t_i > t) = R_i(t)$ and $P(\overline{e}_i) = P(t_i \leq t) = F_i(t)$ represent, respectively, the reliability and the unreliability of the i-th item. Consequently, the probability that the system is successful and the system fails are $P(S) = P(\mathbf{t} > t) = R(t)$ and $P(\overline{S}) = P(\mathbf{t} \leq t) = F(t)$. In the following the effect on the system success of the most important configurations are analyzed from a theoretical point of view; more complex structures can be found in literature [2–4].

9.5.1 Series Configuration

The series configuration can be represented by the RBD shown in Figure 9.7. This configuration, often denoted as *chain* structure, is the simplest one from a theoretical point of view, but, on the other hand, is more critical in terms of reliability performance. The system is operative if and only if all the constituting items are functioning correctly. Consequently the probability that the system is successful is:

$$P(S) = P\left\{\bigcap_{i=1}^{n} e_i\right\} = P(e_1)P(e_2|e_1) \cdot \ldots \cdot P(e_n|e_1 \cap \ldots \cap e_{n-1}), \qquad (9.31)$$

where $P(e_i|e_1 \cap \ldots \cap e_{i-1})$ represents the probability of success of the item E_i evaluated under the condition that items $E1, E2, \ldots E_{n-1}$ are operating. If independent

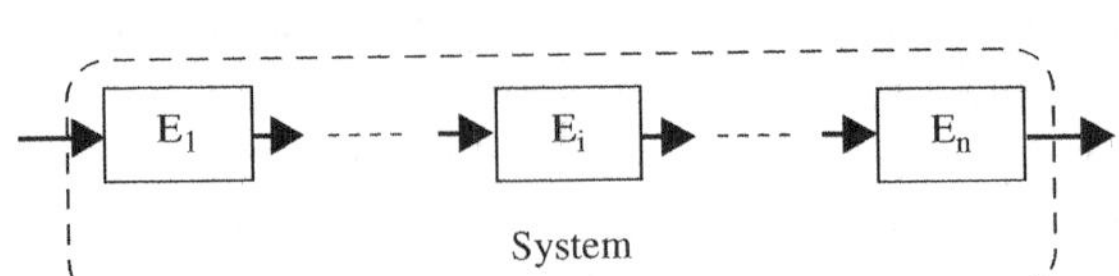

FIGURE 9.7 Chain structure.

failures can be assumed, that is, the functioning or the failure of a given item does not depend on the other items, the system reliability is:

$$R_{SYS}(t) = \prod_{i=1}^{n} R_i(t). \tag{9.32}$$

Assuming random failures and considering λ_i the failure rate of the item E_i

$$R_{SYS}(t) = \exp\left\{-\lambda_{SYS} t\right\} \tag{9.33}$$

being $\lambda_{SYS}(t) = \sum_i \lambda_i$ the failure rate of the whole system. From Equation (9.32) some important considerations can be deduced. First, $R_{SYS}(t) \leq \min_i \left\{R_i(t)\right\}, i = 1, \ldots, n$; so, it appears evident that the chain structure is influenced by the lowest reliable item. Then, the system reliability decreases as the number of the constituting items increases. On the other hand, the series configuration is easy to analyze. Indeed, under random failures assumption, the total failure rate λ_{SYS} is the sum of the single failure rates of all the involved items.

According to Equation (9.29) the mean time between failures is:

$$MTBF = \frac{1}{\lambda_{SYS}} \tag{9.34}$$

Considering the event "item faulty" in the interval $(0, t]$, the probability of failure of the system is given by

$$P(\overline{S}) = P\left\{\bigcup_{i=1}^{n} \overline{e_i}\right\} = \sum_{k=1}^{n} (-1)^{k+1} S_k \text{ with } S_k = \sum_{1 \leq i_1 < \cdots < i_k \leq n} P(e_{i_1} \cap \ldots \cap e_{i_k}). \tag{9.35}$$

Consequently, the cumulative function of failure

$$F_{SYS}(t) = 1 - R_{SYS}(t). \tag{9.36}$$

Obviously $P(S) + P(\overline{S}) = 1$.

9.5.2 Parallel Configuration

The parallel configuration, denoted also as redundant structure, is represented by the reliability block diagram shown in Figure 9.8. The system is operative only if one element E_i, $i = 1 \ldots n$ is operative. Vice versa, the system is not functioning if all the constituting elements are faulty. So, the probability that the system fails is

$$P(\overline{S}) = P\left\{\bigcap_{i=1}^{n} \overline{e_i}\right\} = P\left(\overline{e_1}\right) P\left(\overline{e_2} | \overline{e_1}\right) \cdot \ldots \cdot P\left(\overline{e_n} | \overline{e_1} \cap \ldots \cap \overline{e_{n-1}}\right) \tag{9.37}$$

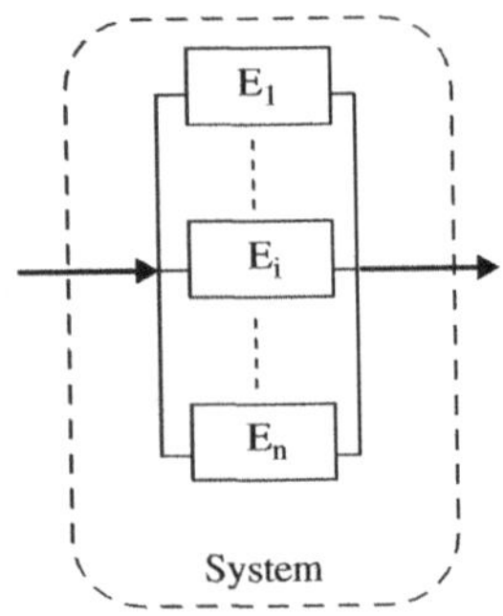

FIGURE 9.8 Parallel structure.

denoting with $\bar{e}_i$ the event "element E_i faulty". Assuming the hypotheses of independent and random failures, Equation (9.37) allows to obtain the system's unreliability as

$$F_{SYS}(t) = \prod_{i=1}^{n} F_i(t) = \prod_{i=1}^{n} (1 - e^{-\lambda_i t}). \tag{9.38}$$

Consequently, the system's reliability is

$$R_{SYS}(t) = 1 - F_{SYS}(t). \tag{9.39}$$

From Equation (9.39) some important properties of the parallel structure can be deduced. First, the system's reliability depends on the highest reliability value of the constituting elements E_i, that is $R_{SYS}(t) \geq \max_i \{R_i(t)\}$, $i = 1, \ldots, n$. Then, the system's reliability increases as n increases. For this configuration the mean time between failure is

$$MTBF = \int_0^{+\infty} R_{SYS}(t)dt \tag{9.40}$$

and under the assumption of independent and random failures

$$MTBF = \sum_{i=1}^{n} \frac{1}{\lambda_i} - \sum_{i=1}^{n-1} \sum_{j=i+1}^{n} \frac{1}{\lambda_i + \lambda_j} + \cdots (-1)^{n+1} \frac{1}{\sum_{i=1}^{n} 1/\lambda_i}. \tag{9.41}$$

Comparing Equation (9.41) with Equation (9.34) it is easy to observe that the evaluation of MTBF for a redundant structure is more difficult than the series

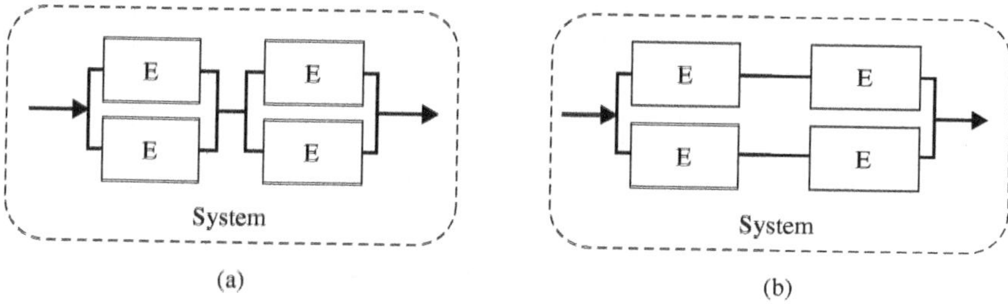

FIGURE 9.9 Different redundancy approaches: (a) component level; (b) system level.

configuration. If the system is constituted by two items with different failure rates λ_A and λ_B, the reliability law R_{sys} and the mean time between failures are

$$R_{SYS}(t) = e^{-\lambda_A t} + e^{-\lambda_B t} - e^{-(\lambda_A+\lambda_B)t} \tag{9.42}$$

$$MTBF = \frac{1}{\lambda_A} + \frac{1}{\lambda_B} - \frac{1}{\lambda_A + \lambda_B}. \tag{9.43}$$

If, for the sake of simplicity, the items are characterized by identical failure rates $\lambda_A = \lambda_B = \lambda$, it is easy to see from Equation (9.43) that the MTBF of the system increases by 50% with respect to the MTBF of the system corresponding to a single item. On the basis of such consideration this type of redundancy can be assumed as a reliability growth technique applicable to any level of design decomposition of Figure 9.6, ranging from the component to a complete sub-system or to the whole system.

9.5.2.1 *Example* In Figure 9.9 two different approaches for reliability growth are proposed for a system constituted by two identical items in series configuration. The design solution (a) takes in to account a component redundancy while a system redundancy is represented in (b).

With respect to the original chain structure with $R_{SYS}(t) = e^{-2\lambda t}$ and $MTBF = 0.5/\lambda$ the reliability and MTBF of the whole systems are:

Solution (a) $R_{SYS}(t) = 4e^{-2\lambda t} + e^{-4\lambda t} - 4e^{-3\lambda t}$; $MTBF \cong 0.9/\lambda$
Solution (b) $R_{SYS}(t) = 2e^{-2\lambda t} - e^{-4\lambda t}$; $MTBF \cong 0.75/\lambda$

Besides this structure others types of redundancy can be considered with different performance [1, 3, 6] such as *warm*, *stand-by*, and so on.

9.5.3 *k*-out-of-*n* Configuration

A particular active redundancy configuration, often implemented for safety systems, is represented by *k*-out-of-*n*, also known as *koon* redundancy, as depicted in Figure 9.10. The system is operative when at least k items out of a total of n are normally

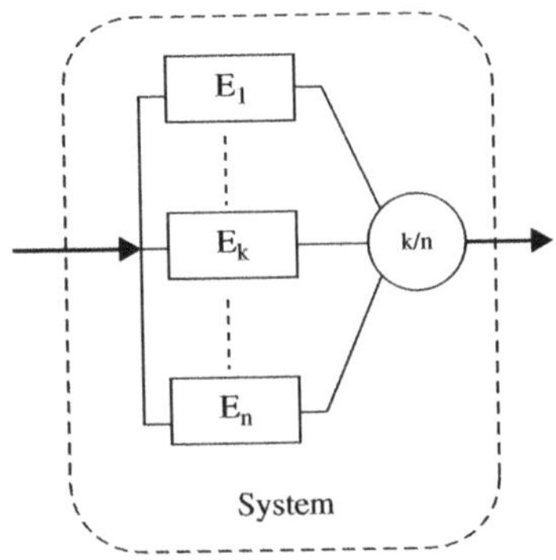

FIGURE 9.10 The *k*-out-of-*n* (*koon*) configuration.

functioning, with $k \leq n$. For this configuration it can be assumed k items in active redundancy and the remaining $(n - k)$ in stand-by.

Assuming the system constituted by identical items with independent failures, the probability of k successes out of n units can be evaluated by considering the binomial distribution

$$P\{k, n, p\} = \binom{n}{k} p^k (1 - p)^{n-k}, \tag{9.44}$$

where p represents the probability of success of any item. On the basis of the definition given for *koon*, the system successes if $k, k + 1, \ldots, n$ succeed. Consequently, the probability of the system success is

$$R_{SYS}(t) = \sum_{i=k}^{n} \binom{n}{i} \left(e^{-\lambda t}\right)^i \left(1 - e^{-\lambda t}\right)^{n-i}. \tag{9.45}$$

Equation (9.45) represents the reliability of the *koon* configuration for random failures, being λ the failure rate of the item and

$$\binom{n}{i} = \frac{n!}{i!(n - i)!} \tag{9.46}$$

the binomial coefficient, with $0! = 1$. It is interesting to observe that Equation (9.45) represents a generalized expression of the series structure if $k = n$ (*noon*) and the parallel structure if $k = 1$(1oo*n*). The system's reliability for some *koon* configurations are summarized in Table 9.3 under the same assumptions of Equation (9.45). It is interesting to observe that 2oo2 and 3oo3 correspond to the series structure, while 1oo2 and 1oo3 to the parallel. The corresponding system reliability reported in Table 9.3 are in accordance with Equation (9.33) and Equation(9.39), respectively.

TABLE 9.3 Reliability evaluation for different *koon* configurations

Koon configuration	System's reliability ($p = e^{-\lambda t}$)	System's MTBF
1oo2	$2p - p^2$	$3/(2\lambda)$
2oo2	p^2	$1/(2\lambda)$
2oo3	$3p^2 - 2p^3$	$5/(6\lambda)$
3oo3	p^3	$1/(3\lambda)$
1oo3	$3p - 3p^2 + p^3$	$11/(6\lambda)$
4oo5	$5p^4 - 4p^5$	$9/(20\lambda)$
Single item	p	$1/\lambda$

More complex systems can be obtained by means of combinations of the elementary structures considered above. Thus, starting from the reliability block diagram and taking into account the related equations, it easy to evaluate the system reliability of *mixed structures* and the corresponding MTBF.

9.5.3.1 *Example* The system depicted in Figure 9.11 represents a mixed structure constituted by a combination of series, parallel and *koon*. It can be observed that two identical sub-systems B are connected together in parallel and, then, in series with a 2oo3 sub-system *A*. The item *C* is in parallel configuration with the combination of *A* and *B*; the sub-system D is in series with the others.

The reliability model is

$$P(S) = \left\{1 - \left[1 - \left(3p_A^2 - 2p_A^3\right)\left(2p_B - p_B^2\right)\right](1 - p_C)\right\} p_D, \qquad (9.47)$$

where $p = e^{-\lambda t}$. Assuming random failures and considering, for sake of simplicity, $\lambda_A = \lambda_B = \lambda_C = \lambda_D = \lambda$, the system's reliability law can be written as

$$R_{SYS}(t) = e^{-2\lambda t} + 6e^{-4\lambda t} - 13e^{-5\lambda t} + 9e^{-6\lambda t} - 2e^{-7\lambda t} \qquad (9.48)$$

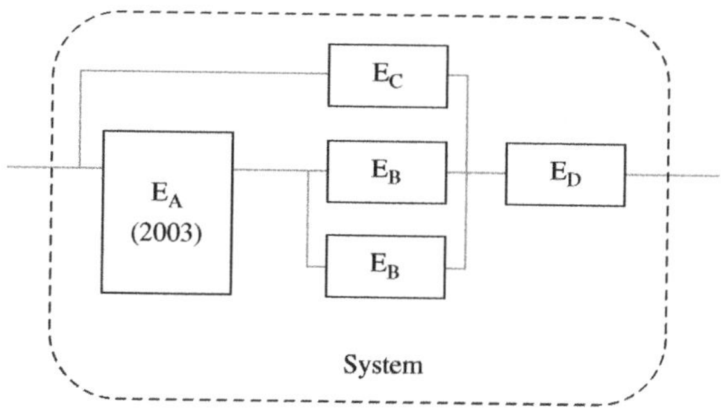

FIGURE 9.11 Mixed structure.

and, consequently, the expression of MTBF is

$$MTBF = \int_{0}^{+\infty} R_{SYS}(t)dt \cong 0.61/\lambda. \tag{9.49}$$

9.6 ANALYSIS TECHNIQUES FOR DEPENDABILITY

In this section a brief presentation of the analysis techniques that can be used for the prediction, evaluation, review and improvement of reliability, maintainability, and availability performance—in a collective term *dependability*—is proposed. Such techniques, also used for the risk and safety assessment, can be applied in different phases of the life cycle of a system, from the design to development, operation and maintenance, and at different levels: from system level to sub-system, from sub-system to component. Independently on the particular approach, it is fundamental to know the system configuration as described in Section 9.5 and to collect information concerning reliability parameters such as the failure rate, the mean time to failure, the mean time between failures, and so on. Dependability techniques can be classified in different categories [8].

1. Methods for fault avoidance
    - parts derating and selection
    - stress–strengh analysis
2. Methods for architectural analysis and dependability assessment
    (a) bottom-up method
    - event tree analysis (ETA)
    - failure mode and effects analysis (FMEA)
    - failure modes, effects, and criticality analysis (FMECA)
    - hazard and operability study (HAZOP)

    (b) top-down method
    - fault tree analysis (FTA)
    - Markov analysis
    - Petri net analysis
    - truth table (structure function analysis)
    - reliability block diagram (RBD)
3. Methods for estimating reliability parameters of basic events
    - failure rate prediction
    - human reliability analysis (HRA)
    - statistical reliability methods
    - software reliability engineering (SRE)

Among these, the bottom-up methods mainly deal with effects of single faults. Top-down, instead, are able to account for effects arising from combination of faults. In particular,

(i) *Bottom-up*: The starting point for such methods is to identify failure modes, that is the manner in which an item fails, at the component level. For each failure mode, the corresponding effect on performance is deduced for the appropriate system level.

(ii) *Top-down*: At first, the undesirable single event or system success at the highest level of interest (denoted as top event) should be identified. The contributory causes of that event at all levels are then defined and analyzed. The starting point of the top-down approach is to proceed from the highest level of interest, that is, the system or sub-system level, to successively lower levels in order to identify undesirable system operations. Only FMEA and FTA methods will be briefly described in the following sections.

9.6.1 Failure Modes and Effect Analysis

FMEA is a systematic procedure for the analysis of a system and identify its potential failure modes, their causes, and effects on system performance. In the method, a system can represent hardware, software (with their interaction) or a process [4, 9]. FMEA is applicable at various levels of system decomposition, from the highest level of block diagram down to the functions of discrete components or software commands. In fact the application of FMEA is preceded by a hierarchical decomposition of the system into its more basic elements. The analysis then starts with lowest level elements. A failure mode effect at a lower level may then become a cause in the failure mode of an item in the upper level. The analysis proceeds in a bottom-up fashion until the end effect on the system is identified.

The FMEA procedure consists of the following four main stages:

(a) establishment of the basic ground rules for the FMEA and planning and scheduling to ensure that time and expertise are available to do the analysis;

(b) executing the FMEA using the appropriate worksheet or other means such as a logic diagrams or fault trees;

(c) summarizing and reporting the analysis results including any conclusions and recommendations;

(d) updating the FMEA as the development activity progresses.

For particular applications and in risk and safety assessment, FMEA is also considered to be a method that allows to identify the severity of potential failure modes and to provide an input to identify solutions for risk reduction. FMEA also includes an estimation of the probability of occurrence of the failure modes. In this case such a technique is denoted as failure modes, effects and criticality analysis (FMECA). In Table 9.4 a classification of severity is proposed.

TABLE 9.4 Classification of severity and effects [4, 9]

Class	Severity level	Consequence to persons or environment
IV	Catastrophic	A failure mode which could potentially result in the failure of the system's primary functions and therefore causes serious damage to the system and its environment and/or personal injury.
III	Critical	A failure mode which could potentially result in the failure of the system's primary functions and therefore causes considerable damage to the system and its environment, but which does not constitute a serious threat to life or injury.
II	Marginal	A failure mode which could potentially degrade system performance function(s) without appreciable damage to system or threat to life or injury.
I	Insignificant	A failure mode which could potentially degrade the system's functions but will cause no damage to the system and does not constitute a threat to the or injury.

In FMECA approach some indices have to be taken into account:

(a) *Severity S* is a dimensionless that stands for severity, that is, an estimate of how strongly the effects of the failure will affect the system or the user on the basis of considerations summarized in Table 9.4;

(b) *Occurrence O* denotes the probability of occurrence of a failure mode for a predetermined or stated time period—even though it may be defined as a ranking number rather than the actual probability of occurrence; occurrence can be correlated with the failure rate of an item.

(c) *Detection D*, an estimate of the chance to identify and eliminate the failure before the system or customer is affected. This number is usually ranked in reverse order from the severity or occurrence numbers: the higher the detection number, the less probable the detection is.

The product of such indices defines the risk priority number, $RPN = S \times O \times D$. This index ranks the identified failure modes of an item according to their resolution priority.

9.6.2 Fault Tree Analysis

FTA is concerned with the identification and analysis of conditions and factors that cause or may potentially cause or contribute to the occurrence of a defined top event [4, 10]. FTA is often applied to the safety analysis of systems (such as transportation systems, power plants, or any other systems that might require evaluation of safety of their operation). Fault tree analysis can be also used for availability and maintainability evaluations. A fault tree is an organized graphical representation of the conditions or other factors causing or contributing to the occurrence of a defined outcome, referred to as the "top event." When the outcome is a success, the fault tree becomes a success tree, where the input events are those that contribute to the top success event.

FTA can be combined with other availability techniques, such as FMEA and FMECA (this approach is often denoted as combined analysis of availability). This analysis combination is often recommended by sector-specific standards, in particular safety standards and transportation standards. The main benefits of the combined analysis are the following:

(i) FTA is a top-down and FMEA a bottom-up analysis method and use of both deductive and inductive reasoning is regarded as a good argument for providing assurance for the completeness of an analysis.

(ii) Safety standards often demand a single failure and, in some cases, a multiple failure analysis, the first requirement being fulfilled by FMEA. Both single and multiple failure analysis are accomplished by FTA.

(iii) FMEA is also a useful method for a comprehensive identification of basic events or hazards, while FTA is a practical method for causal analysis of the undesirable events.

9.7 CONCLUSIONS

In this chapter the importance of reliability assessment has been focused. As we said in the introduction, the reliability measurements represent the starting point for more complex evaluations concerning availability, maintainability, and safety denotes as, in technical terms, with the acronym RAMS. This is justified by the fact that in many fields of application it is fundamental to guarantee and assure high reliability performance for components and systems. Often, in many contexts, the reliability assurance represents a fundamental task for the customer. Assuming such component and system compliance to specification at the output of the design and the production process, the evolution of its performance over time is evaluated by means of reliability measurements. To this aim, some important definitions are recalled in the first part of this chapter and the basic concepts for the evaluation of the reliability low and the failure rate are presented, with the fundamental statistical parameters that are traditionally used in this context: MTTF, MTBF and MTTR. After these, some models of system reliability with details and examples for series, parallel and *koon* structures are presented in the second part. How the knowledge of such models is fundamental for engineers and technicians in the design phase in order to evaluate, and improve with the use of redundant configurations and reliability growth has been highlighted. In conclusion of this chapter, some important techniques used to evaluate the availability of complex systems and equipment are briefly introduced. In particular, a general description of FMEA, FMECA and FTA is proposed for the design review. In References and Further reading, more details and insights on specific topics regarding the broad area of the reliability and reliability measurements can be made.

FURTHER READING

As said in the introduction, the knowledge of the reliability performance represents the starting point of more complex evaluations concerning the RAMS requirements. The references proposed in this chapter cover the topics presented here and can be considered exhaustive. However, the reader can deepen some details and expand their knowledge considering further reading. In particular, the probabilistic aspects of reliability theory can be deepened in Barlow and Proschan (1981), *Statistical Theory of Reliability and Life Testing – Probability Models*, edited by Silver Spring. Always on reliability theory and system reliability, among the various books and publications that are present in literature, can be considered Billinton and Allan (1987), *Reliability Evaluation of Engineering Systems: Concepts and Techniques*, Plenum Press, as well as Elsayed (1996), *Reliability Engineering*, Addison Wesley. More details concerning some availability and dependability techniques, such as FMEA and FMECA, can be found in Stamatis (1995), *Failure Mode and Effect Analysis – from Theory to Execution*, ASQC Quality Press. The topics presented in this chapter are the basic concept for risk assessment and safety evaluation. For these subjects, in addition to the reported references, more details can be found in the Standards, being such topics in evolution. To this aim, it can be suggested [1] and the family of Standard IEC 61508 (parts from 1 to 7), *Functional Safety of Electrical, Electronic, Programmable Electronic Safety–Related Systems*. More specific standards in this field can be found in function of particular industrial contexts.

REFERENCES

1. IEC 60050-191 ed 1.0. International Electrotechnical Vocabulary (IEV), Chapter 191: Dependability and quality of service. IEC International Electrotechincal Commission, Geneve (CH) (1990).
2. Rausand, M., Hoyland, A. *System Reliability Theory: Models, Statistical Methods and Applications*, 2nd edition, John Wiley & Sons, USA (2004).
3. Birolini, A. *Quality and Reliability of Technical Systems: Theory, Practice and Management*, Springer-Verlag Berlin Heidelberg, Germany (1997).
4. Lazzaroni, M., Cristaldi, L., Peretto, L., Rinaldi, L., Catelani, L. *Reliability Engineering – Basic Concepts and Applications in ICT*, Springer-Verlag Berlin Heidelberg, Germany (2011).
5. Modarres, M., Kaminskiy, M., Krivtsov, V. *Reliability Engineering and Risk Analysis – A practical Guide*, 2nd edition., CRC Press, NY (2010).
6. IEC 60068-1. *Environmental Testing*, Part.1-General and Guidance, IEC International Electrotechnical Commission, Geneve (CH) (2011).
7. ISO 3534-1. *Statistics – Vocabulary and Symbols, Part 1: Probability and General Statistical Terms*, International Organization for Standardization, Geneve, CH (1993).
8. IEC 600300-3-1. *Dependability Management, Part.3.1-Application Guide-Analysis Techniques for Dependability-Guide on Methodology*, IEC International Electrotechnical Commission, Geneve (CH).

9. IEC 60812. *Analysis Techniques for System Reliability – Procedure for Failure Mode and Effects Analysis (FMEA)*, IEC International Electrotechnical Commission, Geneve (CH).
10. IEC 61125. *Fault Tree Analysis (FTA)*, IEC International Electrotechnical Commission, Geneve (CH).
11. CEI EN 61703 ed.1.0. *Mathematical Expressions for Reliability, Availability, Maintainability and Maintenance support terms*, CEI-Comitato Elettrotecnico Italiano, Italy (2002).

EXERCISES

9.1 The system represented in Figure 9.12 is a mixed structure where the series of sub-systems A and B is connected in parallel with sub-system C. Sub-system A is characterized by an asymptotic availability equal to 0.99 and MTTR = 168 hours and sub-system B is a 3*oo*4 configuration with reliability for each element equal to 0.5 at 1000 hours. Calculate the system's reliability at 1000 hours assuming that the reliability of the sub-system C is equal to 0.6 at same time.

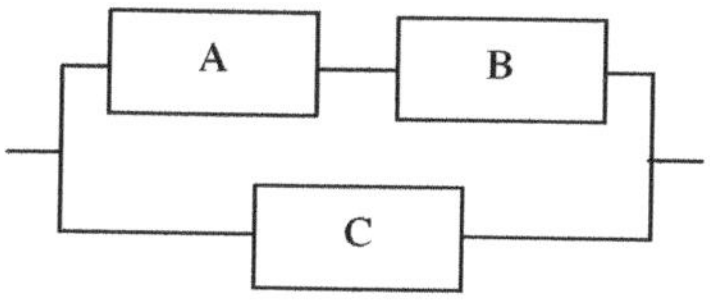

FIGURE 9.12 Mixed structure.

9.2 Evaluate the reliability of a single element at the time equal to MTTF, assuming constant failure rate.

9.3 A system is constituted by 12 identical items, with the same reliability performances, in series configuration. Calculate the reliability of the generic item if the probability of failure of the whole system is equal to 7% at 1 year.

9.4 A system is constituted by four identical items, with same reliability performances, in parallel configuration. Calculate the reliability of the generic item if the probability of operation is equal to 93% at 1 year.

9.5 Assume a failure rate of the single item equal to $3 \cdot 10^{-5}$ h^{-1} and a 2*oo*3 configuration. Calculate the system's reliability versus time. Then, evaluate the probability of failure for the system at the end of the warranty (one year is considered as warranty time).

9.6 For Exercise 9.5, calculate the probability of operation of the 2*oo*3 configuration at time equal to MTBF.

9.7 A power supply with MTBF = 200,000 hours is constituted by five super capacitors. Assuming that all capacitors are functioning successfully, calculate the probability of operation of each item at 8 760 hours. By assuming super capacitors with identical performance, how the MTBF of the power supply can be increased till 300,000 hours?

9.8 A streetlight is constituted by three identical lights with MTTF = 20,000 hours each. It is assumed that the streetlight is in up-state (operating state) if at least one light is in up-state. Evaluate the probability of failure of the streetlight at 8760 hours.

9.9 In a parallel structure evaluate the number of identical items that are necessary to obtain a 50% increase of the mean time between failures assuming, for each item, a failure rate λ.

9.10 A chain structure is constituted by six items so that $R_i = 0.85$, for $i = 1, \dots, 6$. Compare the system's reliability value with that of a parallel configuration designed to operate with the same items.

CHAPTER 10

EMC MEASUREMENTS

CARLO CAROBBI
Department of Information Engineering, Via S. Marta, University of Florence, Firenze, Italy

10.1 INTRODUCTION

By the expression *electromagnetic compatibility measurements* it is here specifically meant *measurements aimed at assessing compliance of a device, equipment, or system to an electromagnetic compatibility requirement*. In this sense electromagnetic compatibility (EMC) measurements are *compliance measurements*. The scope of this chapter is to introduce the reader to the fundamental measuring instrumentation and measurement methods, defined by the International Electrotechnical Commission (IEC), for compliance assessment of commercial products. The essential metrological requirements specified for the EMC measuring instrumentation and test sites are also recalled and the relevant methods of verification of such requirements are briefly described.

The specific measurement techniques, adopted by other than IEC standards organizations, for compliance assessment of special equipment (e.g., designed for military, space, or avionic applications) are not discussed here. However, the reader should keep in mind that the same basic EMC measurement and testing techniques are shared among the different standards organizations. In addition, the measuring instrumentation is often the same, the main deviations being originated by differences in the magnitude of the limits (e.g., higher immunity limits for a mission critical military equipment rather than for a commercial off-the-shelf electronic product) and/or parameters (e.g., the frequency range involved).

The treatment of EMC measurement uncertainty, although of great interest, is beyond the scope of this essay on EMC measurements. Further, the matter is rapidly evolving especially in the field of immunity tests, where also the basic concepts are still under debate among the experts. The reader is therefore referred to the specialized literature available on the subject. In particular, general concepts on the evaluation

Modern Measurements: Fundamentals and Applications, First Edition.
Edited by Alessandro Ferrero, Dario Petri, Paolo Carbone, and Marcantonio Catelani.

of EMC measurement uncertainty are in Reference 1, which is mainly devoted to emission measurements, and Reference 2, where also the basic ideas behind the evaluation of the uncertainty of immunity tests are presented. A quantitative and up-to-date uncertainty analysis including detailed uncertainty budgets of emission measurements is in Reference 3. Examples of uncertainty evaluations of immunity tests are presented and discussed in dedicated informative[1] annexes of the basic immunity standards. The work is still in progress. The immunity standards that currently include the annex on uncertainty are in References 4–8. Further useful references on the subject are the tutorial papers in References 9 and 10: the first deals with measurement uncertainty evaluations when logarithmic quantities are involved, as it frequently occurs in EMC; the second is devoted to the application of the Monte Carlo method to EMC measurement uncertainty calculations. An analysis of the reproducibility of radiated and conducted emission measurements is available in References 11 and 12, respectively.

This introduction section is concluded by a remark concerning the distinction between compliance and *pre-compliance* EMC measurements. Compliance measurements are taken at the final stage of the development process of a product. They are designed for maximizing measurement reproducibility and minimizing the risk of disputes among the interested parties. This is achieved through the adoption of measurement methods that assure metrological traceability of measurement results and a detailed description of the measurement procedure, setup and layout of both the measuring instrumentation and the equipment under test. On the contrary, pre-compliance measurements are performed from the early development stages of the product. They are designed for providing the EMC engineer with sufficient insight into the non-intentional behavior of the product in order to let him identify the simplest, most effective and minimum-cost design solutions for the product's EMC. Pre-compliance measurement methods are flexible, quick, and relatively simple. Metrological traceability is not an issue since, in most cases, measurements are aimed at quantifying the effectiveness of a new design solution by comparing measurement results obtained before and after the implementation of the solution. Pre-compliance measurements are more related to EMC engineering than to measurement science, and this is the main reason why they are not dealt with in this book. The reader interested in pre-compliance measurements is referred to Chapter 18 of Reference 13, where this important subject is presented in a plain and systematic way.

10.2 DEFINITIONS AND TERMINOLOGY

Several of the following definitions are directly taken from the International Electrotechnical Vocabulary (IEV, [14]), others are adapted from IEV.

Electromagnetic compatibility is the ability of an equipment or system to function satisfactorily in its electromagnetic (EM) environment without introducing intolerable EM disturbances to anything in that environment. The opposite of EMC is

[1] Non-normative.

electromagnetic interference (EMI), which is the degradation of the performance of an equipment, transmission channel or system caused by an EM disturbance. The terms "electromagnetic disturbance" and "electromagnetic interference" designate respectively the cause and the effect.

EM *disturbance* is any EM phenomenon which may degrade the performance of a device, equipment, or system, or adversely affect living or inert matter. Note that this is a broad definition, extending the context of EMC to the protection of human beings from exposure to EM fields.

Disturbances are grouped according to their *time-domain* or *frequency-domain* representation. In the time-domain a *sinewave* or *sinusoidal* disturbance is represented by a constant amplitude or modulated (amplitude, frequency modulation, or both) sinewave oscillation. An *impulse* disturbance is represented by an impulse of any shape (oscillatory, unidirectional, more or less regular) which repeats with a constant or random rate. In the frequency-domain a *narrow-band* disturbance has a frequency spectrum enclosed within the bandwidth of the receiver (or victim). A *wide-band* disturbance has a frequency spectrum which occupies a relatively wide frequency range and is flat within the bandwidth of the receiver. It is important to observe that the distinction between narrow- and wide-band depends on the bandwidth of the receiver, it is not ascertained from the intrinsic nature of the disturbance.

A sinewave disturbance may be coherent or incoherent. A *coherent* sinewave disturbance is characterized by a high correlation between the values of the phases at two instants in time. An *incoherent* sinewave disturbance is characterized by a low correlation between the values of the phases at two instants in time. These definitions can be extended to impulse disturbances reasoning in terms of their sinewave spectral components. Therefore an impulse disturbance which regularly repeats at a constant rate (or single shot) is a coherent disturbance. Thermal noise is an incoherent disturbance. Similar to the narrow- and wide-band classification, the distinction between coherent and incoherent depends on the bandwidth of the receiver (or victim). For example, the disturbance generated by the superposition of several uncorrelated impulse disturbances appears as incoherent, when measured with a spectrum analyzer, if the average repetition rate is large with respect to the resolution bandwidth of the spectrum analyzer, otherwise it appears as coherent. In the first case (incoherent) the displayed *power* spectral density of the disturbance is proportional to the spectrum analyzer's resolution bandwidth, in the second case (coherent) it is the displayed *voltage* spectral density to be proportional to the resolution bandwidth.

A test is a technical operation that consists of the determination of one or more characteristics of a given product to a specified procedure. A *compliance test* is a test used to show whether or not a characteristic or a property of an item complies with the stated requirements.

EMC tests are classified as emission or immunity (susceptibility) tests. *Emission* is the phenomenon by which EM energy emanates from a device, equipment, or system. *Immunity* to a disturbance is the ability of a device, equipment, or system to perform without degradation in the presence of an EM disturbance. *Susceptibility* is a lack of immunity.

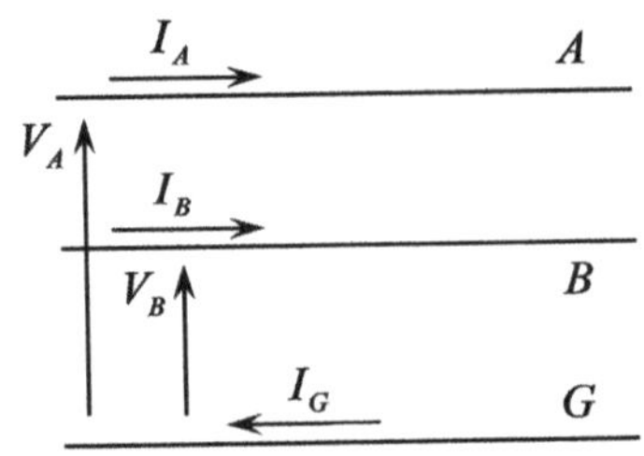

FIGURE 10.1 A connection made of three wires named A, B, and G (ground).

Emission and immunity tests are said to be *conducted* if the disturbance energy is generated in the form of voltage, current, or power. Emission and immunity tests are *radiated* if the disturbance energy is generated in the form of a field quantity (electric, magnetic, EM field, or power density).

The *emission level* of a disturbing source is the level of a given EM disturbance emitted from a particular device, equipment, or system. The *immunity test level* is the level of a test signal used to simulate an EM disturbance when performing an immunity test. The *emission limit* of a disturbing source is the specified maximum emission level of a source of EM disturbance. The *immunity limit* is the specified minimum immunity level of a device, equipment, or system.

Consider a connection between a source and a load which consists of two wires A and B plus a third wire G at ground potential, that is, $V_G = 0$.[2] Let I_A be the current flowing along wire A (see Figure 10.1), I_B flows along B and I_G along G. It is assumed that no current flow can take place outside this three-wires connection system, therefore $I_A + I_B - I_G = 0$. The *differential mode* (DM) or *symmetric* current is given by

$$I_{DM} = \frac{I_A - I_B}{2},$$

and the *common mode* (CM) or *asymmetric* current is

$$I_{CM} = \frac{I_A + I_B}{2}.$$

Therefore the following decomposition applies

$$I_A = I_{DM} + I_{CM},$$

$$I_B = -I_{DM} + I_{CM},$$

and $I_G = 2I_{CM}$. I_{DM} can be interpreted as the *intentional* current supplying the load while I_{CM} may represent an *unintentional* current due to leakage.

[2]Such connection scheme is the one adopted, for example, for the low voltage AC power distribution system. In that case A corresponds to the phase, B to the neutral and G to the safety wire.

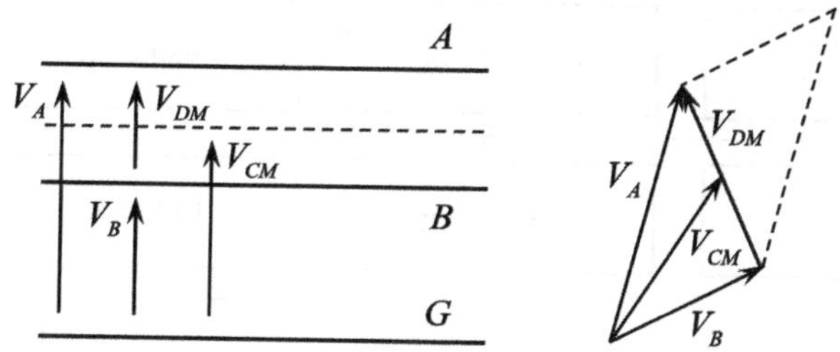

FIGURE 10.2 Voltage decomposition into DM and CM.

DM (symmetric) and CM (asymmetric) voltages can be similarly defined, from the ground referenced voltages V_A and V_B, as

$$V_{DM} = V_A - V_B,$$

$$V_{CM} = \frac{V_A + V_B}{2}.$$

The decomposition of voltages is as follows:

$$V_A = \frac{V_{DM}}{2} + V_{CM},$$

$$V_B = -\frac{V_{DM}}{2} + V_{CM},$$

and it is represented in Figure 10.2. Voltages V_A and V_B are named *unsymmetric* in conducted emission measurements terminology.

10.3 THE MEASURING RECEIVER

Emission tests are indirect measurements. Voltage is the measured quantity. Voltage is measured by using a scalar, frequency selective voltmeter. In the frequency range from 9 kHz to 1000 MHz the standard voltmeter is the so called *EMI receiver*, while in the range above 1 GHz a *spectrum analyzer* is adopted. The most important difference between the two is the detector: the *quasi-peak* detector for the EMI receiver and the *peak* detector for the spectrum analyzer. Both voltmeters have tight metrological performances specified in Reference 15.

10.3.1 Quasi-Peak Measuring Receivers

The building blocks of the quasi-peak measuring receiver, or EMI receiver, are those represented in Figure 10.3. The instrument is a superheterodyne receiver. Its operation consists of frequency conversion through the mixer and local oscillator

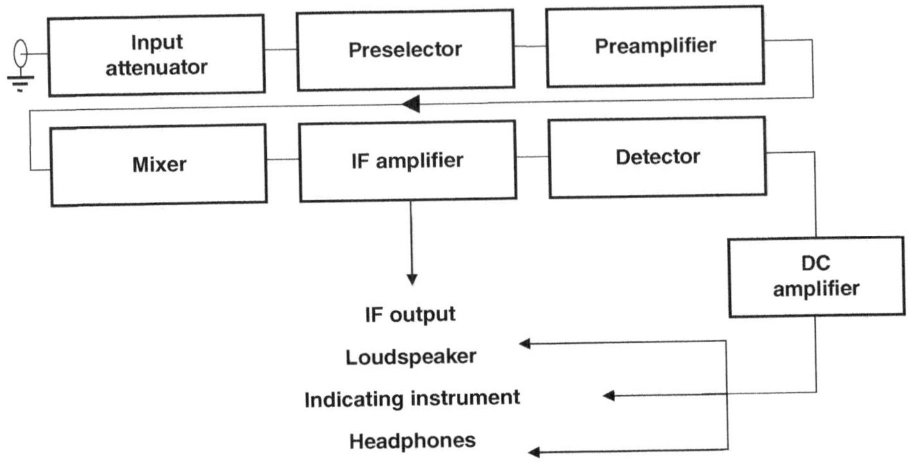

FIGURE 10.3 The EMI receiver.

(local oscillator is not represented in Figure 10.3), amplification and filtering at intermediate frequency (IF) and, finally, detection. Once detected, the disturbance is amplified by a direct coupling (DC) amplifier and sent to a loudspeaker or headphones, in order to listen to its audio characteristic, or quantified through an indicating instrument (usually a digital display).

The input attenuator is resistive, stepped, and broadband. The preselector is a set of manually switched pass-band filters each having a different tuned frequency. The preselector is useful when a low amplitude disturbance must be detected in the presence of a high amplitude disturbance or signal at a different frequency. This is the case when measuring the disturbance generated by an intentional transmitter. The low amplitude disturbance and the high amplitude transmitter carrier are simultaneously present at the input of the receiver. In order to avoid overloading the most sensitive stages of the receiver (i.e., the low noise preamplifier and mixer), the intentional signal can be strongly attenuated by the preselector while the disturbance is left to reach the next stages. Preselector designs which automatically track the tuned frequency of the receiver are also available. The coordinated use of the input attenuator, preselector and low noise preamplifier permits to easily achieve more than 60 dB instantaneous dynamic range[3] and to reliably measure the disturbance even when the difference between the amplitudes of the carrier and the disturbance is of the order of 100 dB.

Tolerances are specified on the metrological characteristics of EMI receivers [15]. The input circuit is unbalanced with nominal 50 Ω input impedance. The voltage standing wave ratio (VSWR) shall not exceed 2.0 to 1 when the radio frequency (RF) attenuation is 0 dB and 1.2 to 1 when the RF attenuation is 10 dB or greater. The specified tolerance on the sinewave amplitude measurement accuracy is ±2 dB, that on the impulse response is between ±1 and ±2 dB, depending on the impulse

[3]The instantaneous dynamic range is the ratio, expressed in decibel, between the largest (spurious-free) and smallest (detectable) signals simultaneously present at the input of the receiver.

repetition rate (the lower the repetition rate the higher the tolerance). The curve of the selectivity[4] of the measuring receiver shall lie within specified limits. The intermediate frequency rejection ratio[5] and the image frequency rejection ratio[6] shall be not less than 40 dB.

Other specifications concern spurious responses (other than intermediate and image frequencies), limitation of intermodulation effects, error due to background noise, screening effectiveness, emissions (both conducted and radiated), intermediate frequency and quasi-peak detector outputs.

The detector is a weighting circuit whose time constants are specified in Reference 15. Four basic types of detector are presently standardized, namely peak, quasi-peak, average, and root-mean-square-average (RMS-average) detector. The peak and average detectors are designed to indicate the peak and average value of the envelope of the voltage at their input, respectively. The quasi-peak detector is aimed at quantifying the subjective average annoyance effect due to a disturbance of radio-reception, in particular, the meter indication increases according to a specific law when the repetition frequency of the disturbance increases. The RMS-average detector is a combination of RMS and average detectors such that it indicates the average value when the repetition frequency is lower than a given corner frequency, while it indicates the RMS value when the repetition frequency is higher than the corner frequency.[7] A different value of the corner frequency is specified depending on the tuning frequency of the receiver: 25 Hz for tuning frequencies between 9 kHz and 150 kHz, 1 kHz above 150 kHz and up to 18 GHz. Most of emission limits are set in terms of quasi-peak, average, and peak detectors. Attention will be mainly drawn here to the quasi-peak detector because peak detection and average detection are accomplished by conventional spectrum analyzers through the maximum-hold and averaging functions of the displayed trace, respectively.

The basic detector circuit is represented in Figure 10.4, where R_C is the sum of the output resistance of the IF amplifier, r_s, and the conduction resistance of the diode D, while R_D is the resistance through which the capacitance C discharges,[8] v is the voltage at the input of the detector, v_d is the voltage at the output.

10.3.1.1 Response to Sinewave Disturbances

The operation of the detector circuit essentially depends on the time constants of the charge and discharge of capacitance C. The *charge* time constant, T_C, is the time needed, after the instantaneous application of a constant amplitude sinewave voltage v at the input of the detector, for the output voltage v_d to reach 63% of its final value V_{qp}. The *discharge* time

[4]Selectivity is quantified by the variation with frequency of the amplitude of the input sinewave voltage that produces a constant indication on the measuring receiver.

[5]Ratio, expressed in decibel, between the amplitude of the input sinewave voltage at the intermediate frequency and that at the tuned frequency that produces the same indication.

[6]The ratio of the input sinewave voltage at the image frequency to that at the tuned frequency that produces the same indication.

[7]Therefore the indicated voltage value increases linearly at frequencies below the corner frequency and proportionally to the square root of the repetition frequency at frequencies above the corner frequency.

[8]It is assumed, as it is in practice, that $R_D \gg R_C$.

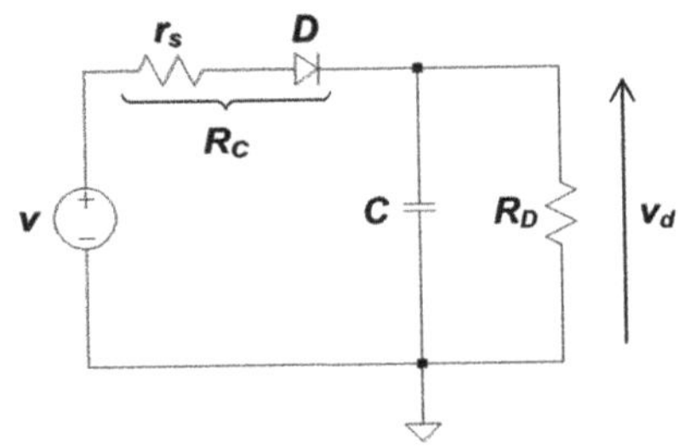

FIGURE 10.4 Detector circuit.

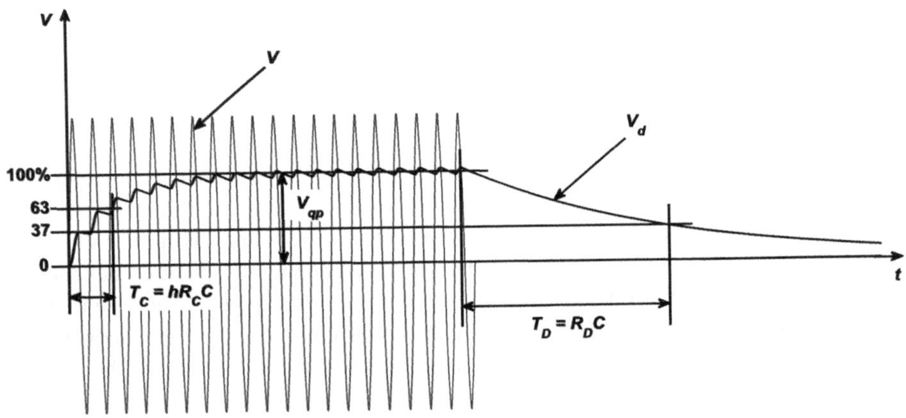

FIGURE 10.5 Charge and discharge time constants.

constant T_D is the time needed, after removal of the input voltage, for the output to fall to 37% of its initial value. A schematic representation of the time constants is in Figure 10.5. It is well known that $T_D = R_D C$. Further, in the case of a step input voltage, $T_C = R_C C$ while in the case of a sinewave input voltage $T_C = hR_C C$, where $h > 1$. The steady state value V_{qp} in Figure 10.5 represents the quasi-peak detector response to a sinewave disturbance. Note that V_{qp} is lower than the amplitude V of the sinewave, then $V_{qp} = k_s V$, and $k_s < 1$.

The time constants specified for the quasi-peak detector are those in Table 10.1, where also the −6 dB resolution bandwidth B_6 and the overload factor, of the IF, OF_{IF}, and DC, OF_{DC}, amplifiers are reported. The value of the parameter k_s can be derived from the analysis of the detector circuit in Figure 10.4 and it results to be quite close to one[9] except in the frequency band from 9 to 150 kHz where its value is 0.81.

Since the time constants of the quasi-peak detector are large, automatic scanning over the relatively wide frequency interval required by tests is slow. Due to this

[9]Hence the name "quasi-peak".

TABLE 10.1 Fundamental characteristics of the quasi-peak receiver

Characteristics	9 to 150 kHz	0.15 to 30 MHz	30 to 1000 MHz
B_6 (kHz)	0.20	9	120
T_C (ms)	45	1	1
T_D (ms)	500	160	550
T_M (ms)	160	160	100
h	3.0	3.9	4.0
k_s	0.81	0.97	0.99
k_p	0.12	0.12	0.011
k_m	0.53	0.27	0.62
k_n	1.3	2.3	2.7
OF_{IF} (dB)	24	30	43.5
OF_{DC} (dB)	6	12	6

reason standard emission measurement methods allow fast pre-scans with a conventional spectrum analyzer or switching the receiver to the peak detection mode,[10] in order to quickly identify the frequencies of the stronger emissions and the worst-case measurement setup.[11] After this preliminary identification phase compliance measurements are performed by using the quasi-peak detector and manually tuning the receiver.

10.3.1.2 *Response to Impulse Disturbances* The response of the quasi-peak detector to recurrent[12] wide-band impulse disturbances is represented in Figure 10.6. The input v to the detector is the impulse response of the IF amplifier. The peak value of the envelope is $V_{max} = V(f_T)B_{imp}$, where $V(f_T)$ is the single-sided voltage spectral density, in V/Hz, of the disturbance at the tuned frequency f_T, and B_{imp} is the impulse bandwidth[13] of the receiver (see References 15–19). The detector output v_d oscillates between the values V_1 and V_2.[14] Due to the inertia of the indicating instrument, the value indicated by the receiver is $V_{qp} = \frac{V_1+V_2}{2}$. The time constant T_M of the critically damped indicating instrument[15] is in Table 10.1.

[10]All types of detectors are implemented in the same measuring receiver owing to analog to digital conversion and numerical processing techniques.

[11]For example, height and polarization of the receiving antenna and angle of the rotating table which maximize radiated emission.

[12]The repetition frequency of the disturbance is assumed to be smaller than the impulse bandwidth of the receiver, so that the transient response of the IF amplifier is nullified in the time interval between each impulse and the next one.

[13]For Gaussian IF filters, which is an acceptable approximation of the actual shape of their frequency response, we have $B_6 = \sqrt{2}B_3$ and $B_{imp} = 1.064B_6$.

[14]The straight lines in Figure 10.6 are an approximation. However, since $T_C \gg 1/Bimp$ then the capacitor charge voltage is approximately linear. Further, also discharge is nearly linear if $T_D \gg T$.

[15]T_M is $\frac{1}{2\pi}$ times the period of free oscillation of the instrument with all damping removed.

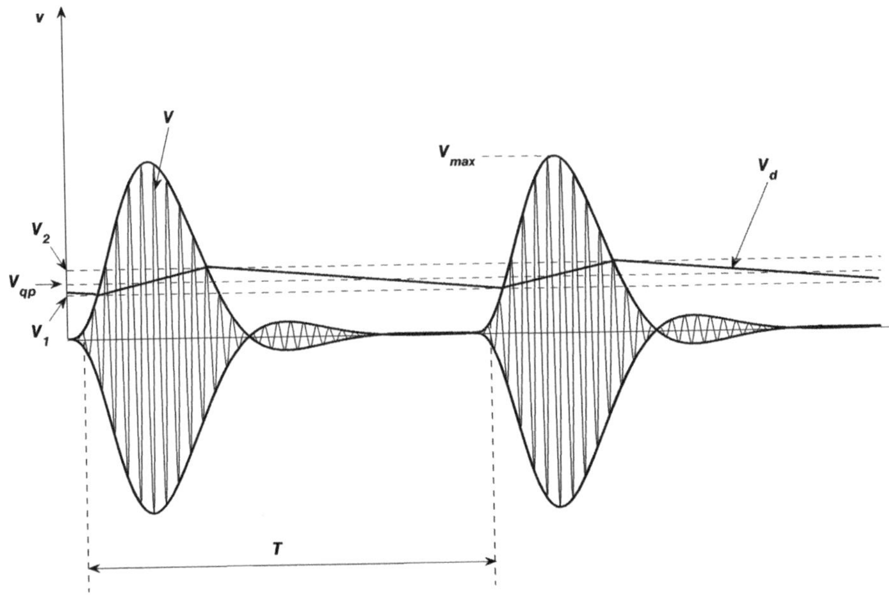

FIGURE 10.6 Response of the quasi-peak detector to a recurrent impulse disturbance.

The response of the EMI receiver to recurrent impulse disturbances is (see References 15 and 16) $V_{qp} = P(\alpha)V_{max}$, where $P(\alpha)$ is the *pulse rectification coefficient*,

$$\alpha = \frac{\pi}{4}\frac{T_C}{T_D}\frac{B_6}{f_R},$$

and $f_R = \frac{1}{T}$ is the impulse repetition frequency. P as a function of α is plotted in Figure 10.7. In order to appreciate the effect of the repetition frequency of the disturbance on the indication of the receiver the plot of P directly as a function of f_R is represented in Figure 10.8 for the specific case of $B_6 = 120$ kHz and $T_D/T_C = 550$. It is seen that the indication of the receiver increases with f_R increasing, the rate of increase being designed to be the same as the annoyance felt by the average listener.

In the case of very low repetition rate impulse disturbances[16] the indication of the receiver oscillates between two extreme values. It is of interest to present the response to the single impulse disturbance, which is the one plotted in Figure 10.9. Since $T_D \gg T_C \gg B_{imp}$ then the impulse response of the quasi-peak detector is approximately given by $V_2 = k_p V_{max} e^{-\frac{t}{T_D}}$, where the k_p value depends on the product $T_C B_{imp}$. The indication of the receiver increases from zero to the peak value $V_{qp,max} = k_p k_m V_{max}$, and then comes back to zero.

[16]Of the order of a few hertz or less.

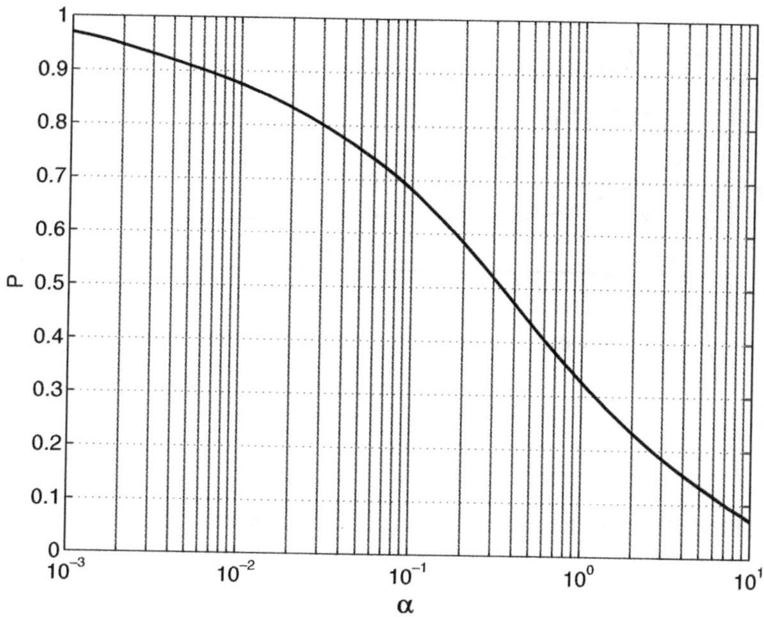

FIGURE 10.7 Parameter P quantifying the response of the EMI reciver to recurrent impulse disturbances.

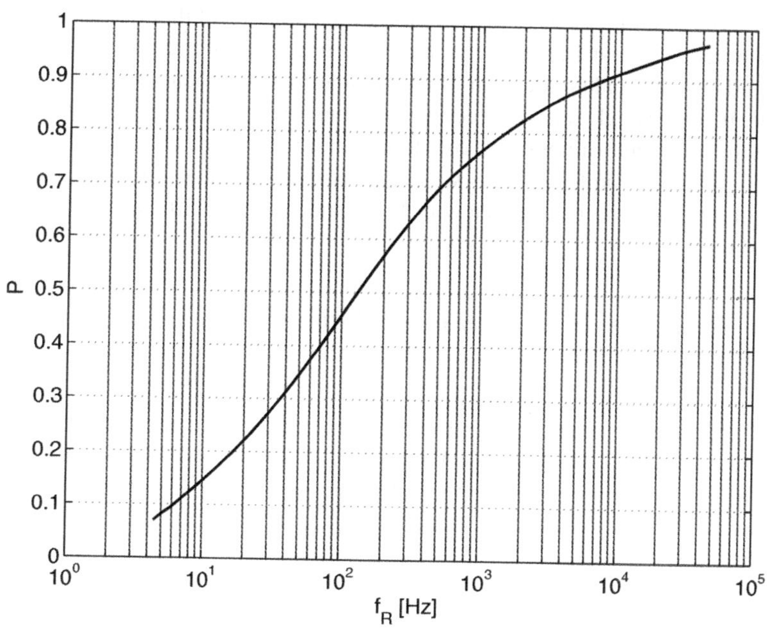

FIGURE 10.8 P as a function of the impulse repetition frequency f_R in the specific case of the 30–1000 MHz frequency band.

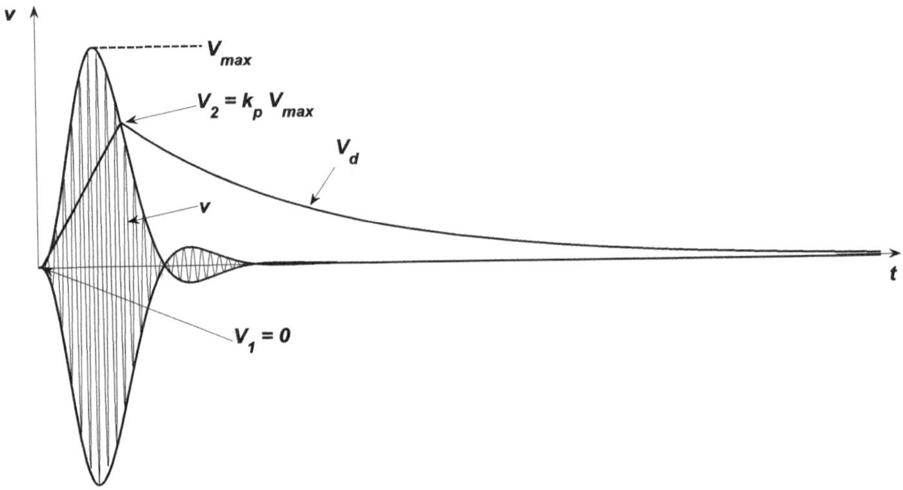

FIGURE 10.9 Response of the quasi-peak detector to a single impulse.

Overload may be an issue when measuring low repetition rate impulses (eventually a single impulse). Some margin of linearity is needed in order to avoid IF and DC amplifiers' saturation especially when the indication is well below the full scale value. The overload factor is defined as the ratio of the level that corresponds to the range of practical linear function[17] of a circuit, or group of circuits, to the level that corresponds to full-scale deflection of the indicating instrument [15]. It is easy to to conclude, from the analysis of the receiver response to impulses of different repetition rate, that the *minimum* values of the overload factor for IF and DC amplifiers are given by $-20\log_{10}(k_p k_m)$, and $-20\log_{10} k_p$, respectively. The values of the parameters k_p and k_m are in Table 10.1, together with the standard values of the overload factors OF_{IF} and OF_{DC}.

10.3.1.3 *Response to Incoherent Sisturbances* We finally consider the response of the quasi-peak detector to incoherent wide-band disturbances (see Reference 16). The level indicated by the receiver is given by $V_{qp} = k_n V_n(f_T)\sqrt{B_n}$, where $V_n(f_T)$ is the voltage spectral density, in $\mathrm{V}/\sqrt{\mathrm{Hz}}$, of the disturbance at the tuned frequency f_T, B_n is the noise bandwidth,[18] and k_n is a numerical coefficient whose value is in Table 10.1. For the sake of comparison, note that the indication in the case of use of the average detector is $V_{ave} = 1.25V_n(f_T)\sqrt{B_n}$, and therefore $V_{qp} > V_{ave}$.

[17] The maximum level at which the steady-state response does not depart by more than 1 dB from ideal linearity.

[18] For Gaussian IF filters $B_n = 0.753B_6$.

10.3.2 Peak Measuring Receivers

A peak measuring receiver is adopted for compliance measurements above 1 GHz [15]. This essentially consists of a spectrum analyzer with maximum-hold function of the displayed trace. The metrological characteristics specified are the same as those for the quasi-peak measuring receiver except for overload, which is not an issue for peak measuring receivers. In particular the input VSWR shall be less than 3.0 to 1 with 0 dB input attenuation, and 2.0 to 1 with 10 dB input attenuation or more, the accuracy of sinewave voltage measurement shall be better than ±2.5 dB and that of impulse measurements shall be better than ±1.5 dB, for any repetition frequency, in the tuning frequency range from 9 kHz to 1 GHz. The accuracy of the impulse response for tuning frequencies above 1 GHz is not specified. The standard resolution bandwidth adopted is 1 MHz (impulse bandwidth). Similar or identical specifications to those of the quasi-peak measuring receiver apply for the rejection of intermediate, image and spurious frequencies, intermodulation effects, screening, and emissions.

10.4 CONDUCTED EMISSION MEASUREMENTS

Conducted emission tests consist in the measurement of the RF disturbance voltage generated by an equipment under test (EUT) connected to a power outlet[19] [20, 21]. Current disturbance measurements, although not frequently required in compliance tests, have practical and conceptual importance.

10.4.1 The Artificial Mains Network

In order that different laboratories obtain reproducible conducted emission test results the EUT should apply the disturbance voltage across the same RF power mains impedance. Unfortunately the impedance seen looking into a power outlet is extremely variable from place to place since it depends on the particular geometrical and electrical configuration of the power distribution system and the local loading conditions. Therefore an artificial mains network (AMN) was standardized providing a normalized mains impedance at RF to EUT [20]. Such impedance is comparable, both in magnitude and phase, to the impedance that would be seen looking into a power outlet connected through a few tens of meter of power cable to the secondary winding of the transformer feeding the unloaded line. The frequency range of interest is comprised between 9 kHz or 150 kHz, depending on the specific product to be tested, and 30 MHz. The AMN simultaneously serves different purposes, namely

- connecting EUT to the mains network at power frequency;
- isolating EMI receiver from disturbances originating from the power-line network;

[19] The important but rather specific case of the measurement of CM disturbance voltage at EUT telecommunications ports is not treated here.

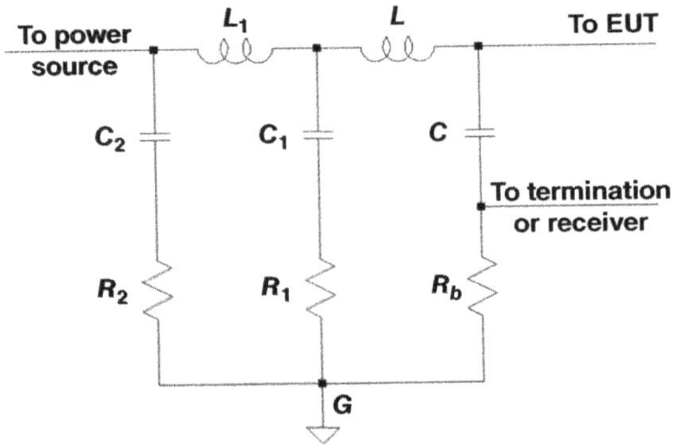

FIGURE 10.10 Basic electrical circuit of the 50 Ω in parallel with 50 μH plus 5 Ω AMN.

- isolating power-line network from disturbances originating from EUT;
- isolating EMI receiver from the power source;
- connecting EUT to the EMI receiver;
- providing EUT with a normalized load impedance.

The circuit representation of the basic network designed for tests in the 9 kHz to 30 MHz frequency range is in Figure 10.10. At power frequency inductances are short circuits while capacitances are open circuits, thus EUT is directly connected to the power source. RF disturbances originating from the power source are attenuated before reaching the input of the EMI receiver by the action of the low- pass filter formed by $C_2 = 4$ μF, $R_2 = 10$ Ω, $L_1 = 250$ μH, $C_1 = 8$ μF, $R_1 = 5$ Ω and $L =$ 50 μH. The same low pass filter prevents the disturbances generated by EUT from reaching the power source. The high pass filter formed by the capacitance $C =$ 250 nF and the input resistance of the receiver, $R = 50$ Ω, limits the amplitude of the power-mains voltage reaching the receiver's input to a safe value.[20] At the same time capacitance C connects EUT to the receiver at RF. Finally, the RF impedance seen by the EUT when looking into the network terminals is dominated by the parallel between the input resistance of the receiver R and the series of R_1 and L, that is, 50 Ω in parallel with 50 μH plus 5 Ω, which is the standard AMN impedance value.

A simplified network is adopted for disturbance measurements in the frequency range between 150 kHz and 30 MHz. Such a network does not include the filter section formed by C_2, R_2, and L_1 since inductance L and capacitance C_1 are deemed to provide adequate isolation in this higher frequency range. Along the same line of reasoning capacitance C_1 is reduced from 8 to 1 μF, resistance R_1 is replaced by a short circuit, because its impedance is negligible with respect to the impedance of the inductor L. Finally, capacitance C is reduced from 250 to 100 nF, because this

[20] Receivers are usually equipped with an internal DC block which is also effective at power frequency. Further, an external RF attenuator can be inserted in order to provide for additional protection.

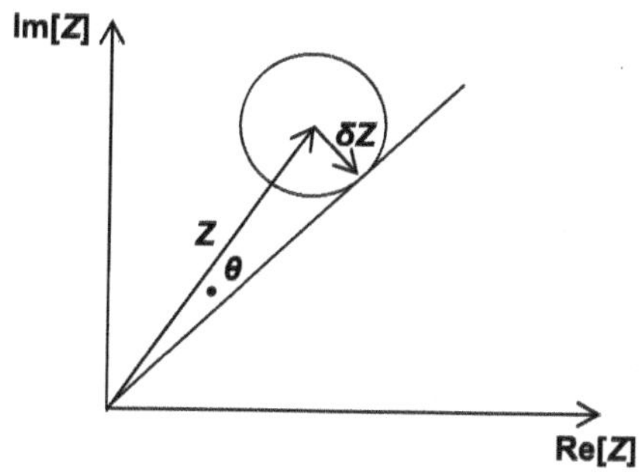

FIGURE 10.11 The AMN impedance tolerance circle.

value still provides adequate coupling of the disturbance generated by EUT to the receiver while further decreasing the amplitude of the unwanted power-mains voltage reaching the input of the receiver.

Since capacitances C_1 and C_2 have large values, for safety reasons the network case should either be solidly bonded to reference earth or an isolating transformer should be used. Further, resistor $R_b = 1000\ \Omega$ has the task of bleeding the charge stored in the capacitors once that the network is disconnected from the power source.

During tests each line supplying the EUT shall be connected to a network as the one in Figure 10.10 (e.g., three networks are used for a three-phase EUT). Disturbance measurements shall be performed connecting the receiver to one network at a time while the others shall be terminated into a 50 Ω load. Commercial AMNs may include up to three networks with a switch that permits to select the line to which the receiver is connected and to terminate the other ones.

Tolerances on magnitude and phase of AMN impedance are specified and they are ±20% and ±11.5 deg, respectively. Such tolerances were established by using the concept of the *impedance tolerance circle*, see Figure 10.11. If Z is the standard value of the complex impedance and $\delta Z = 0.2Z$ then $\theta = \arcsin(0.2) = 11.5$ deg. This means that all the complex impedance values within the circle in Figure 10.11 are compliant with the impedance tolerance specification. The impedance requirement shall be satisfied with both the power source port open and short-circuited.

A minimum value of isolation between EUT port and power source port is also specified. The residual attenuation between EUT port and receiver port shall be corrected for by using an appropriate correction factor obtained through calibration.

The most critical component of the network is the inductor L because it is interested, at the same time, by the flow of the large current supplying the EUT,[21] and must not deviate too much from its ideal behavior, especially in the higher frequency range.[22] A special design was conceived for this purpose and described in Reference 20.

[21] For example, 100 A, which implies the use of a large air-coiled inductor.

[22] Because otherwise the impedance tolerance requirement would not be satisfied.

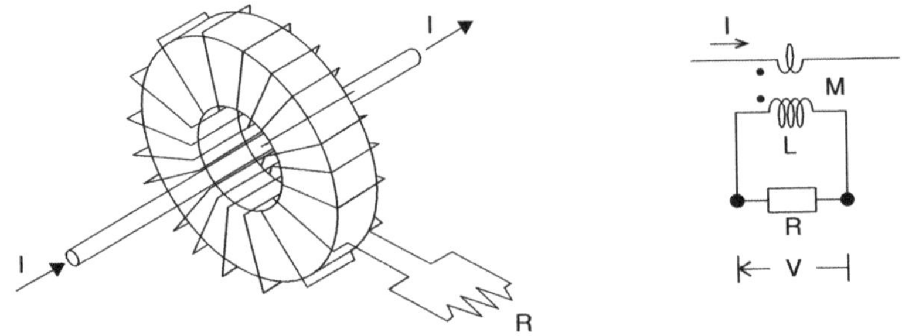

FIGURE 10.12 Basic structure of the current probe (left), and its equivalent circuit model (right).

10.4.2 The Current Probe

The current probe is a very popular device, although mainly used for pre-compliance testing. It essentially consists of an N-turns coil wound around a ferrite toroidal core and closed by the load resistance R (see Figure 10.12, left portion). The probe operation is that of a transformer whose primary circuit is the one where the current I under measurement flows, and the secondary circuit is the coil (right portion of Figure 10.12). The self-inductance of the secondary circuit is L and the mutual-inductance between the primary and secondary circuits is $M = L/N$. The current I is obtained from the measured voltage V through the *transfer impedance* of the probe, Z_T, given by

$$Z_T = \frac{V}{I} = \frac{f/f_c}{\sqrt{(f/f_c)^2 + [(f/f_r)^2 - 1]^2}}\frac{R}{N},$$

where f is frequency, $f_c = \frac{R}{2\pi L}$, $f_r = \frac{1}{2\pi\sqrt{LC}}$ and C is the distributed capacitance of the secondary winding. The frequency range of operation of the probe can start well below the corner frequency f_c, where the magnitude of Z_T is sufficient to provide for adequate sensitivity, and it usually stops just around the resonant frequency f_r. The larger the ratio f_r/f_c the wider the frequency interval where Z_T is frequency-independent ($Z_T = R/N$).

The loading effect of the current probe is characterized in terms of an impedance in series with the flow of the current under measurement. The complex insertion *insertion* impedance of the probe, $\dot{Z}_{ins}$, results from the superposition of two effects [22]: mutual coupling, which causes the impedance of the secondary circuit to be reflected to the primary circuit, and intrusion of the ferromagnetic material into the primary circuit. The first effect is described by the *reflected* impedance[23] $\dot{Z}_{refl} = \frac{(\omega M)^2}{R+j\omega L}$, the second one by the *intrusion* impedance $\dot{Z}_{intr} = \frac{j\omega(L-L_0)}{N^2}$, where $\omega = 2\pi f$,

[23]The presence of the stray capacitance C is neglected.

$L_0 = L/\mu_r$ and μ_r is the relative permeability of the ferrite material of the core. Since $\dot{Z}_{ins} = \dot{Z}_{refl} + \dot{Z}_{intr}$ we have, after calculations,

$$\dot{Z}_{ins} = \frac{jf/f_c}{1+jf/f_c}\frac{R}{N^2} - j\omega\frac{L_0}{N^2}. \tag{10.1}$$

It is seen from (10.1) that, in the wide frequency range comprised between f_c and $\mu_r f_c$, $\dot{Z}_{ins}$ is frequency-independent and equal to $\frac{R}{N^2}$. The maximum value of the magnitude of the insertion impedance is specified to be 1 Ω [20].

Commercial current probes are *clamp* devices provided with a *selective shield*, that is, a shield capable to reduce the unwanted coupling of the internal coil with the stray electric field and the axial magnetic field present outside but that, at the same time, permits coupling to the wanted magnetic field produced by the current under measurement. This is achieved through a cut in the shield along the whole perimeter of the hole. Current probe designs are available that can cover several frequency decades. Low frequency models can work from some tens of hertz while high frequency models can reliably operate up to about 1 GHz.

10.5 RADIATED EMISSION MEASUREMENTS

Field measurements can be conveniently classified according to the measurement frequency. Reactive magnetic and electric (near) fields are separately measured from 9 kHz to 30 MHz at a distance not exceeding, or comparable with, one wavelength λ from EUT. Electromagnetic (far) fields are measured at a distance not less than, or comparable with, the greater between λ and D^2/λ, where D is the EUT size, above 30 MHz. The type of EM environment where measurements are performed determines a further subdivision of the frequency range: field measurements below 1 GHz are done over a reflecting surface (metal ground plane), measurements above 1 GHz are in free space. This last distinction has important consequences especially on the different measurement methods adopted in the 30–1000 MHz and above 1 GHz frequency ranges [23].

The radiation characteristic of transmitting antennas is here characterized in terms of the *realized gain*, G, while the parameter adopted for receiving antennas is the *antenna calibration factor*, briefly ACF, given by

$$ACF = \frac{E}{V_r},$$

where E is the electric field impinging on the antenna and V_r is the voltage received across a 50 Ω load connected to antenna terminals. Due to reciprocity, gain and ACF are related through the following equation:

$$ACF = \sqrt{\frac{4\pi\zeta}{R}}\frac{1}{\lambda\sqrt{G}} = \frac{9.73}{\lambda\sqrt{G}}, \tag{10.2}$$

where ζ is free space impedance ($\zeta = 377\ \Omega$) and $R = 50\ \Omega$.

The *magnetic* antenna calibration factor, ACF_H, is sometimes preferred for characterizing receiving loop antennas. ACF_H is defined as $ACF_H = \frac{H}{V_r}$, where H is the magnetic field strength which induces the voltage V_r across the loop terminals (loop loaded by resistance R).

10.5.1 Antennas for the 9 kHz to 30 MHz Frequency Range

Magnetic fields are measured by using a screened loop antenna [24]. Practical designs employ a low noise amplifier to provide increased sensitivity and nearly flat frequency response. Care is required to avoid amplifier's overloading. The cut in the shield is necessary to not incur in magnetic flux cancellation. The shield is responsible for balanced-to-unbalanced (balun) conversion [25, 26]. Typical loop diameter is 56 cm (22 inch), which represents a compromise between sensitivity and the basic requirement that the loop behaves as a short (small perimeter with respect to wavelength) magnetic dipole (high rejection of electric field).

The elementary loop is an example of *calculable antenna*, this means that its radiation properties are predicted from the antenna geometry by using analytical formulas. In the case of a loop antenna whose shield has circular cross section of radius a and whose inner radius is b, we have

$$ACF_H = \frac{\sqrt{1 + (f/f_c)^2}}{2\pi f \mu_0 S},$$

where f is frequency, $f_c = \frac{R}{2\pi L}$, $R = 50\ \Omega$, $S = \pi b^2$ is the loop surface area, μ_0 is the magnetic permeability of vacuum and L is the loop inductance given by $L = \mu_0(b + a)[\ln \frac{8(b+a)}{a} - 2]$.

Electric field measurements are performed by using a vertical rod antenna [24]. The height of the rod is $h = 1$ m and a square metallic counterpoise, whose typical side dimension is 50 cm, is attached to the bottom of the base of the antenna. A low noise amplifier is usually employed for flat frequency operation and high sensitivity. Passive rod antennas have selectable frequency ranges. In order to increase antenna sensitivity an inductor is placed in series with the rod's terminal which resonates with the capacitance C of the rod at the center frequency of each range. The capacitance of the rod can be calculated as

$$C = \frac{2\pi\varepsilon_0 h}{\ln \frac{h}{2a} - 1}, \tag{10.3}$$

where a is the radius of the rod (average radius for telescopic rods) and ε_0 is the permittivity of vacuum. C results to be about 10 pF for practical rod diameters.

Calibration of the rod antenna is performed by substituting the actual rod with a lumped capacitor (dummy rod), whose value is given by (10.3), in series with the output of a 50 Ω RF generator. If V_0 is the open circuit voltage generated by the RF generator and V_r is the voltage across antenna terminals, as measured by a

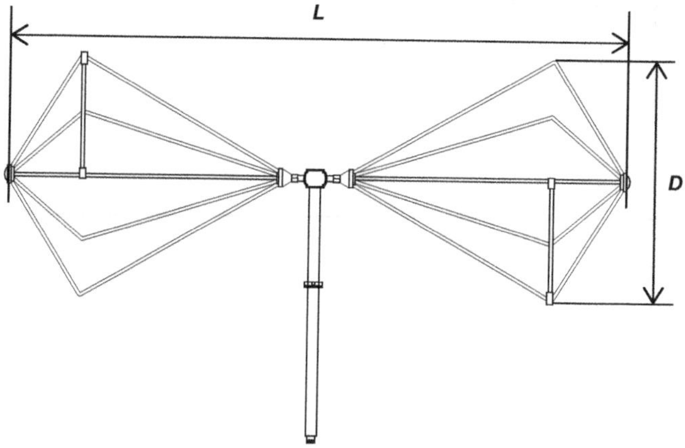

FIGURE 10.13 Biconical antenna (reproduced by courtesy of Schwarzbeck - Mess Elektronik).

50 Ω receiver, then we have $ACF = \frac{V_0/l_e}{V_r}$, where $l_e = h/2$ is the *equivalent length* of the rod.[24]

Both in the case of the shielded loop antenna and the rod antenna a passive transmitting and active receiving couple is used for measurements of screening effectiveness of shielded enclosures to magnetic and electric fields, respectively.

10.5.2 Antennas for the Frequency Range Above 30 MHz

The reference antenna in this frequency range is the nearly half-wave, *tuned dipole* [24]. Tuning is required above 80 MHz. $ACF = 2.6\pi/\lambda$ in a 50 Ω system not including balun losses (less than 0.5 dB for well-designed baluns). Dipole length is slightly smaller than $\lambda/2$ to achieve tuning. Tuned-length is specified by the manufacturer (more shortening is required as the dipole radius increases). E-plane half-power beamwidth is 78 deg (frequency independent). The tuned dipole is a precision antenna, mainly used for metrological applications, such as antenna calibration through the substitution method. VSWR is less than 1.5:1.

Broadband antennas are preferred to tuned dipoles for testing (no need for tuning). The most popular broadband antenna used in the VHF frequency range (30–300 MHz) is the biconical antenna, see Figure 10.13. L and D are about 1.3 m and 0.5 m, respectively. ACF has a bell-shaped frequency behavior with a minimum at the resonant frequency (about 70 MHz). The values attained by ACF typically fall between 5 and 20 dB(1/m). E-plane half-power beamwidth ranges from 60 to 80 deg. VSWR spans from 1.4:1 at resonance (antenna input resistance is about 35 Ω) to more than 60:1 at 30 MHz (the input is essentially a capacitance of 15 to 20 pF).

[24] It is assumed that, due to the large impedance of capacitor C, the voltage drop across the 50 Ω output impedance of the generator is negligible.

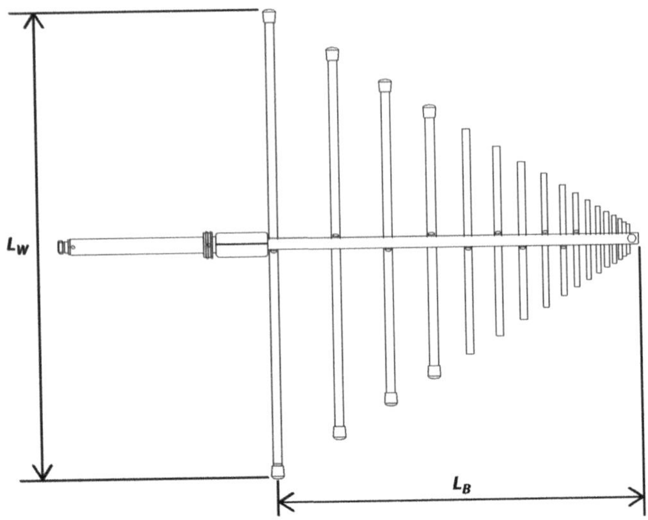

FIGURE 10.14 Log-periodic antenna (reproduced by courtesy of Schwarzbeck - Mess Elektronik).

Due to the large input VSWR at low frequency a matching pad (attenuator) is usually connected at its terminals in order to reduce mismatch uncertainty [27].

Compact size ($L < 0.5$ m) biconical antennas are available that are calibrated through the full 30 to 1000 MHz frequency range. Obviously they have reduced sensitivity when compared with the larger models in the overlapping frequency range (ACF comprised between 20 and 40 dB(1/m)). Small biconical antennas are particularly suitable for measurements of exposure to EM fields, that is, when a relatively large field strength is involved.

The logarithmic periodic dipole array (or log-periodic antenna) antenna, see Figure 10.14, is widely used in the frequency range from 100 to 2000 MHz. It is a broadband antenna having good match (VSWR< 1.5:1) and relatively high power handling capability (1 kW or more). It is frequently adopted for immunity tests from 80 to 1000 MHz. Typically the realized gain is 7 dB, half-power beamwidth is about 50 deg in the E-plane and 100 deg in the H-plane. Front-to-back ratio may be 20 dB or more. The cross-polar rejection is relatively poor (i.e., not greater than 20 dB), especially at high frequency. This is due to the fact that, in practical realizations, the poles on one side are not aligned with the poles on the opposite side, along the direction perpendicular to the poles themselves and to the antenna axis (see Figure 10.15). This negatively impacts the ability to resolve vertical from horizontal polarization, particularly when the shorter dipoles are active. The dimensions of a model covering the 200 to 2000 MHz frequency range are $L_B = 0.5$ m and $L_W = 0.65$ m.

The most significant complexity associated with the use of log-periodic antennas is the *shift of the phase center* position with frequency, being close to the longer radiating elements at low frequencies and to the shorter ones at high frequencies. If

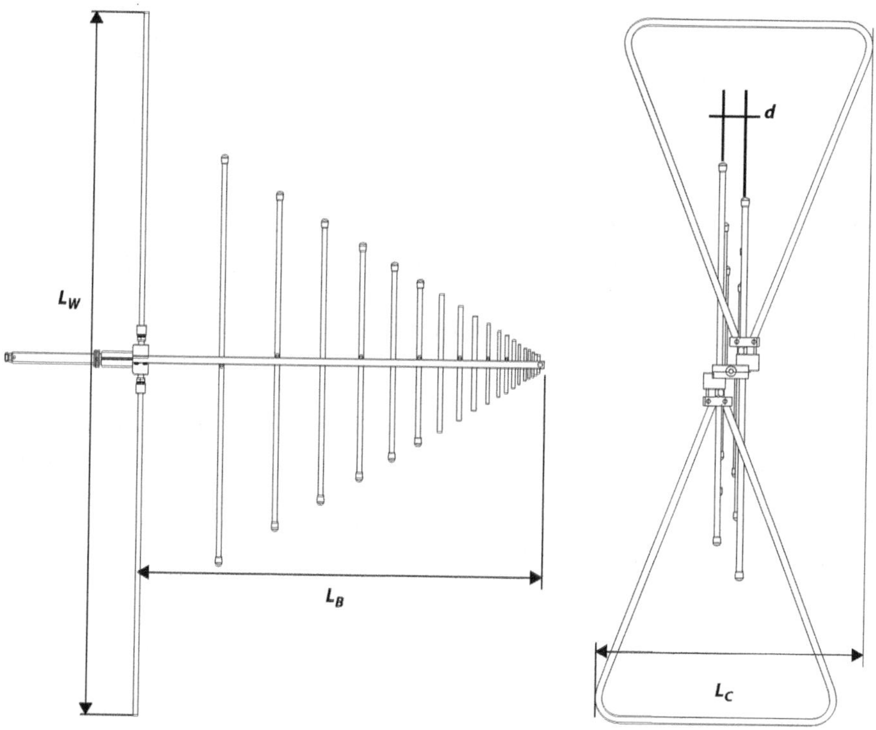

FIGURE 10.15 Hybrid biconical-logarithmic-periodic antenna (reproduced by courtesy of Schwarzbeck - Mess Elektronik).

one takes the tip of the antenna as the geometrical reference position then the phase center position can be described in terms of its distance from the tip along the boom. With fairly good approximation the phase center displacement is proportional to the wavelength when the profile of the antenna is triangular (as the one in Figure 10.14, the most common shape). This property permits to obtain a feasible correction of measurements and calibrations in a free space environment. More problematic is the case of height scanning over a conducting ground (as is usually the case in compliance tests in the 30 to 1000 MHz frequency range).

The hybrid biconical-logarithmic-periodic (or bi-log) antenna is the preferred choice for radiated emission testing up to 1 GHz since it permits to cover the full 30 to 1000 MHz frequency range in one sweep, with adequate sensitivity. Its appearance is that of a biconical and a log-periodic antennas stitched together, see Figure 10.15. It takes on the limitations of both antennas, that is, poor match at low frequency (biconical) and phase center shift and poor cross-polar performance at high frequency (log-periodic), and adds its own (presence of sharp resonances at the frequencies where the operation of the antenna switches from biconical to log-periodic). It is a pretty large antenna: typical dimensions are $L_W = 1.7$ m, $L_B = 0.8$ m, $L_C = 0.5$ m. Note in Figure 10.15 the displacement d of the radiating elements of the log-periodic

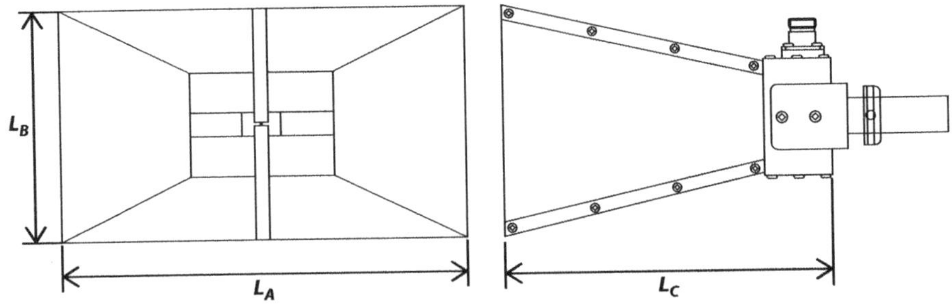

FIGURE 10.16 Double ridged horn antenna (reproduced by courtesy of Schwarzbeck - Mess Elektronik).

portion of the bi-log antenna responsible for poor cross-polar discrimination, especially at high frequencies, where d is comparable with the size of the active radiating elements.

Double ridged horn antennas, see Figure 10.16, are commonly used for emission and immunity testing above 1 GHz. They have broadband operation: commercial models are available covering the frequency ranges 1–18 GHz and 18–40 GHz. The realized gain is frequency-dependent and falls between 5 and 20 dB. Half-power beamwidth decreases from 90 deg at 1 GHz to 10 deg at 18 GHz (both E- and H-planes). Front-to-back ratio is 25 dB or more, the same value applies to cross-polar rejection. Power handling is several hundreds of watts. The dimensions of a model covering the 1–18 GHz frequency range are $L_A = 24.5$ cm, $L_B = 14.2$ cm and $L_C = 19.5$ cm.

Standard gain horns are used as reference antennas above 1 GHz. They are the above 1 GHz counterparts of nearly half-wave tuned dipoles below 1 GHz. The frequency range spanned by each standard gain horn is less than one octave. A popular set of ten standard gain horns covers the frequency range comprised between 1 and 40 GHz. Each element has a constant ACF in its frequency range of operation[25] and it has nearly the same gain, between 15 and 18 dB. The VSWR of the horn itself is 1.1:1 while the VSWR of the horn equipped with a waveguide to coaxial adapter is 1.5:1.

Several methods of calibration of EMC antennas are available for the frequency range above 30 MHz [28]. Two identical transmit and receive antennas are involved in the *two-antenna method*. Antenna separation d must be such that each antenna is in the far field of the other one. Boresight alignment is required. If P_T is the power that the generator feeding the transmitting antenna delivers to a matched (50 Ω) load and P_M is the power measured by a 50 Ω receiver connected to the terminals of the receiving antenna, then the realized gain G is obtained as

$$G = \frac{4\pi d}{\lambda}\sqrt{\frac{P_M}{P_T}}. \tag{10.4}$$

[25] Hence its gain increases quadratically with frequency, see (10.2).

In the *three-antenna method* three antennas are involved with three combinations in pair (one transmitting and one receiving). The realized gain of each antenna is obtained through the simultaneous solution of three equations of the form $P_M = P_T G_T G_M (\frac{\lambda}{4\pi d})^2$.

The *standard site method* [29] is a variation of the three antenna method conceived for taking into account the ground reflected contribution (height scanning is performed). It is used for the calibration of dipole, biconical, and log-periodic antennas. The accuracy of the method relies upon the quality of the measurement site (see Reference 30] for a description of the characteristics and validation methods of the sites for antenna calibration).

In the *standard antenna method* the antenna factor of the antenna under calibration is determined through substitution with a standard antenna whose antenna factor is already known (e.g., a calculable dipole or a standard gain horn). The accuracy of the calibration obtained by using this method is independent of the quality of the measurement site, provided that the antenna under calibration and the standard antenna have the same radiation pattern [31].

Standards impose several requirements on antenna performances, such as calibration accuracy, directivity, cross-polar discrimination and VSWR. One important aspect to check is balun DM/CM conversion. It consists in verifying that, when a vertically polarized and stable field impinges on the antenna, the rotation of the antenna through 180 deg does not produce a significant variation[26] of the voltage at its terminals. The balun DM/CM conversion check is performed over a ground plane and it is particularly important for dipole, biconical and bi-log antennas in the VHF range (antennas conceived for use above 300 MHz do not exhibit significant unbalance). Variations of the order of 10 dB were found especially with old and high power balun designs. Antennas with such poor balance cannot even be calibrated. Many respectable test sites may have appeared[27] to fail to comply with the applicable field uniformity requirements due to the fact that they were verified by using poorly balanced antennas [32].

10.5.3 Measurement Sites

The ideal reference test site for radiated emission measurements in the 30–1000 MHz frequency range is a semi-infinite empty space above a perfectly conducting, perfectly flat, infinite ground plane [24]. The practical realization of the ideal test site is the *open area test site*. The minimum size and shape of the obstruction free area is specified by standards. If the receiving antenna is fixed and the EUT is rotated through a turntable[28] then the boundary of the minimum size obstruction free area is elliptic. EUT and receiving antenna are at the foci of the ellipse, along the major axis and at distance d. The length of the major axis shall be $2d$, while that of the minor axis $\sqrt{3}d$. The rational behind is that an EM field wave leaving the EUT, reflected by an

[26]The maximum allowed variation is 1 dB.
[27]Especially in the past, when this problem with VHF antennas was not identified.
[28]The most common situation.

obstacle on the boundary and finally reaching the receiving antenna is attenuated at least by 6 dB with respect to the direct wave traveling directly from EUT to antenna because it travels a distance two times larger. It is assumed that obstacle and antenna are in the far field of the EUT, where field decreases proportionally to the reciprocal of distance. The distance d between antenna and EUT can be 3, 10, or 30 m. The ground plane may be composed of a wide range of materials from earth to highly conductive metallic material.

Due to the exposure to ambient noise and weather conditions radiated emission tests are actually performed in screened, semi-anechoic chambers (alternative test sites) whose walls and ceiling are lined with EM absorbing (no echo) material while the floor is reflecting (metallic). The absolute majority of chambers are designed for measurements at distance $d = 10$ or 3 m. Due to the presence of the ground plane radiated emission measurements from 30 to 1000 MHz are performed through height scanning of the antenna from 1 to 4 m in order to search for constructive superposition of the direct wave with the ground reflected wave. Further, EUT is placed on a motor-driven turntable and rotated through 360 deg in order to capture maximum radiation from all its sides. Larger size chambers, for 10 m measurements, are equipped with a special floor, having a motor-driven, turning portion capable to carry on and rotate heavy floor standing EUTs. The rotating portion is in electrical contact with the fixed one through sliding contacts in order to preserve the electrical continuity of the ground plane at RF. Radiation from EUT, both at vertical and horizontal polarization, is investigated.

Test site validation in the range between 30 and 1000 MHz is obtained through the *normalized site attenuation* (NSA) measurement [33], which consists in measuring the EM field attenuation between a pair of antennas, one transmitting (T) and one receiving (R), placed in the site under evaluation. This *site attenuation*, SA in dB, is normalized subtracting the sum of the antenna calibration factors, ACF_T and ACF_R in dB(1/m). Therefore NSA is defined as

$$NSA = SA - ACF_T - ACF_R.$$

The measurement unit of NSA is dB(m^2). Measurements are performed both at vertical and horizontal polarization and at different positions of T and R antennas around the test site.[29] Measured NSA values shall not differ from the calculated ones by more than ±4 dB at each frequency between 30 and 1000 MHz. Calculated NSA values are obtained as (in linear units)

$$NSA_H = \frac{\lambda d}{k_H}\frac{R}{\zeta}, \text{ where } k_H = d\left|\frac{e^{-j2\pi\frac{r}{\lambda}}}{r} - \frac{e^{-j2\pi\frac{r'}{\lambda}}}{r'}\right|_{\mathrm{MAX}}, \tag{10.5}$$

[29]This is a specific requirement for alternative test sites, such as semianechoic chambers. It does not apply to open area test sites.

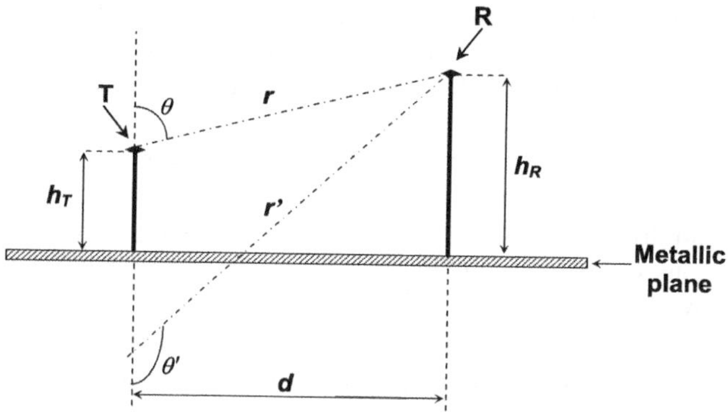

FIGURE 10.17 Schematic representation of NSA measurement method.

for horizontal polarization, and

$$NSA_V = \frac{\lambda d}{k_V}\frac{R}{\zeta}, \text{ where } k_V = d\left|\frac{e^{-j2\pi\frac{r}{\lambda}}}{r}\left(\frac{d}{r}\right)^2 + \frac{e^{-j2\pi\frac{r'}{\lambda}}}{r'}\left(\frac{d}{r'}\right)^2\right|_{\text{MAX}}, \quad (10.6)$$

for vertical polarization. Equations (10.5) and (10.6) can be easily derived considering the schematic representation in Figure 10.17, where the effect of the image antenna is taken into account. The distance between the transmitting and receiving antennas is $r = \sqrt{d^2 + (h_R - h_T)^2}$ while that from the image antenna to the receiving one is $r' = \sqrt{d^2 + (h_R + h_T)^2}$. Further, $\zeta = 377\ \Omega$ and $R = 50\ \Omega$. The currents on the radiating elements of the image antenna are in phase opposition in horizontal polarization (minus sign in the formula for k_H), while in phase in vertical polarization (plus sign in the formula for k_V). The subscript MAX means that the maximum value, obtained by varying h_R between the limits required for height scanning of the R antenna,[30] is taken. It is assumed that antennas, including image antenna, have negligible dimensions with respect to their distance[31] and that they have the same radiation pattern as that of the short dipole, that is, $\sin\theta$ pattern.[32]

The rationale for the ±4 dB acceptability criterion is 1 dB maximum error for each antenna factor, plus 1 dB for receiver, plus 1 dB for test site. In other words NSA measurements in a test site which deviates from ideal by 1 dB can result in an absolute deviation of up to 4 dB from the values calculated by using (10.5) and (10.6). Other validation methods, much more suitable to resolve the deviation from ideal of the test site, have been recently proposed by standards. For example, SA measurements by using *calculable dipoles* [34], where measured SA is compared with that calculated

[30] h_R ranges from 1 to 4 m, h_T is 1 or 2 m in horizontal polarization, 1 or 1.5 m in vertical polarization.

[31] Thus excluding mutual coupling effects, which in practice, however, are present and significant, in particular at lower frequencies.

[32] A poorly verified assumption, especially at higher frequencies

assuming that the dipoles are used in a perfect site, or the *reference site method* which consists in comparing the results of SA measurements performed, with the same instrumentation, both in a reference, high quality site and in the test site under evaluation.

It is important to observe that in free space[33] the process of test site validation would be much more simple, both from the conceptual and practical point of view. In free space, no distinction exists between horizontal and vertical polarization, no ground reflections or mutual coupling effects between an antenna and its image take place. Provided that $h_R = h_T$, NSA measurements can be compared with the unique value[34] $\lambda d \frac{R}{\zeta}$. Radiated emission measurements in a free space environment are permitted for small equipment (table-top or whose size is within the validated test volume).

The ideal reference test site for radiated emission measurements above 1 GHz is free space. Height scanning is needed only if the radiation pattern of the receiving antenna does not encompass the EUT vertical size h at the measurement distance d. If $h \geq 2d \tan(\theta_3/2)$ where θ_3 is the minimum -3 dB antenna beamwidth (in both E-plane and H-plane and in the whole test frequency range) then height scanning is required. The usual measurement distance is $d = 3$ m.[35] EUT is rotated through a turntable. Both vertical and horizontal polarizations are investigated.

Test site validation from 1 to 18 GHz is performed through the measurement of the *site voltage standing wave ratio*, or SVSWR. It is assumed that the dominant imperfection of the site originates from limited EM absorption from the walls, ceiling, and floor. The receiving (R) antenna used for site validation is the same one used for emission measurements. The transmitting (T) antenna shall have a short dipole-like radiation pattern (detailed requirements on radiation pattern are given in Reference 24). Compact size biconical antennas usually satisfy such requirements. Measurements are made placing T antenna at different distances from R antenna, which is in a fixed position. Distance d between antennas is increased in steps of 2, 10, 18, 30, and 40 cm from a starting position of the T antenna. Antennas are kept aligned. Two heights and both vertical and horizontal polarizations are investigated.[36] SVSWR is the peak-to-peak variation of the voltage at the output of the R antenna, as obtained by moving the T antenna through the six distances from the R antenna. The effect of free space attenuation with increasing distance is corrected for before calculation of the SVSWR. In an ideal site SVSWR would be 0 dB. SVSWR of the site under evaluation shall not exceed ± 6 dB in the whole frequency range, for each starting position of the T antenna, polarization and height. While NSA validation method is *absolute*, the SVSWR validation method is *relative* and, therefore, weakly dependent on the characteristics of the measuring instrumentation (e.g.,

[33]The practical realization of free space is the fully anechoic chamber

[34]This result is easily obtained from (10.5), or (10.6), in the case $r = d$ and $r' \to \infty$, hence $k_H = k_V = 1$.

[35]Recall that at high frequency the far field condition requires $d \geq D^2/\lambda$, where D is EUT size.

[36]The size of the validated test volume shall depend on the maximum EUT size, typically 1.5 m × 1.5 m.

SVSWR measurement results do not depend on the calibration factors of antennas and linearity of the receiver, not its absolute calibration, is required).[37]

10.6 IMMUNITY TESTS

Differently from emission tests, the result of immunity tests is not a measured value. In other words, immunity tests are not measurements. Immunity test results are classified in terms of the loss of function or degradation of performance of EUT, relative to an expected minimum performance level (or normal performance). The recommended classification is as follows:

- Normal performance;
- Temporary loss of function or degradation of performance which ceases after the disturbance ceases, and from which the equipment under test recovers its normal performance, without operator intervention;
- Temporary loss of function or degradation of performance, the correction of which requires operator intervention;
- Loss of function or degradation of performance which is not recoverable, owing to damage to hardware or software, or loss of data.

A sample of the equipment used to generate conducted and radiated disturbance phenomena and the relevant immunity test methods adopted are briefly presented.

10.6.1 Conducted Immunity Tests

Impulse conducted disturbances can be categorized into high energy/low frequency or low energy/high frequency.[38] *Surge* disturbance is an example from the high energy/low frequency disturbance category [7]. Surge disturbances are mainly present in the AC power mains and they may be originated by switching operations within the power system (switching of capacitor banks or loads) or by indirect effects of lightning strokes. An ad hoc test generator capable of simulating surge disturbance, named *combined* test generator, is specified by standards. The impulse waveforms generated are specified in terms of both open-circuit voltage and short-circuit current.[39] The generator must meet its specifications in the referenced open and short-circuit conditions, that is, specifications do not concern the actual surge wave-shape applied to

[37] Relative test site validation methods were proposed also for the 30–1000 MHz frequency (see Reference 35) but the old NSA measurement is presently, by far, the most used site validation method.

[38] For an analysis of the metrological issues connected with the characterization of impulse disturbance generators, see References 36 and 37.

[39] Therefore two standard sources are "combined" in one: a voltage source and a current source. See Reference 38 for an equivalent circuit model of the combined generator.

EUT.[40] The combined generator is designed to generate a unidirectional pulse having an open-circuit voltage rise time (30 to 90%) of 0.72 μs, an open-circuit voltage time width at half value of 50 μs, a short-circuit current rise time (10 to 90%) of 6.4 μs and a short-circuit current time width at half value of 16.9 μs. The peak of the open-circuit surge voltage shall be adjustable in the range between 0.5 and 4 kV according to the test level required for the specific EUT. The peak of the short circuit current shall be that of the open circuit voltage divided by 2 Ω.[41] All the specifications have a corresponding tolerance. For example, time width is specified having a 20% tolerance. Other characteristics of the combined test generator are polarity (positive and negative), phase shifting (in a range between 0 to 360 deg relative to the phase angle of the AC power-line voltage feeding the EUT), pulse repetition rate (1 per minute or faster), and generator output (floating). The surge disturbance shall be applied to EUT through a *coupling/decoupling network*, or CDN. The CDN shall, at the same time, couple the impulse to EUT with the appropriate impedance, couple the AC power source to EUT and decouple the surge generator output from the AC power source. The CDN shall also permit to couple the disturbance line-to-line (DM coupling) and line-to-ground (unsymmetrical coupling).

Electrical fast transient/burst (briefly EFT/burst) is an example of impulse disturbance from the low energy/high frequency category [6]. Fast transients are generated in the AC and DC power mains by interruption of inductive loads or bounce of a relay contact. Due to their high frequency content, these disturbances can easily couple also to control and signal ports of an equipment. The EFT/burst consists of bursts of unidirectional impulses. The waveform of the impulses forming each burst has a specified rise time, width, and peak value. Also the repetition frequency of the impulses within each burst and the burst repetition period are specified. The open-circuit output voltage of the EFT/burst generator shall be adjustable at least within 0.25 to 4 kV. The generator shall be capable of operating under short-circuit conditions. The polarity of the impulse must be selectable by user. The generator is provided with a coaxial, single ended output and a DC blocking capacitor of 10 nF. Output impedance is 50 Ω. Pulse repetition frequency within each burst can be selected between 5 or 100 kHz. The burst relation to power supply is asynchronous. The burst duration shall be 15 ms at 5 kHz repetition frequency and 0.75 ms at 100 kHz repetition frequency.[42] The burst period is 300 ms. The waveform of the pulse shall have 10 to 90% rise time of 5 ns and duration at 50% of 50 ns. Tolerances on these specifications and characteristics are defined by standards. Similar to surge disturbance, also the EFT/burst disturbance is coupled to the power supply ports (both AC and DC) of EUT through an appropriate CDN. Coupling to signal and control lines is obtained through a *capacitive coupling clamp* which injects a bulk CM current along the cable under test. The typical coupling capacitance between cable and clamp is 100 to 1000 pF.

[40]This is the general approach within IEC: The actual stimulus applied to the EUT is not probed during testing, quite the opposite is true for US EMC military standards.

[41]Note that this does not imply that the output impedance of the generator is 2 Ω.

[42]The number of impulses per burst is 75.

The *bulk current injection*, or BCI, immunity test is another example of conducted immunity test [8]. CM currents are induced on cables and connections through near-field magnetic coupling by using a BCI probe. A BCI probe is essentially a current probe whose operation is reversed, that is, a generator feeds the secondary winding of the current probe (the primary circuit of the BCI probe) and a load is connected to the primary (the secondary circuit of the BCI probe). The main differences between current and BCI probes are power-handling capability (50 to 200 W for BCI probes, few watt for current probes) and number of turns (one to three turns in the case of BCI probes, 10 turns or more for current probes). The motivation behind this test is that in many cases, and particularly at low frequency, the disturbance is coupled to the victim through the undesired CM current induced on cables and connections by the disturbance EM field. This is because direct coupling between the victim and the disturbing field through apertures and slots at low frequency is negligible, since their dimensions are small with respect to wavelength. A very large transmitted power would be needed in order to radiate a low frequency electric field strength whose magnitude is comparable with actual threats, therefore it is convenient (i.e., relatively low power is needed) to directly induce CM currents through the BCI probe. During the test the disturbance is injected onto one cable at a time. This is a crude approximation of the real situation where the disturbing source acts on all the victim's cables and connects simultaneously and with different amplitudes and phases.

BCI probes are characterized in terms of *insertion loss* (IL). Insertion loss is measured by using a calibration fixture which consists of a round stripline whose dimensions are (approximate) 15 mm diameter, 70 mm length and 50 mm distance from ground planes. The two ports of the stripline are loaded by $R = 50\,\Omega$ and the BCI is clamped around the strip. The schematic representation of the IL measurement is in Figure 10.18. IL measurement is in two steps. During the first step (see Figure 10.18a) voltage V_1 is measured directly connecting an RF generator to the $50\,\Omega$ receiver. Since the output resistance of the generator is $R_G = R$ we have $V_1 = V_G/2$. During the second step (see Figure 10.18b) the generator is connected to the BCI probe input and the voltage V_2 across the resistance R loading one of the two ports of the jig is

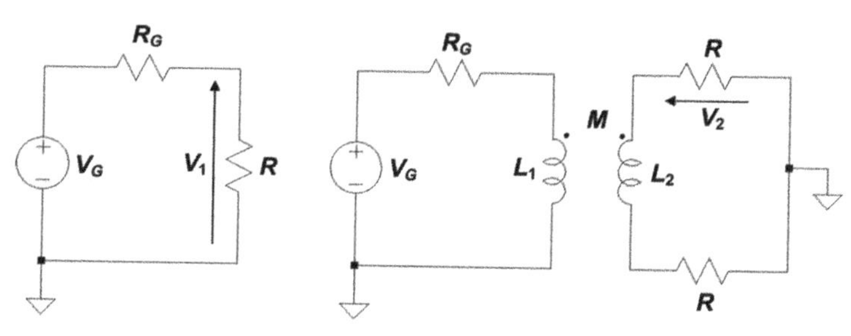

FIGURE 10.18 Schematic representation of the two-steps measurement, (a) and (b), of the insertion loss of a BCI probe.

measured. IL can be obtained in terms of the parameters of the lumped equivalent circuit model of the BCI probe (see again in Figure 10.18b). We have

$$IL = 20\log_{10}\frac{V_1}{V_2} = 20\log_{10}\frac{\sqrt{R^2 + \left[2\pi f\left(L_1 + \frac{L_2}{2}\right)\right]^2}}{2\pi f M},$$

where L_1 is the inductance of the primary circuit (the BCI probe coil), L_2 is the inductance of the secondary circuit (the calibration fixture), M is the mutual coupling and f is the frequency. It is interesting to observe that IL decreases with increasing frequency, it is frequency-independent at high-frequency and equal to $20\log_{10}\frac{L_1+L_2/2}{M}$. Further, the lower the IL the lower the generator power needed to induce the same voltage V_2. If we assume perfect coupling between the primary and secondary circuits we have $M^2 = L_1 L_2$. If N is the number of turns forming the probe coil we have $M = \frac{L_1}{N} = NL_2$. Therefore $\frac{L_1+L_2/2}{M} = N + \frac{1}{2N}$ which is minimum when $N = 1$ and the corresponding value of IL is 3.5 dB.

BCI immunity tests are performed generating narrow band disturbances,[43] whose amplitude is modulated by a 1 kHz signal with 80% modulation depth, and the carrier frequency is in the range between 9 kHz and 80 MHz. The test level is set with the same calibration fixture used for IL measurement. Voltage V_2 of the unmodulated carrier shall be in the range between 0.5 and 5 V (RMS), depending on the product to be tested.

10.6.2 Radiated Immunity Tests

The test site where radiated immunity tests take place is the anechoic chamber. The frequency range is 80 to 1000 MHz for general purpose protection and up to 6 GHz for protection against RF emissions from digital radio telephones and other RF emitting devices [5]. The test field strength is the RMS value of the unmodulated carrier signal and ranges between 1 and 30 V/m. This carrier signal is 80% amplitude modulated with a 1 kHz sinewave[44] to simulate actual threats. Hence the actual peak-to-peak electric field strength amplitude is $2\sqrt{2}(1.8) = 5.1$ times the test field strength. In the 80 to 1000 MHz frequency range the field must be uniform within −0 to +6 dB of the stipulated test level in 12 out of 16 equally spaced points over a 1.5 m × 1.5 m square surface, named uniform field area (UFA) [39]. Test frequencies are chosen according to a geometric sequence whose ratio is 1.01 (i.e. 1% increment). The 12 points of the UFA satisfying the −0 to +6 dB uniformity requirement have not necessarily to be the same at each frequency. Test level setting and evaluation of field uniformity are performed in absence of EUT. Non-uniformity is due to the directivity of the transmitting antenna, reflections from walls, ceiling and floor, spherical wave-front.

[43] Wide band impulse disturbances (unidirectional or oscillatory) are induced by using the BCI probe when testing an EUT against US military standards.

[44] The same as in the case of BCI tests.

The distance between the transmitting antenna and the UFA is usually 3 m, below 1 GHz, and 1 m (minimum allowed distance), above 1 GHz.[45] Above 1 GHz it is allowed to divide the whole area occupied by the face of the EUT into 0.5 m × 0.5 m windows (4 points) where the field is uniform within −0 to +6 dB. The test level is separately set for each window. Test level setting and immunity test must be performed both in horizontal and vertical polarization. A linearity check of the source is required. The maximum compression allowed is 2 dB when the selected test level is increased by 5.1 dB.[46] Each side of the EUT shall be illuminated and the minimum dwell time at each frequency shall be not less than 0.5 s.

The field generating antennas are usually log-periodic antennas (below 1 GHz) and double ridged horns (above 1 GHz). Antennas for radiated immunity tests must tolerate high power levels and have relatively high gain.[47]

10.7 CONCLUSIONS

EMC measurements are aimed at quantifying the non-intentional electromagnetic emissions from an equipment under test and at assessing its immunity to electromagnetic disturbances. EMC measurements deal with the non-intentional behavior of an equipment. The electromagnetic quantities associated with such non-intentional behavior are not available at (emission) or applied to (immunity) a port supplied by the manufacturer of the equipment for this purpose. For example, in EMC testing the power supply port of an equipment under test is seen as the port through which the radio frequency conducted disturbances generated by the equipment itself are applied to the power supply network. Similarly, the power cord is seen as an antenna along which common mode radio frequency currents flow and radiate unwanted electromagnetic energy. Impulses of different energy and spectral content are applied to the power supply in order to simulate the real world threats that the equipment will face in its electromagnetic environment. Such peculiarity of EMC measurements is reflected by the variety of measurement techniques that were and are developed for the purpose of capturing the non-intentional behavior of the equipment. Those who are involved in the standardization process and in research in this area have the difficult task of improving the present measurement techniques or designing new ones while balancing the ability to capture or reproduce real-world phenomena while preserving measurement reproducibility (which usually implies a rigid measurement setup, poorly representative of the actual operation and use of the equipment under test). We hope that the reader of this chapter appreciated the effort of making EMC measurements as rigorous as possible at the same time as effective.

[45] Distance is taken from antenna tip or phase center to UFA.
[46] $20 \log_{10} 1.8 = 5.1$.
[47] An increase of 3 dB of the maximum required output power of the amplifier may correspond to double its cost.

FURTHER READING

The content of this chapter cannot obviously cover the whole set of EMC test and measurement techniques implemented by the relevant standards organizations. Further, different test methods apply to different products, such as domestic appliances, lighting, medical, telecommunication, automotive, railway, avionic, space, military equipment, and so on. Finally, EMC test equipment and measurement methods are rapidly changing in order to keep pace with the technological evolution of the devices to be tested and the new telecommunication techniques. The recent EMC standards are a major source of information for those involved in EMC measurements, and the following standards are therefore recommended for further reading. Within the IEC standards framework the CISPR 16–1–X and CISPR 16–2–X basic standards series set the requirements applicable to measuring equipment and (mainly) emission measurement methods, respectively. Similarly, the IEC 61000–4–X basic standards series defines the requirements applicable to immunity test equipment and measurement methods. Harmonic current, voltage fluctuations and flicker measurement methods, and associated instrumentation are the subject of the standards IEC 61000–3–2, IEC 61000–3–3, IEC 61000–3–11 and IEC 61000–3–12. A comprehensive source of information about both EMC emission and immunity test methods is provided by the United States military standard MIL-STD 461. Many of the test methods described in this standard are also adopted for automotive, avionic, and space equipment. Noteworthy, an application guide is included in MIL-STD 461 as an appendix, providing for background information. Far from being a textbook, the standard is a rich source of information about EMC measurement techniques.

REFERENCES

1. CISPR/TR 16-4-1. *Specification for Radio Disturbance and Immunity Measuring Apparatus and Methods – Part 4-1: Uncertainties, Statistics and Limit Modelling – Uncertainties in Standardized EMC Tests*, Ed. 2.0 (2009).
2. IEC/TR 61000-1-6. *Electromagnetic Compatibility (EMC) – Part 1-6: General – Guide to the Assessment of Measurement Uncertainty*, Ed. 1.0 (2012).
3. CISPR 16-4-2. *Specification for Radio Disturbance and Immunity Measuring Apparatus and Methods – Part 4-2: Uncertainties, Statistics and Limit Modelling – Measurement Instrumentation Uncertainty*, Ed. 2.0 (2012).
4. IEC 61000-4-2. *Electromagnetic Compatibility (EMC) – Part 4-2: Testing and Measurement Techniques – Electrostatic Discharge Immunity Test*, Ed. 2.0 (2008).
5. IEC 61000-4-3. *Electromagnetic Compatibility (EMC) – Part 4-3: Testing and Measurement Techniques – Radiated, Radio-Frequency, Electromagnetic Field Immunity Test*, Ed. 3.1 (2008).
6. IEC 61000-4-4. *Electromagnetic Compatibility (EMC) – Part 4-4: Testing and Measurement Techniques – Electrical Fast Transient/Burst Immunity Test*, Ed. 3.0 (2012).

7. IEC 61000-4-5. *Electromagnetic Compatibility (EMC) – Part 4-5: Testing and Measurement Techniques – Surge Immunity Test*, Ed. 3.0 (2014).
8. IEC 61000-4-6. *Electromagnetic Compatibility (EMC) – Part 4-6: Testing and Measurement Techniques – Immunity to Conducted Disturbances, Induced by Radio-Frequency Fields*, Ed. 4.0 (2014).
9. Carobbi, C.F.M. The use of logarithmic units in the uncertainty evaluations of EMC measurements, *IEEE-EMC Newsletter*, (224), 46–50 (Winter 2010).
10. Carobbi, C.F.M. The GUM Supplement I and the uncertainty evaluations of EMC measurements, *IEEE-EMC Newsletter*, (225), 53–57 (Spring 2010).
11. Carobbi, C.F.M., Cati, M., Panconi, C. Reproducibility of radiated emissions measurements in compact, fully-anechoic, rooms – the contribution of the site-to-site variations, *IEEE Transactions on Electromagnetic Compatibility*, **51** (3) 574–582 (August 2009).
12. Carobbi, C.F.M., Stecher, M. The effect of the imperfect realization of the artificial mains network impedance on the reproducibility of conducted emission measurements, *IEEE Transactions on Electromagnetic Compatibility*, **54** (5) 986–997 (October 2012).
13. Ott, H.W. *Electromagnetic Compatibility Engineering*, John Wiley & Sons, Inc., Hoboken (NJ) (2009).
14. IEC 60050-161. *International Electrotechnical Vocabulary (IEV) – Chapter 161: Electromagnetic Compatibility.*
15. CISPR 16-1-1. *Specification for Radio Disturbance and Immunity Measuring Apparatus and Methods – Part 1-1: Radio Disturbance and Immunity Measuring Apparatus – Measuring Apparatus*, Ed. 3.0 (2010).
16. Geselowitz, D.B. Response of ideal radio noise meter to continuous sine wave, recurrent impulses, and random noise, *IRE Trans. Radio Freq. Interference*, **RFI-3** (1), 2–10 (May 1961).
17. ANSI/IEEE Standard 376-1975. *IEEE Standard for the Measurement of Impulse Strength and Impulse Bandwidth*, Corrected Edition (June 1992).
18. Military Standard. *Measurement of Electromagnetic Interference Characteristics*, U.S. Department of Defense, MIL-STD-462 (July 31, 1967).
19. Carobbi, C.F.M., Cati, M., Panconi, C. A double inequality for equivalent impulse bandwidth, *IEEE Transactions on Electromagnetic Compatibility*, **52** (3) 516–520 (August 2010).
20. CISPR 16-1-2. *Specification for Radio Disturbance and Immunity Measuring Apparatus and Methods – Part 1-2: Radio Disturbance and Immunity Measuring Apparatus – Ancillary Equipment – Conducted Disturbances*, Ed. 1.2 (2006).
21. CISPR 16-2-1. *Specification for Radio Disturbance and Immunity Measuring Apparatus and Methods – Part 2-1: Methods of Measurement of Disturbances and Immunity – Conducted Disturbance Measurements*, Ed. 2.1 (2010).
22. Carobbi, C.F.M., Millanta, L.M. Circuit loading in radio-frequency current measurements: the insertion impedance of the transformer probes, *IEEE Transactions on Instrumentation and Measurement*, **59** (1) 200–204 (January 2010).
23. CISPR 16-2-3 *Specification for Radio Disturbance and Immunity Measuring Apparatus and Methods – Part 2-3: Methods of Measurement of Disturbances and Immunity – Radiated Disturbance Measurements*, Ed. 3.0 (2010).

24. CISPR 16-1-4. *Specification for Radio Disturbance and Immunity Measuring Apparatus and Methods – Part 1-4: Radio Disturbance and Immunity Measuring Apparatus – Ancillary Equipment – Radiated Disturbances*, Ed. 3.0 (2010).
25. Carobbi, C.F.M., Millanta, L.M., Chiosi, L. The high-frequency behavior of the shield in the magnetic-field probes, in *Proc. 2000 IEEE Intern. Symp. on EMC*, pp. 35–40, Washington, DC (August 21–25, 2000).
26. Carobbi, C.F.M., Millanta, L.M. Analysis of the common-mode rejection in the measurement and generation of magnetic fields using loop probes, *IEEE Transactions on Instrumentation and Measurement*, **53** (2), 514–523 (April 2004).
27. Carobbi, C.F.M., Cati, M., Panconi, C. Note on the expected value and standard deviation of the mismatch correction, *IEEE Transactions on Electromagnetic Compatibility*, **53** (4) 1098–1099 (November 2011).
28. ANSI C63.5-2006. *American National Standard For Electromagnetic Compatibility – Radiated Emission Measurements in Electromagnetic Interference (EMI) Control – Calibration of Antennas (9 kHz to 40 GHz).*
29. Smith, A.A. Standard Site Method for Determining Antenna Factors,*IEEE Transactions on Electromagnetic Compatibility*, **24** (3) 316–322 (August 1982).
30. CISPR 16-1-5 *Specification for Radio Disturbance and Immunity Measuring Apparatus and Methods – Part 1-5: Radio Disturbance and Immunity Measuring Apparatus – Antenna Calibration Test Sites for 30 MHz to 1000 MHz*, Ed. 1.0 (2003).
31. Betta, G., Capriglione, D., Carobbi, C. F. M., Migliore, M.D. The accurate calibration of EMC antennas in compact chambers – measurements and uncertainty evaluations, *Elsevier, Computer Standards and Interfaces*, **33** (2), 201–205 (February 2011).
32. Measurement Good Practice Guide No. 73. *Calibration and use of Antennas, Focusing on EMC Applications*, National Physical Laboratory, Teddington (UK) (December 2004).
33. Smith, A.A., German, R.F., Pate, J.B. Calculation of site attenuation from antenna factors, *IEEE Transactions on Electromagnetic Compatibility*, **24** (3) 301–316 (August 1982).
34. Alexander, M., Salter, M., Loader, B., Knight, D. Broadband calculable dipole reference antennas, *IEEE Transactions on Electromagnetic Compatibility*, **44** (1) 45–58 (February 2002).
35. Carobbi, C.F.M., Millanta, L.M., Cati, M. A New Procedure for Evaluating the Performance of the Sites for Radiation Tests or Antenna Calibration, in *Proc. Intern. Symp. EMC Europe 2004*, pp. 702-706, Eindhoven, The Netherlands (September 6–10, 2004).
36. Carobbi, C.F.M., Bonci, A., Stellini, M., Borsero, M. Time-domain characterization of the surge, EFT/Burst and ESD measurement systems, *IEEE Transactions on Instrumentation and Measurement*, **62** (6) 1840–1846 (June 2013).
37. Carobbi, C.F.M. Measurement error of the standard unidirectional impulse waveforms due to the limited bandwidth of the measuring system, *IEEE Transactions on Electromagnetic Compatibility*, **55** (4) 692–698 (August 2013).
38. Carobbi, C.F.M., Bonci, A. Elementary and ideal equivalent circuit model of the 1,2/50 8/20 μs combination wave generator, *IEEE-EMC Magazine*, **2** (4), 51–57.
39. Carobbi, C.F.M., Borio, G., Festa,D., Zanobini, A. Field non-uniformity over the uniform field area as per IEC 61000-4-3 radiated immunity test, *IMEKO TC4 Int. Symp.*, Kosice, Slovakia (September 8–10, 2010).

EXERCISES

10.1 The disturbance level displayed by an EMI receiver at 815 MHz is 65 dB(μV/m). Quasi-peak detector is set. Calculate the spectral density of the impulse disturbance assuming that its repetition frequency is 5 kHz.

10.2 An incoherent disturbance whose spectral density is 80 $\mu V/\sqrt{Hz}$ is applied at the input of an EMI receiver whose tuned frequency is 20 MHz. Calculate the level displayed by the receiver assuming that the average or the quasi-peak detector are set.

10.3 Calculate the complex impedance seen looking into the EUT port of the circuit in Figure 10.10 assuming that the receiver port is terminated into a 50 Ω load and that the power source port is open- or short-circuited to ground G. Verify that the impedance requirements on magnitude and phase are satisfied.

10.4 Calculate the transfer impedance of a current probe which consists of a 50-turns coil wound around a ferrite core whose relative permeability is 1000, inner radius is 2.5 cm, outer radius is 5 cm and thickness is 2 cm. The distributed capacitance of the winding is 5 pF and the load resistance is 50 Ω.

10.5 Calculate the realized gain of an antenna whose calibration factor is 13 dB (1/m) at 100 MHz (load resistance is 50 Ω).

10.6 Calculate the magnetic antenna calibration factor at 10 MHz of a loop antenna whose inner radius is 50 cm, the radius of the shield is 6 mm and the load resistance is 50 Ω.

10.7 A rod antenna is connected, through a coaxial cable of negligible attenuation, to the input of an EMI receiver. Calculate the electric field impinging on the rod antenna at 1 MHz assuming that the reading of the receiver is 45 dB (μV), the length of the rod is 1 m and the radius of the rod conductor is 3 mm.

10.8 Calculate the radio frequency power needed to generate a field strength of 30 V/m at 3 m distance from the aperture of a double ridged horn antenna whose realized gain is 10 dB.

10.9 Calculate the normalized site attenuation (in horizontal and vertical polarization) of a site for measurements at 3 m distance from the equipment under test in the frequency range comprised between 30 and 1000 MHz. Assume that the height of the transmitting antenna is 1 m and the height of the receiving antenna scans the range between 1 and 4 m.

10.10 Calculate the radio frequency power needed to generate 10 V across the 50 Ω loads connected at the two ports of the calibration jig of a BCI probe, assuming that the insertion loss of the probe is 5 dB.

PROBLEM SOLUTIONS

SOLUTIONS FOR CHAPTER 1

1.1 d)

1.2 b)

1.3 c)

1.4 d)

1.5 c)

1.6 a)

1.7 a)

1.8 b)

1.9 a)

1.10 c)

1.11 a)

1.12 d)

1.13 c)

Modern Measurements: Fundamentals and Applications, First Edition.
Edited by Alessandro Ferrero, Dario Petri, Paolo Carbone, and Marcantonio Catelani.

1.14 b)

1.15 b)

1.16 a)

1.17 c)

1.18 b)

1.19 c)

1.20 a)

SOLUTIONS FOR CHAPTER 2

2.1 From (2.11) and (2.13) it results

$$m = \frac{c_0/\Delta\nu_{Cs}}{\{c_0\}/\{\Delta\nu_{Cs}\}}.$$

Thus, the very reference quantity of the metre is $c_0/\Delta\nu_{Cs}$, a length corresponding to the wavelength of a radiation of frequency $\Delta\nu_{Cs}$, whose numerical value is

$$Q_R(\text{m}) = \frac{299\,792\,458}{9\,192\,631\,770}\ \text{m} \cong 0.0326122557\ \text{m},$$

where the numerical value can be exactly expressed only by a fraction, being a rational number.

2.2 The force is due to interaction of the current I flowing in one of the two parallel conductors with the magnetic induction B produced by an equal current flowing in the other conductor. Let r be the distance between the conductors, l the conductor length to which the force is referred, and μ_0 the permeability of vacuum (magnetic constant). Then it is $B = \mu_0 H$, where the magnetic field H is evaluated as the magnetomotive force, equal to I, acting along the circumference of radius r lying in a plane perpendicular to the conductors and centered on the other conductor, divided by the length of the same circumference. That is $B = \mu_0 I/2\pi r$, from which

$$F = B\,I\,l = \frac{\mu_0\,I}{2\,\pi\,r}\,I\,l = \frac{\mu_0\,A^2}{2\,\pi},$$

being $I = 1\text{A}$, $r = 1$ m, and $l = 1$ m.

Note that the force produced by interaction of each current with the magnetic induction generated by itself is correctly considered null, in the absence of any magnetic material in the surrounding space.

2.3 Expressing the newton in terms of the base units (cf. Table 2.3), the defining Equation (2.15) can be rewritten as

$$A = m^{1/2}\,kg^{1/2}\,s^{-1/2}\,\frac{\mu_0^{-1/2}}{\{\mu\}^{-1/2}}$$

$$= \frac{c_0^{1/2}\,\Delta\nu_{Cs}^{1/2}\,m(\mathcal{K})^{1/2}\,\mu_0^{-1/2}}{\{c_0\}^{1/2}\,\{\Delta\nu_{Cs}\}^{1/2}\,\{m(\mathcal{K})\}^{1/2}\,\{\mu_0\}^{-1/2}}, \tag{10.7}$$

where the last equation is obtained by substituting the base units with their respective reference quantities resulting from Problem 2.1, (2.12), and (2.13).

Then, for the very reference quantity it results

$$\mathcal{Q}_R(A) = \sqrt{\frac{c_0\,\Delta\nu_{Cs}\,m(\mathcal{K})}{\mu_0}} \cong 1.480\,898\,70\ A. \tag{10.8}$$

2.4 From equation (2.30) it is

$$\mathcal{Q}_R(kg) = \frac{h\,\Delta\nu_{Cs}}{c_0^2} = \frac{\{h\}m^2\,kg\,s^{-1}\ \{\Delta\nu_{Cs}\}\,s^{-1}}{\{c_0\}^2\,m^2\,s^{-2}} = \frac{\{h\}\,\{\Delta\nu_{Cs}\}}{\{c_0\}^2}\,kg, \tag{10.9}$$

from which it results $\mathcal{Q}_R(kg) = 6.\,777\,264\ldots \times 10^{-41}$ kg.

The same Equation (2.30), rewritten as

$$1kg\,c_0^2 = \frac{\{c_0\}^2}{\{h\}\{\Delta\nu_{Cs}\}}\,h\,\Delta\nu_{Cs}, \tag{10.10}$$

relates the equivalent energy of the kilogram to the photon energy of frequency $\Delta\nu_{Cs}$.

A possible definition based on the equation $1kg\,c_0^2 = h\nu$ could be "The kilogram is the mass whose equivalent energy corresponds to the energy of a photon of frequency $\nu = c_0^2/h = 1.\,475\,521\,66\ldots \times 10^{50}$ Hz". Of course, a photon with such a huge frequency has no physical significance and should be replaced by a collection of photons with lower frequencies (a similar wording was first proposed for the kilogram redefinition).

2.5 From Table 2.3 it results $F = m^{-2}\,kg^{-1}\,s^4\,A^2$. Then, from the respective reference quantities of the base units given in Section 2.5, one finds

$$\mathcal{Q}_R(F) = \frac{\Delta\nu_{Cs}^2\,A^2}{c_0^2\,m(\mathcal{K})\,\Delta\nu_{Cs}^4} = \frac{1}{\Delta\nu_{Cs}\,c_0\,\mu_0}, \tag{10.11}$$

where the last expression is obtained introducing the reference quantity of the ampere obtained in Problem 2.3.

From 2.36 and 2.25, considering the calculable capacitor as a primary standard, one can write

$$\{C_{PS}\} = \frac{(\ln 2/\pi)\,\varepsilon_0\, l}{1/(\Delta\nu_{Cs}\, c_0\, \mu_0)}\, \frac{1}{\{\Delta\nu_{Cs}\}\{c_0\}\{\mu_0\}} = \frac{l}{c_0/\Delta\nu_{Cs}}\, \frac{\ln 2/\pi}{\{\Delta\nu_{Cs}\}\{Z_0\}}, \tag{10.12}$$

where the relations $\varepsilon_0\,\mu_0 = 1/c_o^2$ and $c_0\,\mu_0 = Z_0$ have been recalled. Thus, C_{PS} can be considered as a primary capacitance standard, whose value depends only on a length measurement. Such a measurement is usually performed with an interferometric method, which determines the ratio between l and the wavelength λ_I corresponding to the frequency ν_I of the laser radiation used. Therefore, the wavelength $c_0/\Delta\nu_{Cs}$, of impractical length, can be substituted with $\lambda_I\,(\nu_I/\Delta\nu_{Cs})$, so that the value of the primary standard can be determined as

$$\{C_{PS}\} = \frac{l}{\lambda_I}\,\frac{\Delta\nu_{Cs}}{\nu_I}\,\frac{\ln 2/\pi}{\{\Delta\nu_{Cs}\}\{Z_0\}}, \tag{10.13}$$

with no involvement of any unit.

2.6 From Table 2.3 it results $\mathrm{J} = \mathrm{m}^2\,\mathrm{kg}\,\mathrm{s}^{-2}$ and then, from the reference quantities of the relevant base units,

$$\mathcal{Q}_R(\mathrm{J}) = \frac{c_0^2}{\Delta\nu_{Cs}^2}\, m(\mathcal{K})\,\Delta\nu_{Cs}^2 = m(\mathcal{K})\, c_0^2. \tag{10.14}$$

Incidentally, the joule reference quantity coincides with the equivalent energy of the kilogram prototype and this could appear as a surprising derivation of the relativity equation from pure classical mechanics relations. In reality, the coincidence depends on being the joule homogeneous with a kinetic energy (kilogram times the squared velocity unit) and on having assigned c_o as the reference quantity of the unit of velocity.

With electrical quantities, it is $\mathrm{J} = \mathrm{W\,s} = \mathrm{V\,A\,s}$, but from Table 2.3 it results $\mathrm{V} = \mathrm{m}^2\,\mathrm{kg}\,\mathrm{s}^{-3}\,\mathrm{A}^{-1}$ and substituting in the expression of the joule one finds the same expression in terms of mechanical units leading to the already found reference quantity.

With the system references proposed for the revised SI (see Section 2.6), that is with fixed values for h, e, k, and N_A, the kilogram has the reference quantity found above (cf. Problem 2.4) and then

$$\mathcal{Q}_R(\mathrm{J}) = \frac{c_0^2}{\Delta\nu_{Cs}^2}\,\frac{h\,\Delta\nu_{Cs}}{c_0^2}\,\Delta\nu_{Cs}^2 = h\,\Delta\nu_{Cs}. \tag{10.15}$$

In this case, the expression of $Q_R(\mathrm{J})$ has really a quantum-related meaning, as the quantum and relativistic relations are implicitly introduced by the kilogram definition.

2.7 The present defining equation for the second (2.13) would become

$$\mathrm{s} = \frac{1/\nu_C}{1/\{\nu_C\}} = \frac{\{m_e\}\{c_0\}^2}{\{h\}} \frac{h}{m_e\, c_0^2} = 1.235\,589\,8 \ldots \times 10^{20} \frac{h}{m_e\, c_0^2}. \tag{10.16}$$

The possible future-defining equation for the kilogram (2.30) would become simply

$$\mathrm{kg} = \frac{m_e}{\{m_e\}} = 1.097\,769\,3 \ldots \times 10^{30}\, m_e. \tag{10.17}$$

2.8 Replacing in Equation (2.19) the derived unit watt with its expression in terms of base units and those with their reference quantities, one obtains $\mathrm{cd} = \mathrm{m}^2\ \mathrm{kg}\ \mathrm{s}^{-3}\ \mathrm{sr}^{-1}\ K_m/683$ and then

$$\begin{aligned}\mathrm{cd} &= \left(\frac{\{\Delta\nu_{Cs}\}}{\{c_0\}} \frac{c_0}{\Delta\nu_{Cs}}\right)^2 1\, m(\mathcal{K}) \left(\frac{\{\Delta\nu_{Cs}\}}{\Delta\nu_{Cs}}\right)^{-3} \\ &= \frac{c_0^2\, m(\mathcal{K})\, \Delta\nu_{Cs}\, K_m}{\{c_0\}^2 1 \{\Delta\nu_{Cs}\}\{K_m\}}\end{aligned} \tag{10.18}$$

where the numerator is the reference quantity of the unit expressed in terms of the SI reference constants.

SOLUTIONS FOR CHAPTER 3

3.1 A casual reader might be misled into thinking that 10 samples per period could be a high enough sampling rate in any case. Actually, it should be remembered that the spectrum of any periodic waveform, other than a sinewave, is made up of several components, at the frequency $f_0 = 1/T_0$ (the fundamental frequency) as well as at integer multiples of f_0 (harmonic frequencies). Hence, the answer depends on the spectral composition of the analyzed waveform. Another issue concerns what is a "correct" acquisition, or, what could be an acceptable sampling interval "for practical purposes". In general, time-domain signal measurements have less demanding requirements, whereas frequency-domain measurements are often aimed at analyzing the detailed spectral composition of the signal. In the former case, accuracy to within few percents of the actual waveform is usually fully satisfactory.

For a triangular waveform, mathematical analysis shows that amplitudes of harmonic components at frequencies nf_0 decrease as $1/n^2$, but extend indefinitely (it may be remembered that, for a perfectly symmetric waveform, only odd-order harmonics are present). The proposed sampling rate is 100 samples per second, corresponding to a Nyquist frequency of 50 Hz. The triangular waveform fundamental frequency is 10 Hz, that means the Nyquist frequency corresponds to the fifth order harmonic, whose amplitude is 4% (=1/52) of that of the fundamental component. Then, acquisition at 100 samples per second can be "correct" as long as component amplitudes below 4% of the fundamental are considered negligible. It should as well be remembered that, in practice, higher order harmonics would be folded back into the 0–50 Hz range unless an anti-aliasing filter is employed. In the latter case the fifth order harmonic would be significantly attenuated, since no actual filter can be an ideal "brick-wall."

3.2 The Nyquist frequency is 800 Hz, therefore aliasing affects the third signal component. For the range 0–800 Hz, the sampled signal has $f_1 = 324$ Hz; $f_2 = 648$ Hz; and $f_3 = 1600 - 972 = 628$ Hz.

3.3 The sample set covers an observation interval of 400 ms or, approximately, 20 periods of the sinewave. If the quasi-coherent estimate was obtained using just one period of the sinewave, this would correspond to using about $N_P = 50$ samples in the algorithm, with an uncertainty (proportional to $1/N_P$) in the order of 2%. If the algorithm can select NP so that it matches the largest integer number of waveform periods within the acquisition interval, then either 19 or 20 periods (depending on the actual sinewave frequency) could be considered and the approximate value of N_P would be between 950, and 1000. Thus, uncertainty drops to order of 0.1.

3.4 The specific values of signal frequencies and sampling interval given in this exercise provide the opportunity to discuss several aspects of digital RMS measurement.

It can be checked that $x(t)$ is indeed a periodic waveform with a period of 20 ms, since the largest common divider of the three component frequencies is 50 Hz. Harmonic indexes of the three components are then $k_1 = 4$, $k_2 = 7$, and $k_3 = 8$. Continuous-time periodicity does not in general carry over to the sampled waveform and, in this particular case, it turns out that samples of the digitized waveform repeat with a period of 140 ms. As the acquisition interval is exactly 140 ms, processing the whole set of 400 samples would allow exact determination of the RMS value, that is equal to $[\frac{1}{2}(10^2 + 3^2 + 1^2)]^{-\frac{1}{2}} =$ 7.42.

As already noted, unless a perfectly coherent sampling condition is obtained by design, occasional occurrence cannot be relied upon for a practical implementation of the RMS algorithm. Period counting based on zero-crossing is employed instead as a simple way to determine NP for the quasi-coherent sampling approach. Unfortunately, in this specific case it is somewhat inaccurate, with zero-crossing being actually sensitive to the lower frequency 200 Hz component of the signal, that has the largest amplitude. Direct inspection of the figure shows that, from the analysis of zero-crossings, the digital algorithm would suggest that 27 periods occur during the observation interval. N_P is then about 385 samples or, approximately, 135 ms. Thus, in the digital computation of the RMS value at least some of the spurious terms will not be canceled, affecting the accuracy of the measured value. Closer inspection of formulae on digital RMS estimation shows that critical contributions come from cross-terms involving $k_2 = 7$, nevertheless, impact on overall accuracy is still limited to the order of a few percent.

3.5 Observation interval length is 40 ms, its reciprocal yielding frequency granularity $F = 25$ Hz. Computing the ratios f_i/F yields normalized frequencies $\nu_1 N = 730.56$, $\nu_2 N = 830$, and $\nu_3 N = 867.52$. Therefore, only the amplitude of the second component can be estimated accurately. For the first and third ones, magnitudes are attenuated by a factor approximately equal to *WCSL*, that is, about 35%. Separation among the three components is large enough ($\nu_3 N - \nu_2 N = 37.52$ in the worst case) to make spectral interference negligible.

3.6 The minimum observation interval, under coherent sampling conditions, is the same as the signal period. In this case, the signal has a fundamental frequency $f_f = 5$ Hz, obtained as the maximum common divider of the frequencies of the three given spectral components. Therefore, the minimum required observation interval is $T_o = T = 1/f_f = 0.2$ s. The maximum signal frequency is 75 Hz, that corresponds to harmonic $N = 75/5 = 15$. Therefore, the minimum sampling frequency, to avoid aliasing, is $\hat{f}_s = (2N + 1) \cdot f_f = 155$ Hz. This sampling frequency, however, will return 31 samples/period, which is not a power of 2, as required to implement an FFT algorithm. Therefore, it is necessary to acquire 32 samples, at least, that is the nearest power of 2 that provides a number of samples greater than 31. The required sampling frequency is, therefore: $f_s = 32 \cdot f_f = 160$ Hz.

3.7 To answer this question, the frequency resolution f_r with which the frequency domain components are evaluated must be determined. The frequency resolution is the inverse of the observation interval $T_o = \frac{512}{f_s}$. Considering that the kth harmonic of the signal is located at frequency $f_k = k \cdot f_1$, the required position h is the one that satisfies the following equation: $h = \mathrm{INT}(\frac{k \cdot f_1}{f_r}) = \mathrm{INT}(\frac{55 \cdot 50 \cdot 512}{13500}) = 104$.

3.8 The frequency of the highest harmonic component in the continuous-time signal is $f_M = 133 \cdot f_1 = 6650$ Hz. Since this frequency is higher than f_s, and not only higher than $f_s/2$, the shifted replica of the Fourier transform of the continuous-time signal are totally overlapping. Therefore, all harmonic components of the continuous-time signal are contributing to aliasing, and all harmonic components are affected by aliasing.

SOLUTIONS FOR CHAPTER 4

4.1 Solutions are

1. The Nyquist frequency, f_N, is 100 Hz. The lowest frequency, 25 Hz, is below f_N but the highest frequency, 110 Hz, will be shifted down to 110–f_N=90 Hz by alias. Thus, the answer is 25 and 90.
2. The 175 Hz tone will be shifted down to 25 Hz and fall on top of the lowest or original frequency component. There is no phase information in the problem text but we know that the magnitude is only 0.5. Thus, the original frequency component will not be completely canceled out.

4.2 The common frequency components for both sampling frequencies are 10 Hz and 117 Hz. These are then most likely frequencies in the first Nyquist band. The remaining frequencies must have been folded down from higher Nyquist bands. If they are from the second Nyquist band they will be

$$f_s = 320 \text{ Hz: } 4 \rightarrow 316, 158 \rightarrow 162, 159 \rightarrow 161$$

$$f_s = 312 \text{ Hz: } 12 \rightarrow 300, 150 \rightarrow 162, 153 \rightarrow 159$$

As can be seen two more matches can be made. The frequency 162 Hz and the frequency 159 Hz. The latter is folded down for the sampling frequency 312 Hz but not for the sampling frequency 320 Hz. To find the last match we continue to investigate if they are from the third Nyquist band: f_s=320 Hz: $4 \rightarrow 324$, f_s=312 Hz: $12 \rightarrow 324$. Thus, the frequencies are 10, 117, 159, 162, and 324 Hz.

4.3 The frequency components will be $f_{2n+1} = 60 + 160n$ and $f_{2n+2} = 160 - 60 + 160n, n = 0, 1, 2, \ldots$. The amplitude will be $sinc(f_i/160), i = 1, 2, \ldots$. Figure 4.7 illustrates the output spectrum but for other frequencies.

4.4 Solutions are

1. Let the signal be sinewave with peak value FS. The signal RMS value will then be $FS/\sqrt{2}$. The quantisation noise has an RMS value $Q/\sqrt{12}$, where $Q = (V_{\max} - V_{\min})/2^B = 2FS/2^B$.

$$\begin{aligned}
SINAD &= 20\log\left(\frac{FS}{\sqrt{2}}/\frac{2FS}{2^B\sqrt{12}}\right) \\
&= 20\log\left(2^B\sqrt{\frac{12}{8}}\right) \\
&= 20B\log 2 + 20\log\sqrt{\frac{12}{8}} \\
&\approx 6.02B + 1.76
\end{aligned}$$

2. If the signal peak value is μFS, where μ is 0..1 then

$$SINAD = 20\log\left(\frac{\mu FS}{\sqrt{2}} / \frac{2FS}{2^B\sqrt{12}}\right)$$
$$= 20\log\left(2^B\sqrt{\frac{12}{8}}\mu\right)$$
$$= 20\log\mu + 20B\log 2 + 20\log\sqrt{\frac{12}{8}}$$
$$\approx 6.02B + 1.76 + 20\log\mu.$$

The last term will be negative due to that μ is lower than one. When $\mu < \sqrt{\frac{8}{12}}/2^B$ SINAD will be negative.

3. $SINAD = 6.02ENOB + 1.76$ $71 = 6.02ENOB + 1.76 \Leftrightarrow ENOB \approx 11.5$ bits.
4. When oversampling with two times the required sampling frequency, the noise will be spread over twice the frequency range. After filtering the noise level has been reduced by two or 3 dB. Consequently, SINAD has increased by 3 dB. $71 + 3 = 6.02ENOB + 1.76 \Leftrightarrow ENOB \approx 12$ bits.

SOLUTIONS FOR CHAPTER 5

5.1 a. −3.4%, b. 1000 Ω, c. 333 kΩ

5.2 −4.2%, maximum.

5.3 −0.26

5.4 a. 90 MΩ, b. 1.25 MΩ

5.5 ±0.78 V

5.6 a. 11%, b. −29%

5.7 The elimination of lead and contact resistances.

5.8 In very low dc voltage precision measurements, typically at microvolt voltage level.

5.9 Depending on the accuracy and use of the DMM, typically between 1 and 3 years.

SOLUTIONS FOR CHAPTER 6

6.1 Analog scopes.

6.2 Memory, powerful measurements, communication capability, bandwidth.

6.3 AC coupling blocks low frequency signals. DC coupling allows a direct connection of the signal to the input amplifier.

6.4 Set coupling to AC, vertical scale to 1 volt/division to see the AC signal using the screen full scale.

6.5 When measuring signals synchronized with power line voltage.

6.6 Adjust trigger level to around half value of the signal amplitude and select positive slope triggering.

6.7 When measuring the 2040 Hz signal you should be sampling at least at 8 to 10 kHz but you are really sampling at 2 kS/s so you see the 40 Hz component.

6.8 Auto triggering. It will always trigger even if no valid trigger condition is met.

6.9 Connect the probe to a 1 kHz square generator and adjust the control until you see an almost perfect square shape.

6.10 Input amplifiers, probes, cathode ray bandwidth.

6.11 Analog scopes cannot show pre-trigger information unless they have phosphor memory in the screen. Digital scopes are always sampling so you only need to set the pre-trigger time to be displayed on the screen.

6.12 You need to use the holdoff control until you see a stable waveform. You can also trigger on any of the four pulses only adjusting the holdoff control.

SOLUTIONS FOR CHAPTER 7

7.1 Possible answers are

1. The general property is temperature expressed with the Celsius scale; the individual property is the temperature of the body; the property values are 20°C and 40°C.
2. The statement is meaningless because the Celsius scale is an interval scale, not a ratio scale (zero of the scale is established in a conventional way).
3. The ratio of two temperature measurements expressed using the absolute scale has a metrological meaning since that scale is a ratio scale.
4. The statement has metrological meaning because ratios of temperature differences can be performed in an interval scale as the Celsius scale

7.2 Possible answers are

1. The described process is not a measurement for different reasons. In particular, a provable level of objectivity and inter-subjectivity of the assigned values is not provided.
2. Multiple assessments by independent experts of the subject is expected to increase the inter-subjectivity of the assigned values.

3. Uncertainty of the assigned value can be evaluated by multiple assessments performed by independent experts of the subject and by applying type A evaluation methods (see Chapter 1).
4. Many features may affect the validity of this process of value assignment. For instance, the questions submitted to the students could not properly cover all the different topics in the subject (this can be considered as a source of model uncertainty). Also, the assigned value is expected to depend on the evaluator (this can be considered as a source of instrumental uncertainty). Moreover, the answers provided by students could depend on the way the questions are written (this can be considered as a source of interaction uncertainty).

7.3 Possible answers are

1. The described process is not a measurement mainly because a provable level of objectivity and inter-subjectivity of the assigned values is not provided.
2. The repeatability of judges can be improved, for instance, with suitable education and experience provided, for example, by panel judgments harmonization sessions; also environment conditions ensuring that wines do not change their characteristics are important to ensure repeatability of measurements.
3. Intersubjectivity of results can be ensured by considering the ranking returned by a sufficient high number of independent experts.
4. Assessment uncertainty can be evaluated by considering multiple assessments and applying type A evaluation methods (see Chapter 1).

7.4 Possible answers are

1. According to the definition provided in Section 1.2.8, to obtain a measurement process we should assure both a morphic assignment and a specified and provable level of objectivity and inter-subjectivity, that is, a specified value for measurement uncertainty. It seems quite hard to satisfy these constraints for each considered dimension of the service quality. Moreover, any admissible composition of the obtained grades does not satisfy the representation condition (see the example in Section 1.2.10). Thus we can conclude that the proposed assessment is not a measurement.
2. Repeatability of assessment related to each single question can be quantified by using a dispersion statistic when considering repeated answers provided by a single customer. Similarly, assessment reproducibility can be quantified in a similar way but considering, for instance, different customers.
3. For instance, results obtained from a representative set of customer population could be used to select the most important (or critical) aspects of service quality, so that subsequent assessment could be focused only on these aspects, improving the process efficiency.
4. The overall service quality defined by this process is not a measurement so talking about the related uncertainty is meaningless. However, instrumental uncertainty associated with results related to each question separately can

be evaluated by considering a representative set of customer population. Conversely, modeling contribution to uncertainty is hardly quantified since the concept of "service quality" underlying the measurand is not clearly defined.

7.5 Before starting modeling this measurement process the *ethical* issues associated with this task should fully be explored. Thus, what follows is only to be considered as an hypothetical approach. Moreover, the maturity in measuring pain tolerances and thresholds is currently of limited nature and does not enable the setup of a reliable and consistent measurement process with a predictable and modeled measurement uncertainty. Nevertheless, the presented framework helps in reasoning about the directions to be taken if advancements in this level of maturity are sought.

According to the presented measurement framework the goal-setting phase could serve to clarify the purpose of this measurement: scientific goals or information collection for pain relieving procedures may be two examples of purposes. Following the framework, the first difficulty to overcome would be that of defining the *general property*, as pain involves personal emotional experiences that are difficult to make inter-subjective. This, in turn, will result in a large *definitional uncertainty* that can be possibly reduced only by modeling, as precisely as possible, the measurement context, for example, by confining it to small application domains. Thus, pain measurement associated with for example, dentistry, is an easier task to address than pain measurement associated with a larger domain, for example, generic diseases. Again, the need to make measurement results as much as possible *repeatable* and *reproducible* would result in a limited set of pain inducing mechanisms, whose outcomes may tentatively be represented using an ordinal scale. To *calibrate* the measurement process volunteers can be involved: pain-inducing mechanisms are applied and responses in terms of pain magnitude registered using the defined ordinal scale. Then, dispersion statistics applied to collected data can be exploited to assess repeatability and reproducibility of measurement results.

The role of the pain assessor is also of major importance: procedures must be defined and harmonization sessions managed to reduce subjective interpretation of experimental outcomes, when attributing the position on the scale. This would help in controlling *instrumental* and *interaction uncertainties*.

SOLUTIONS FOR CHAPTER 8

8.1 The fact that the mean value converges asymptotically towards its expected value implies that

$$plim_{N\to\infty}\left[\frac{1}{N}\sum_{i=1}^{N}u^2(i)\right] = \lim_{N\to\infty}\frac{1}{N}\sum_{i=1}^{N}E[u^2(i)]$$

$$= \sigma_u^2 + \lim_{N\to\infty}\frac{1}{N}\sum_{i=1}^{N}u_0^2(i).$$

For the least-squares estimator (8.2), one needs to take into account that the input and the output additive noise sources are uncorrelated, hence

$$plim_{N\to\infty}\alpha_{LS} = plim_{N\to\infty}\frac{\sum_{i=1}^{N} y(i)u(i)}{\sum_{i=1}^{N} u^2(i)}$$
$$= \alpha_0\frac{\lim_{N\to\infty}\frac{1}{N}\sum_{i=1}^{N} u_0^2(i)}{\sigma_u^2 + \lim_{N\to\infty}\frac{1}{N}\sum_{i=1}^{N} u_0^2(i)}.$$

Note that there is a systematic bias for the least-squares estimator in an errors-in-variables framework. The least-squares estimator is only unbiased in the absence of input noise ($\sigma_u^2 = 0$).

The asymptotic value of the errors-in-variables solution equals

$$plim_{N\to\infty}\alpha_{EV} = \frac{plim_{N\to\infty}\frac{1}{N}\sum_{i=1}^{N} y(i)}{plim_{N\to\infty}\frac{1}{N}\sum_{i=1}^{N} u(i)}$$
$$= \frac{\lim_{N\to\infty}\frac{1}{N}\sum_{i=1}^{N} y_0(i)}{\lim_{N\to\infty}\frac{1}{N}\sum_{i=1}^{N} u_0(i)}$$
$$= \alpha_0,$$

showing that the errors-in-variables solution converges towards its real value, even when both input and output are perturbed with additive noise.

Similarly, one can compute the expected value of the instrumental variable estimator, assuming that the reference signal $r(i)$ is known. Hence, $r(i)$ can be considered as a deterministic value with respect to the expectation operator

$$plim_{N\to\infty}\alpha_{IV} = \frac{\lim_{N\to\infty}\frac{1}{N}\sum_{i=1}^{N} E[y(i)]r(i)}{\lim_{N\to\infty}\frac{1}{N}\sum_{i=1}^{N} E[u(i)]r(i)}$$
$$= \frac{\lim_{N\to\infty}\frac{1}{N}\sum_{i=1}^{N} y_0(i)r(i)}{\lim_{N\to\infty}\frac{1}{N}\sum_{i=1}^{N} u_0(i)r(i)}$$
$$= \alpha_0.$$

8.2 The MATLAB code given below generates the (noisy) input and output signals u and y given the reference signal r, which equals a vector of size N. Performing a statistical analysis on (a large number of independent) realization gives the Monte-Carlo analysis.

MATLAB code

```
u=(R1+R2)/(R1+R2+Rg)*r + std_u*randn(size(r));

y=R1/(R1+R2+Rg)*r + std_y*randn(size(r));
```

```
alpha_avg = mean(y./u); % simple averaging
alpha_LS = mean(y.*u)/mean(u.*u); % Least-squares
alpha_EV = mean(y)./mean(u); % Errors-in-variables
alpha_IV = mean(y.*r)./mean(u.*r); % Instrumental variables
```

8.3 The MATLAB code given below computes the errors-in-variables estimates in an explicit way, computes the uncertainty on the estimates, and then gives an example on how to compute the uncertainty when applying a nonlinear transformation (computing R_1 starting from α).

MATLAB code

```
alpha_EV = mean(y)./mean(u);
% Error vector
e = (alpha_EV*u-y)/sqrt(var_u*(alpha_EV)^2 + var_y);
% The cost function
K=e*e';
% Jacobian of the cost function (derivative of e wrt parameters)
J = u/sqrt(var_u*(alpha_EV)^2 + var_y) - ...
var_u*(alpha_EV)*(alpha_EV*u-y)/(var_u*(alpha_EV)^2 + var_y)^(1.5);
% Uncertainty on the estimate
std_alpha_EV=1/sqrt(J*J');
% Nonlinear transformation of variable(s) towards R1
R1_EV = alpha_EV * R2 / (1-alpha_EV);
% Derivative of the transformation of variable(s)
Jacob = alpha_EV * R2 / (1-alpha_EV)^2 + R2 / (1-alpha_EV);
% Estimated uncertainty on estimate R1
std_R1_EV=sqrt(Jacob^2/(J*J'));
```

8.4 The ML estimator is simply given by a least squares solution since only output noise is present. Assuming that the output noise has a noise variance of $\sigma_n^2(t_i)$, the ML estimators can be written as $\sum_{i=1}^{K} e_i^2$ with

$$e_{i1} = \frac{A\sin(2\pi f t_i + \phi) - y(t_i)}{\sigma_n(t_i)}$$

and

$$e_{i2} = \frac{A\sin(2\pi f t_i) + B\cos(2\pi f t_i) - y(t_i)}{\sigma_n(t_i)}.$$

To minimize these cost functions, one needs to take the first derivative of the cost with respect to the parameters. In the former case, one ends up with a set of equations which is nonlinear in the parameter ϕ

$$\sum_{i=1}^{K} \frac{(A\sin(2\pi f t_i + \phi) - y(t_i))\sin(2\pi f t_i + \phi)}{\sigma_n^2(t_i)} = 0$$

$$\sum_{i=1}^{K} \frac{(A\sin(2\pi f t_i + \phi) - y(t_i))\cos(2\pi f t_i + \phi)}{\sigma_n^2(t_i)} = 0.$$

Hence, a nonlinear solver is needed to determine the solution. In the latter case, the set of equations,

$$\sum_{i=1}^{K} \frac{(A\sin(2\pi f t_i) + B\cos(2\pi f t_i) - y(t_i))\sin(2\pi f t_i)}{\sigma_n^2(t_i)} = 0$$

$$\sum_{i=1}^{K} \frac{(A\sin(2\pi f t_i) + B\cos(2\pi f t_i) - y(t_i))\cos(2\pi f t_i)}{\sigma_n^2(t_i)} = 0$$

is linear in A and B and can be solved in an non-iterative and numerical stable way.

8.5 For a given ϕ, one can compute the amplitude A using

$$A(\phi) = \frac{\sum_{i=1}^{K} \frac{y(t_i)\sin(2\pi f t_i + \phi)}{\sigma_n^2(t_i)}}{\sum_{i=1}^{K} \frac{\sin^2(2\pi f t_i + \phi)}{\sigma_n^2(t_i)}}.$$

This solution can be fed back into the second equation, resulting in a single equation which only depends on the phase ϕ :

$$\sum_{i=1}^{K} \frac{(A(\phi)\sin(2\pi f t_i + \phi) - y(t_i))\cos(2\pi f t_i + \phi)}{\sigma_n^2(t_i)}.$$

This shows that the amplitude A can be eliminated from the set of equations.

SOLUTIONS FOR CHAPTER 9

9.1 The reliability model for a mixed structure is obtained by combining equations referred to the canonical configurations: series, parallel, and *k*oo*n*. In this case, looking at the system depicted in Figure 9.12, the series between A and B is evaluated by the product of $R_A(t)$ and $R_B(t)$, considering the sub-system B a k*oo*n configuration with $k = 3$ and $n = 4$. Then, the parallel structure can be taken into consideration as a product of the unreliability between the upper branch A and B, $F_{AB}(t)$, and the lower C, $F_C(t)$. The final model of the system reliability $R_{sys}(t)$ can be evaluated as $1 - F_{sys}(t)$, with $F_{sys}(t) = F_{AB}(t)F_C(t)$ obtained by the previous step. In this evaluation the simplified conditions with constant failure rate and independent failures are assumed.

9.2 If the failure rate λ can be considered constant in time (useful life), the reliability model is $R(t) = e^{-\lambda t}$. Considering $\lambda = 1/MTTF$ for the useful life of the component, the reliability value at the time equal to MTTF is 0.37.

9.3 The reliability of the system, R_{sys}, is given by 93% at 1 year. Assuming independent conditions among the items, $R_{sys}(t) = \prod_{i=1}^{12} R_i(t)$ and, consequently, $R_i(t)|_{1year} = 0.9934$.

9.4 The reliability model referred to mentioned system is

$$R_{sys}(t) = 1 - F_{sys}(t) = 1 - \prod_{i=1}^{4} (1 - R_i(t)) = 1 - (1 - R_i(t))^4. \tag{10.19}$$

If $R_{sys}(t)|_{1year} = 0.93$ it is immediate from the previous equation the evaluation of the reliability referred to the generic item.

9.5 The reliability model for the redundant configuration k*oo*n is

$$R_{sys}(t) = \sum_{i=k}^{n} \binom{n}{i} (R(t))^i (1 - R(t))^{n-i}. \tag{10.20}$$

Consequently, if $k = 2$ and $n = 3$ we obtain $R_{sys}(t) = 3e^{-2\lambda t} - 2e^{-3\lambda t}$. Replacing the value of the failure rate λ and considering a warranty time 8 760 hours (medium value for 1 year), the probability of failure is 13.55%.

9.6 Considering that the probability of operation of the 2*oo*3 configuration is given by $R_{sys}(t) = 3e^{-2\lambda t} - 2e^{-3\lambda t}$, remembering the relationship between the failure rate λ and the mean time between failure MTBF (or the mean time to failure MTTF, for a component), the probability of operation is equal to 0.3064.

9.7 For the chain structure constituted by n independent and identical items $MTBF = 1/(n\lambda)$. Consequently, $\lambda = 10^{-6}\text{h}^{-1}$ and the reliability of a single item at 8 760 hours is 0.99. Six identical capacitors are needed in order to obtain the required performance.

9.8 The system under consideration is an 1*oo*3 structure. Consequently, the reliability model is $R_{sys}(t) = 3e^{-\lambda t} - 3e^{-2\lambda t} + e^{-3\lambda t}$, and $MTBF = 11/(6\lambda)$. Considering that $\lambda = MTTF = 5 \cdot 10^{-5}\text{h}^{-1}$, $R_{sys} = 0.96$ and the probability of failure of the streetlight is 0.04.

9.9 The reliability of a parallel structure constituted by two independent items with constant failure rate λ_1 and λ_2 is $R_{sys}(t) = e^{-\lambda_1 t} + e^{-\lambda_2 t} - e^{-(\lambda_1+\lambda_2)t}$ and, assuming the failure rate $\lambda_1 = \lambda_2 = \lambda$, $R_{sys}(t) = 2e^{-\lambda t} - e^{-2\lambda t}$. Consequently, the mean time between failure referred to two identical items is $3/(2\lambda)$. It is easy to observe an increase of 50% with respect to the single item.

9.10 If six elements are connected in series configuration, the reliability of the system is $R_{sys} = 0.85^6 = 0.377$; if the same items are in parallel, $R_{sys} = 1 - (1 - 0.85)^6 = 0.99998$. Considering, for instance, a mixed configuration constituted by three items in series, in parallel with a series of the other three items—two branches in parallel, each of three items—the system reliability is $R_{sys} = 1 - (1 - 0.85^3)^2 = 0.851$. It can be observed that any possible mixed configuration constituted by six identical items so that $R_i = 0.85$, for $i = 1, \ldots, 6$ is characterized by a reliability value in the range [0.377 ; 0.99998].

SOLUTIONS FOR CHAPTER 10

10.1 Refer to Section 10.3.1.2. Obtain the value of P corresponding to $f_R = 5$ kHz from Figure 10.8. Then calculate the impulse bandwidth as $B_{\text{imp}} = 1.064B_6$. The spectral density S of the impulse disturbance is given by $S = \frac{L}{PB_{\text{imp}}}$ (all quantities in linear units) where L is the displayed disturbance level. The numerical result (in log units) is $S = 84.0$ dB(μV/m/MHz)

10.2 Refer to Section 10.3.1.3. In case of use of quasi-peak detector we obtain $V_{qp} = 15.1$ mV. In case of use of average detector $V_{ave} = 8.2$ mV.

10.3 The simplest way to obtain the solution is by using a circuit simulator. For the purpose of verification take into account that impedance magnitude is 14.1 Ω and phase is 54 deg at 45 kHz, when the power source port is open.

10.4 Use the formula for the transfer impedance Z_T in Section 10.4.2. In addition take into account that for a toroidal core $L = \frac{\mu_r \mu_0 \Sigma}{2\pi \hat{R}} N^2$, where Σ is the cross-section and $\hat{R}$ is the average radius of the core.

10.5 Use Equation (10.2) (all quantities in linear units). Then convert to log units (as usual in EMC). The numerical result is $G = -2.8$ dB. Note that, due to mismatch, the realized gain can be negative.

10.6 Use the equation for ACF_H in Section 10.5.1. The numerical result is (in log units) $ACF_H = -24.5$ dB(S/m).

10.7 Refer to Section 10.5.1 where the the ACF of the rod antenna is given as a function of the length of the rod, its radius, and frequency. Calculate the rod capacitance through Equation (10.3). The numerical result is $E = 98.5$ dB(μV/m).

10.8 The electric field E radiated at distance d in the far-field of a transmitting antenna whose realized gain is G and fed by the power P is $E = \sqrt{\frac{GP\zeta}{4\pi}}\frac{1}{d}$ (all quantities in linear units). The numerical solution is $P = 27$ W.

10.9 Refer to Equations (10.5) and (10.6). Some simple numerical code is needed in order to calculate the maximum. For verification purpose take into account that at 100 MHz we have $NSA_H = -2.0$ dB(m^2) and $NSA_V = -0.7$ dB(m^2).

10.10 Refer to Section 10.6.1. The radiofrequency power P needed to generate the voltage V given the insertion loss IL is $P = IL\frac{V^2}{R}$, where $R = 50\ \Omega$ (all quantities in linear units). Hence the numerical result is $P = 6.3$ W.

INDEX

Modern Measurements: Fundamentals and Applications, First Edition.
Edited by Alessandro Ferrero, Dario Petri, Paolo Carbone, and Marcantonio Catelani.

Printed and bound by CPI Group (UK) Ltd, Croydon, CR0 4YY
16/09/2026
14963423-0002